Concepts, Methodologies and Technologies for Domestic Retrofit

Other related titles:

You may also like

- PBBE009 | Matin | Enabling Technologies for Sustainable Smart Cities: Design, development and management
- PBBE002 | Clarke | Applied Building Performance Simulation | 2024
- PBBE006| Lam | Empowering Smart Cities through Community-Centred Public Private Partnerships and Innovations | 2023
- PBBE003 | Al-Kodmany | Sustainable High-Rise Buildings: Design, technology, and innovation | 2022
- PBBE001| Cao | Handbook of Ventilation Technology for the Built Environment | 2021

We also publish a wide range of books on the following topics:
Computing and Networks
Control, Robotics and Sensors
Electrical Regulations
Electromagnetics and Radar
Energy Engineering
Healthcare Technologies
History and Management of Technology
IET Codes and Guidance
Materials, Circuits and Devices
Model Forms
Nanomaterials and Nanotechnologies
Optics, Photonics and Lasers
Production, Design and Manufacturing
Security
Telecommunications
Transportation

All books are available in print via https://shop.theiet.org or as eBooks via our Digital Library https://digital-library.theiet.org.

IET BUILT ENVIRONMENT SERIES 04

Concepts, Methodologies and Technologies for Domestic Retrofit

Edited by
William Swan and Richard Fitton

The Institution of Engineering and Technology

About the IET

This book is published by the Institution of Engineering and Technology (The IET).

We inspire, inform and influence the global engineering community to engineer a better world. As a diverse home across engineering and technology, we share knowledge that helps make better sense of the world, to accelerate innovation and solve the global challenges that matter.

The IET is a not-for-profit organisation. The surplus we make from our books is used to support activities and products for the engineering community and promote the positive role of science, engineering and technology in the world. This includes education resources and outreach, scholarships and awards, events and courses, publications, professional development and mentoring, and advocacy to governments.

To discover more about the IET, please visit https://www.theiet.org/.

About IET books

The IET publishes books across many engineering and technology disciplines. Our authors and editors offer fresh perspectives from universities and industry. Within our subject areas, we have several book series steered by editorial boards made up of leading subject experts.

We peer review each book at the proposal stage to ensure the quality and relevance of our publications.

Get involved

If you are interested in becoming an author, editor, series advisor, or peer reviewer please visit https://www.theiet.org/publishing/publishing-with-iet-books/ or contact author_support@theiet.org.

Discovering our electronic content

All of our books are available online via the IET's Digital Library. Our Digital Library is the home of technical documents, eBooks, conference publications, real-life case studies and journal articles. To find out more, please visit https://digital-library.theiet.org.

In collaboration with the United Nations and the International Publishers Association, the IET is a Signatory member of the SDG Publishers Compact. The Compact aims to accelerate progress to achieve the Sustainable Development Goals (SDGs) by 2030. Signatories aspire to develop sustainable practices and act as champions of the SDGs during the Decade of Action (2020–30), publishing books and journals that will help inform, develop, and inspire action in that direction.

In line with our sustainable goals, our UK printing partner has FSC accreditation, which is reducing our environmental impact to the planet. We use a print-on-demand model to further reduce our carbon footprint.

Published by The Institution of Engineering and Technology, London, United Kingdom

The Institution of Engineering and Technology (the "**Publisher**") is registered as a Charity in England & Wales (no. 211014) and Scotland (no. SC038698).

First published 2026

The Institution of Engineering and Technology
Futures Place,
Kings Way, Stevenage,
Herts, SG1 2UA,
United Kingdom

www.theiet.org

British Library Cataloguing in Publication Data
A catalogue record for this product is available from the British Library

ISBN 978-1-83953-404-1 (hardback)
ISBN 978-1-83953-405-8 (PDF)
ISBN 978-1-80705-314-7 (EPUB3)

Typeset in India by MPS Limited

Cover image: Jill Jennings/Energy House 1 via The Energy House 2.0, University Salford Manchester, UK

Contents

Preface

Retrofit is a compelling research topic. It covers a wide range of disciplines from building physics to social sciences. It directly affects people's lives, their finances, health, comfort and wellbeing. This makes retrofit research both intellectually challenging and directly impactful. These varied perspectives of retrofit as a research or policy problem also make it problematic. Ideas that seem effective from one perspective can have unintended consequences that arise in another. If there was a subject area that required a genuine interdisciplinarity, retrofit is clearly that problem.

This book presents multiple perspectives of the problem, ranging from the materials we choose, how we measure and evaluate retrofit performance, how we model, and how we assess the impact on individuals. It considers the problem at national and international scales, as well as reflecting on the compromises of the multiple aspects of sustainability that need to be addressed. This book takes a largely practical perspective, addressing our understanding of what retrofit can deliver, how we can deliver it, how we evaluate it, and the lessons that can be learned from understanding the physical application of retrofit. As such, the work touches on policy, occupants, supply chains, measurement, modelling, and physical retrofit. The authors are both academic and industry-based, recognising the need for a constant dialogue between the research and innovation base and the industry tasked with delivering a sustainable existing built environment.

The built environment is subject to the impacts of the wider world. The cost-of-living crises, economic development, and net zero are all driving factors when we talk about retrofit. The energy transition, the emergence of new technology, and better understanding of issues such as indoor air quality and the links between housing and health are all shaping the policy and technical responses year-on-year. However, the fundamental issue we are addressing is people's homes. This should be a major consideration for all the individuals working in the retrofit field. While some issues might be technical or scientifically abstract, we can never lose sight of the critical responsibility we all have to occupants. People's homes are central to their lives and underpin their ability to engage in society fully. Health, education, work, and well-being are all inextricably linked to a warm, safe, secure home. This is what makes retrofit a compelling problem with which to engage.

Professor William Swan and Professor Richard Fitton
Energy House Labs
University of Salford
UK

About the editors

William Swan is a professor at and the director of the Energy House Laboratories at the University of Salford, Manchester, UK, an externally facing research unit that works with industry and the public sector around issues concerning energy and buildings, from building physics to the application of smart meters to digital energy services. These include Energy House 2.0 and the Salford Energy House, which test retrofits and new builds at the whole house scale. He has advised governments at both regional and national levels around issues of retrofit and energy systems, including a recent appointment to the national Energy Efficiency Taskforce, chaired by Lord Callanan. He is deputy chair of the Greater Manchester Low Carbon Buildings Challenge Group, chair of the GM Retrofit Accelerator and one of the founder members of the National Building Performance Network.

Richard Fitton is a professor in building performance and technical director of Energy House Laboratories at the University of Salford, UK. He leads several international projects on the energy performance of homes and the development of UK standards for retrofit assessment. He is also active at the International Energy Agency, where he is studying the use of smart meter data to provide energy efficiency data for dwellings. He holds a place on the SAP Scientific Integrity Group at the Building Research Establishment (BRE), which oversees the domestic energy model used in the UK. He is the chairman of the upcoming British Standard for Retrofit Assessment. He is also the technical lead for the new Energy House 2 project, a building physics test lab. He holds a PhD in Building Physics and is also a chartered building surveyor and a Fellow of the RICS.

Chapter 1

Domestic retrofit as a wicked problem

Will Swan[1]

1.1 Introduction

Both in the UK and internationally, retrofitting has been identified as a central building block of many national decarbonisation strategies. It is identified as a 'good thing' for policymakers, bringing with it, not only a pathway to decarbonisation, but also co-benefits, such as reductions in fuel poverty, creation of employment as part of a post-COVID recovery strategy [1], health and comfort benefits for occupants, and improvements to, sometimes, low-quality built assets [2]. Given its importance and identified capacity to deliver these benefits, why have many nations failed to deliver retrofits at a meaningful scale? After more than a decade of UK policy action to deliver sustainable retrofit, we have not made the inroads we promised ourselves. In 2010, policy identified we would retrofit seven million homes and create 65,000 jobs in the sustainable retrofit sector by 2015 [3], helping the UK to meet its climate goals as laid out in the Climate Change Act (2008) [4]. This ambition was reiterated, more than a decade later, in the UK's Heat and Buildings Strategy [5], which targeted thermal upgrades for homes as well as a transition to low-carbon heating. The Green Finance Institute [6] indicated the retrofit market would need to grow by a multiple of ten to meet these aspirations, suggesting the market was slow to develop.

The Household Energy Statistical release for 2022 identified a total of 3.7 million energy efficiency measures installed in 2.6 million properties between 2013 and 2022, with 78% through supplier obligation schemes, such as the Energy Company Obligation. Of these measures, 24% were loft and cavity wall insulation, often defined as 'low hanging fruit' [7], ostensibly easy to install and delivering clear savings. These data take a measures-based approach and do not directly relate to individual households, which may have received multiple measures, making the number of homes difficult to fully establish. The EU Building Data Observatory [8] identifies that 12% of the EU residential stock had been renovated by 2020, with data being sourced through Energy Performance Certificates (EPC). EPCs can have issues with reliability in terms of understanding the actual performance of

[1]Energy House Labs, University of Salford, UK

buildings, with significant measurement error identified by Crawley *et al.* [9]. Galvin and Sunikka-Blank [10] highlighted a 30% underperformance of measured against modelled performance of retrofit in Germany, using their EPC equivalent as the model. These data highlight two issues, firstly, that the measures-based approach can create difficulty in understanding how many individual homes have been retrofitted and, secondly, the reliance on modelled data can mean we lose sight of the actual performance outcomes of the delivered projects.

One of the fundamental difficulties of retrofitting is the multiple lenses through which it can be viewed. It takes different hues as it shifts through the economic perspective of a policy problem, to the social reality of an occupant, to the technical perspectives of building physics and building performance, through to an investment or purchasing decision. Often, policy and economic perspectives fail to take account of the commercial realities of developing and financing retrofit projects, the physical realities of delivering retrofits on the ground, and the concrete perspective of an occupant who must live through, and with, a retrofit. Actors as diverse as planning officers, materials manufacturers, energy companies, sustainability champions, and financial institutions may all look for their own individual outcomes or perspectives. Does this mean a major national priority becomes so complex that it becomes virtually impossible to deliver?

This brings us squarely into the realm of the ‘wicked problem’, a policy problem that eludes traditional problem-solving [11]. It is not just a hard problem that can be resolved over time following traditional approaches; rather, it is complex, far-reaching, and has an interconnectedness that creates a ‘whack-a-mole’ response, where one part of the problem may be solved, but other new issues are created, something that is familiar to academics and professionals working in the field of retrofit. Additionally, the boundaries and constraints of the system may be viewed as fluid. Viewing the delivery of retrofit as a decarbonisation problem, we see the changes in the carbon intensity of the grid and options for renewables, storage, and energy trading at the domestic level, potentially changing the ‘exam question’ from an energy efficiency problem to an energy systems problem [12]. We might consider retrofitting as a systemic change to the energy system or view energy efficiency as infrastructure [13]. The framing of the problem matters, as it shapes the response to it.

Here, we explore how sustainable retrofit has developed over the last 15 years, considering the changes in perception from the perspective of multiple actors, what the problem might look like now, and how some of the issues are being addressed. We also consider how theoretical perspectives of wicked problems and socio-technical transitions might provide insight into how we might address these issues.

When we think of some of the early conceptions of retrofit, they are driven very much as a technical response to a policy issue. Over the last decade, as mistakes have been made and lessons have, hopefully, been learned, our concept of retrofit has become more complex. To tell this story effectively, it is useful to look at the interplay of policy, technology, actors, and delivery. These complex relationships mean we are clearly located within a socio-technical space in terms of retrofit as a theoretical problem [14–16].

1.2 Defining retrofit

Prior to engaging in a detailed debate as to the scope of the problem, it is useful to consider what we mean when we talk about retrofit, or sustainable retrofit. Placing boundaries around a retrofit can be complex. The debates around defining a 'deep' retrofit, where a higher percentage of energy savings, often 60–80%, is from the baseline performance of the property prior to retrofit [17], or the interrelationship of the retrofit with the reduction of carbon emissions, based on the interaction of the home with the energy system, create different perspectives, as highlighted by Saffari and Beagon [18]. Many of these views are driven by the policy imperatives and the associated instruments that have largely driven the retrofit market in the UK and EU. In the Energy White Paper, Powering Our Lives [19], three issues were highlighted: carbon dioxide emissions reduction, fuel poverty, and energy security. Of these, the retrofit agenda was largely driven by fuel poverty [20] and carbon emissions reduction ambitions, which were embedded in law and regulation, such as the Climate Change Act in the UK, or EU directives, such as the Energy Performance of Buildings Directive. While the two issues are complementary – energy efficiency will theoretically lead to lower bills and reduced emissions – they do drive two slightly different views of retrofit: one emissions-centred and one people-centred. The third issue of energy security and resilience creates an additional systems-focused view of retrofit. This is one that cannot be ignored as we move significant heating and transport loads from fossil fuels to the electricity grid, creating issues of electrical grid capacity [21], while also creating opportunities for an increase of renewables and flexible storage to better balance this demand [22].

Galvin and Sunnika-Blank [2] identify the debates that highlight the complexity of these issues. The conflicting perspectives around 'deep' retrofit against 'light' retrofit and additional considerations, such as indoor air quality, occupant comfort and utility, the protection of heritage, and aesthetic concerns. These have developed over the last decade and, to some extent, remain. The complexity of defining the desired outcomes for, and technical responses to, retrofits contributes to its 'wickedness' as a problem.

1.2.1 Developing views of retrofit

Many early definitions of retrofit give a largely technical definition of the problem, which may be related to measures or in terms of performance outcomes. Swan *et al.* [23] define retrofit as 'sustainable retrofit', considering it as improvements to a property's fabric, systems, or controls that would improve its energy efficiency performance. This tends to have a technologically based view built on early refurbishment literature [24–28] that, while recognising retrofit as an issue that has socio-technical components, considers the physical delivery of a low-carbon housing stock and the potential technical solutions to get there. There were large-scale research projects such as CALEBRE [29] and Retrofit for the Future [30] in the UK and significant investment in retrofit research through the EU Framework and Horizons Programme, with a strong focus on low carbon and energy efficiency [31]. This research often focused on the technical issues, with strong modelling and technical demonstration components, but issues of the consumer were also being considered [32,33], particularly in the context of fuel poverty. Additionally, retrofitting was also considered in terms of issues of delivery, supply chains, and operation and maintenance of retrofits in the long term.

There is a next phase of learning programmes such as the European Regional Development Fund [34] and programmes supported by central governments, such as the UK's Warm Front Scheme [35], with similar responses in Germany, and France [36]. These focused on the deployment of retrofits at scale and developing learning around delivery expertise, often with a focus on municipalities or social housing. In addition, many countries use supplier obligation programmes, where energy suppliers are required to fund retrofit measures against carbon emissions reduction or energy savings [37]. In the UK, there was the Carbon Emissions Reduction Target, focused on emissions reduction, and the Communities Energy Savings Programme, focused on place-based retrofit. The supplier obligation model has existed since the 1970s but became increasingly focused on housing during the last 20 years. This gave more opportunity to deliver retrofits at scale. This larger-scale practical delivery phase moved the debate into more concrete discussions of delivery, quality, and consumer experience. Issues around the performance of external wall insulation, driven by quality issues [38], or the acceptance of new technologies by consumers [33], as well as the performance of these technologies [23,39], were driven by problems of, not only consumer experience, but also specification and installation, creating a more sceptical view of what retrofit could achieve.

Energy efficiency solutions, particularly around fabric design and installation, often failed due to a poor understanding of wider building performance issues [6]. The performance of the retrofitted building may have been good 'on paper', with the outcomes often established using Energy Performance Certificates as a proxy for performance, particularly in grant funding contexts. However, design and installation issues often meant that, not only did retrofits not perform as intended, but pathologies were also introduced to the building, creating environmental issues such as air quality, condensation, or mould [40] that could be hazardous to the occupants and the building. This gave rise to delivery standards in the UK such as PAS2035, designed to manage this risk [41].

While these problems were not universal, they were sufficiently common to position retrofit as a risk. This also gave rise to early discussions around the 'performance gap' [42,43] in retrofit. This was identified as being driven by a number of different factors, including poor design, installation or commissioning; comfort taking by occupants, known as the rebound effect [44]; or a gap between the model and what was ultimately constructed. The relationship between modelling and measurement became an important area of work. If we established a national programme of retrofit, how could we be sure it would deliver outcomes in terms of energy saving or carbon emissions if we did not measure? Reliance on the EPC to drive building performance at the individual and policy levels was subject to criticism [45,46].

Technologically, we did not see major shifts in terms of the components of domestic retrofit solutions. Swan *et al.* [39] showed an adoption trajectory for different technologies from 2010 to 2015 in social housing. The changes in adoption of different technologies, while becoming more widespread, did not bear witness to major changes in technology types. There were issues around some types of new technologies, such as micro-wind, air source heat pumps, and ventilation systems, where complexities of installation and use, led to slower adoption. However, the technologies that were being adopted did not change. Retrofit was driven by a fabric-first, whole-house approach [30], which looked to minimise energy consumption, largely through fabric energy efficiency, followed by low-carbon technology. Renewable energy generation, mainly photovoltaics, funded through subsidy programmes such as the Feed-in Tariff in Germany and the UK, made some moves towards zero-carbon energy production at the household level. The model for both of these phases is, generally, a passive consumer of energy paying for energy consumed. The home is viewed as a demand load on the end of the gas and electricity networks. However, in the background, technological and systemic changes were being undertaken to potentially change the view of what a retrofit might be in terms of both technical solution and how it might change the home's role in the energy system.

In the UK [47], but also internationally [48], there were technological shifts that changed the way that retrofit might be delivered. The smart meter infrastructure, adopted in a number of countries, created an opportunity to bring the home within the energy system rather than purely a demand load. Initially, much of the work around smart meters focused on the visibility of energy data for consumers [49], which could be regarded as important in terms of creating a transparency that was not there before to better support decision-making around energy consumption using the in-home display. However, the purpose of the infrastructure was to allow better use of data and control and better manage grid capacity [50]. Taking an energy systems approach could identify the energy efficiency interventions as part of the energy systems infrastructure, where investments in energy efficiency have the benefit of making the wider electrical energy system more resilient, requiring lower investment in energy transmission infrastructure due to the reduced electricity demand at peak times. This allows us to see retrofitting as an investment whose co-benefits are seen more widely than just energy savings [51]. Renewables, often criticised for their intermittent nature, were paired with storage, such as batteries or V2X from electric vehicles, and an emerging Home Energy Management System (HEMS) market allowed consumers to use, generate, store, and export energy, making them complex energy actors [52]. This created thinking around issues such as microgrids, peer-to-peer trading, and network services that might be considered as part of a retrofit solution. Controls moved from simple thermostatic models to smart self-learning devices, with new entrants to the market looking to create a smart home infrastructure. Challenger energy companies offered new models of consumption with time-of-use tariffs, linked to the adoption of electric vehicles. These new models created the opportunity to deliver new modes of retrofit, such as Energiesprong [53], which linked energy performance guarantees with systems, controls, and monitoring to create a heat-as-a-service model for retrofit.

These changes have seen recent shifts away from the fabric-first model that previously dominated thinking to a more systems-based model that may not require deep retrofit, as identified by Eyre *et al.* [54]. Creating the capacity for load shifting at the home and system level creates opportunities for decarbonisation that were previously unavailable. The discussion moves from deep fabric retrofit as a gold standard to a wider view of energy system decarbonisation. When combined with the decarbonisation of the grid, the solutions for retrofit may become more complex and varied. This can be seen in the UK with the emerging concept of 'heat pump ready', a designation that considers heat pump performance in the context of low-level retrofit, designed to ensure effective performance but reduces the costs and risks around deep retrofit. Lamb and Elmes [55] identified that approximately 35% of UK homes were technically heat pump ready, defined as those with an Energy Performance Certificate rating of C or above. This definition should be considered in the light of previous discussions concerning the actual performance of homes as opposed to their EPC performance. It also does not necessarily track that these homes are as poorly performing as the EPC might suggest, as studies such as Rye [56] have identified that some assumptions in the model, in this case around traditional buildings, can also underestimate energy performance when compared to measurement. However, both under- and overestimations of performance have implications for correctly specifying heating systems.

This final phase also placed another policy imperative on retrofit. Concepts such as Clean Growth in the UK or Build Back Better in the US identified the nexus between *decarbonisation and post-COVID recovery*. Retrofit is seen as an opportunity to drive economic growth and create new 'green' businesses. This, again, creates a need for a more systemic focus that incorporates skills, supply chains, finance, and market-driven solutions. From this perspective, we could equally see the health benefits of retrofitting as having a wider impact on the health system [57] through reduced impact of cold and improved indoor air quality. The 'boiler on prescription' projects undertaken in the Northeast and Northwest of England, where GPs prescribed heating systems on a public health basis, also identified clear health benefits of improved housing quality, particularly for those with chronic conditions [58].

These three phases should not be viewed as clearly demarcated, but there is a clear shift in the shared view of what retrofit might achieve and how it might be delivered. The changing nature of how we define retrofit is further complicated by the number of 'moving parts' that are required to address the development of a retrofit market that is less reliant on grant funding, which remains a significant market driver for retrofit in the UK.

1.3 Retrofit as a wicked problem

Up to this point, we have perhaps fallen into the trap of stating the obvious. That retrofit is complicated and difficult to deliver at scale. If we are to solve the problem, we need an appropriate theoretical lens to frame the problem. The various approaches of socio-technicality to understand a problem that reaches down from policy into people's homes may provide some insight, but do they necessarily deliver a retrofit? The concept of wicked problems was a theoretical response to complex policy problems. Wicked problems, as defined by Rittel and Webber [11], identify ten qualities of wicked problems that define them as such. If we consider the delivery of a single retrofit or the delivery of a multi-home retrofit project, these are plainly solvable issues at the project management and technical delivery level. However, the problem at hand is the large-scale delivery of a national domestic retrofit market that will move the housing stock to a very low- or zero-carbon position across all tenures, while also addressing fuel poverty and energy security.

The ten rules of Rittel and Weber applied to the problems of retrofitting we have discussed so far.

1. *They are difficult to define.* As highlighted in our definition of retrofit, issues of scale, lens, definition, and desired outcome appear to shift over time and between stakeholders.
2. *Wicked problems have no stopping rule.* A stopping rule might be defined as when the problem is solved. We could perhaps argue this definition – when all homes meet all outcomes at an optimal level for occupants, it is likely that we could say the problem is solved. The reality is retrofit is likely to be an ongoing shifting and repeating process, as retrofit installations, such as heating systems and insulation, pass through their warranty periods, meaning the need for replacement and upgrade.
3. *Solutions to wicked problems are not true or false but good or bad.* This is clearly outlined in the discussion above. The solutions are changing and contingent; the challenge is to do no harm, something that is not always achieved. This is clearly reflected in the shift of retrofit perspectives over time that have previously been discussed. Technological strategies and a better understanding of future needs to address issues such as overheating, for example, will mean the space will remain contested.
4. *There is no immediate or ultimate test for solutions.* As stated previously, a retrofit project or programme is not a wicked problem; construction projects are solvable engineering problems. These may be, and often are, assessed and evaluated [17]. However, at the national scale, understanding how to deliver is more complex and, as we have identified, potentially difficult to evaluate at the scale that we need.
5. *All attempts at solutions have effects that may not be reversible or forgettable.* Retrofitting of homes clearly falls into this category; individuals who are left with the technology they do not understand [33] or that does not perform [59] are plainly impacted in terms of utility and, sometimes, health.
6. *The problems have no clear solution and perhaps not even a possible set of solutions.* Again, at scale, this issue is clear. The failure of policy vehicles, such as the Green Deal [60], and, most recently, the slow uptake of the UK Green Homes Grant, shows the difficulty in policy design [61]. Policy design in this space generates issues and unintended consequences, showing that the solutions can be elusive for policymakers.

7. *Every wicked problem is essentially unique.* Retrofit is unique in that its components and perspectives are specific to retrofit as a problem. There are few problems that are as multi-layered, embracing a wide range of socio-technical problems, ranging from design to the impact on the occupant. Issues such as health, climate justice, and fuel poverty are balanced against issues such as supply chains, hygrothermal risk, and data systems in a complex melange that is difficult to view from a single perspective.
8. *Every wicked problem is a symptom of another problem.* This is a more complex question to answer. The national delivery of retrofits was identified as part of a programme to address net zero, as raised in the Climate Change Act (2008), as well as addressing fuel poverty. These are the underlying reasons for undertaking a retrofit, but they do not reflect underlying symptoms. The difficulty of delivering retrofit might be attributed to other underlying problems, such as ageing housing stock, an ongoing issue of skills in the construction industry, and a lack of financial or regulatory models that encourage retrofit, but whether the issues of delivering retrofit identify a wider underlying problem that exists outside the system, as identified above, is difficult to state.
9. *There are multiple explanations for the wicked problem.* This is a clear problem for retrofit. The underlying issues that make its delivery difficult, ranging from the contested technical approaches to its social dimensions and its place in a complex institutional framework, do mean that the failure of retrofit has multiple explanations. However, does this merely denote a multi-faceted problem rather than one that might be defined as 'wicked'?
10. *The planner has no right to be wrong.* Individuals whose homes are retrofitted may perceive that they are in a worse place than they were before. We are dealing with people's homes, which places a significant responsibility on those who make the decisions around retrofits. A planning decision must be lived with at the personal level for many individuals. As we see with the Fishwick scheme in the UK, or other failed projects, this can have health, personal and financial impacts on households if we get it wrong [62].

It might be argued that there is a degree of overlap between these ten rules. Koppenjan and Klijin [63] summarise this in three themes, complexity, uncertainty and divergence. Complexity is viewed as the number of variables that need to be considered; uncertainty is reflected in the risks that may be identified, while divergence is the existence of multiple perspectives. It might be viewed that this does address the main elements of a 'wicked problem'; it might be considered that uncertainty does not adequately reflect the consequences of failure. However, it does provide a useful distillation of the ten points (Figure 1.1).

These ten points might be viewed as a pushback against the rational policy-making of the early 1970s. The debates around retrofitting over the last decade do show that policymaking does recognise this complexity, even if the policy instruments have not always been able to overcome it. Rational technical and engineering approaches to problems often fail [64].

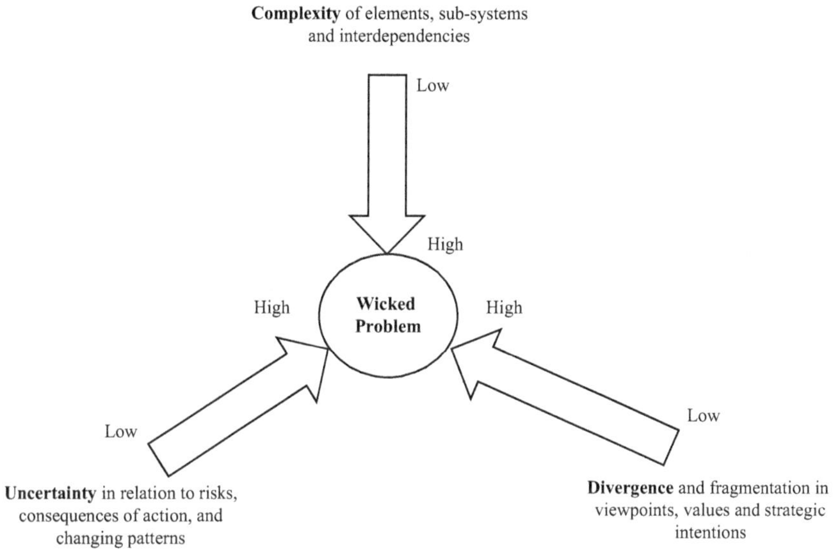

Figure 1.1 Complexity, uncertainty and divergence of wicked problems. Adapted from Koppenjan and Klijin [63].

> 'This preference is reinforced by administrative and budgeting processes which are predisposed towards tangible outputs and the measurement of incremental changes. Political and financial accountability systems encourage the specification of discrete and finite items rather than large amorphous interlinked outputs.'

This criticism could be attributed to many of the policy instruments that have been used to drive retrofit over the past few years. The supplier obligation, which focused on carbon savings, is specifically output-driven, generally relying on the EPC as the model that determines the level of those carbon savings. The distribution of low-energy light bulbs did not guarantee their savings would be realised, just as the presence of solid wall insulation does not guarantee its performance. Wicked Problem Theory takes the view that policymakers are locked in a particular 'rational' policymaking and planning model. However, the period in which Rittel and Weber [11] outlined the problem in the early 70s might not fully be a fair reflection of how the debates among planners are currently shaped.

Head identifies the alignment of wicked problems with a socio-technical perspective.

> 'Wickedness is not just about a clash of ideas and values; it is also implicated in structures, processes and institutional arrangements including power, authority, and procedural rules.' [64]

Power is the ability to shape the debate, either by determining what issues are debated or addressed in the policy space. This can be seen through the increasing political discussion around net zero. It is also the ability to change or hold elements of the socio-technical system through rules or institutional structures. However, Head's summation of wicked problems bears a close similarity to socio-technical perspectives. Positioning this view in the context of retrofit, Galvin and Sunnika-Blank [2] highlight the complexity that is reflected in the two theoretical positions.

> '...it must also be emphasized that thermal retrofitting is very diverse throughout the world. It is influenced by the material features of local building construction; cultural norms regarding thermal comfort and household practices; local climate conditions; levels and distribution of economic wealth; regulatory frameworks and other factors.' [2]

However, if you see the problem as wicked, can you knit together components of a potential solution that relate to one another in a way that recognises that complexity?

1.4 Socio-technical framing of the retrofit domain

The wicked problem approach helps us identify that retrofit is a challenging policy space. However, this may not be sufficient to help us fully reflect on the problem. The socio-technical perspective looks at the interplay between artefacts, rules, and actors [65]. Artefacts can be viewed as the physical and technical elements of the system. Rules can range from explicit policy, regulations, and institutional frameworks to 'rules of thumb', or heuristics applied by site practitioners, each of which can have an impact on the delivery of retrofit. The actors are the individuals, market actors, and institutions that make the decisions about the adoption, delivery, and financing of retrofits. Ince and Marvin [16] conceive of hard and soft elements, with technological elements being denoted as hard, while rules, values and expectations are identified as 'soft' in four major categories: technology, production, user practices, and government regimes. Figure 1.2 identifies a possible perspective of how these various elements that make up the retrofit regime might be framed [15].

These models are particularly concerned with 'transitions': how do we move from one state to another? However, understanding where we are transitioning from relies on reliable data, which, as we have seen, presents a challenge. The exact nature and mode of the transition are often shifting or contested. As we see from the previous section, retrofit as a concept embraces people and technology, policy and economics, engineering and the social sciences. This systemic view of retrofit requires a theoretical framework that embraces the physical and technical, the various influential stakeholders and the policies and practices that organise them. Geels [65] develops a framework for understanding socio-technical systems issues that, while criticised as being vague [66], has created a useful body of work to understand energy systems issues [67–69]. The multi-level perspective defines three layers, the landscape, the socio-technical system and the innovation layer, where new ideas are developed that may become part of, or disrupt, the socio-technical system.

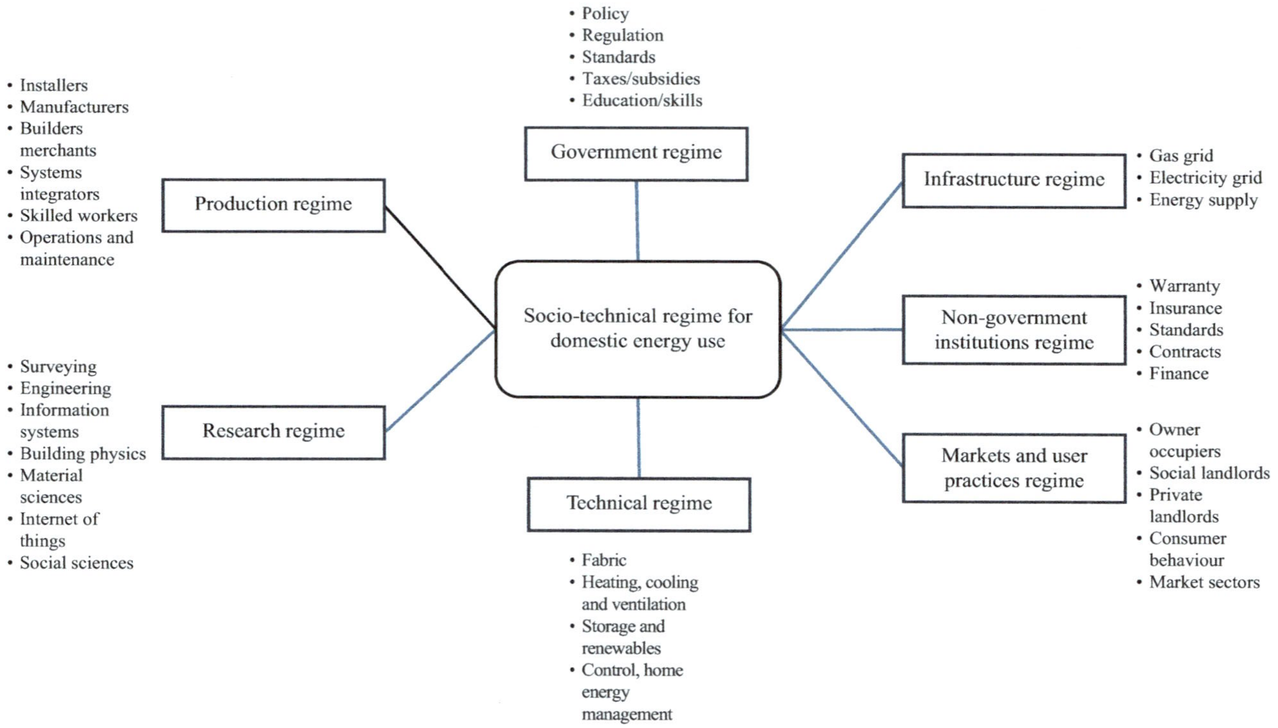

Figure 1.2 Socio-technical regime for domestic energy use (Adapted from Rittel and Webber [11])

A useful way of considering the landscape is to determine what a society or group of people might determine as 'important' external factors that need to be addressed by a specific system. Within Geels' model, he considers these as something similar to the economic principle of exogenous shocks, such as an oil crisis. In terms of retrofits, these may include issues such as climate change, energy security, or fuel poverty, as have been outlined by the Energy White Paper, Powering Our Lives or the Net Zero Strategy. A society, or government, may take a more economic developmental view, where the issues of Clean Growth [70] and the recently withdrawn Building Back Better [71] are brought to the fore. Recent pushback against previous net-zero policies in the UK and the US clearly identifies how political power can mediate the landscape. While there may be wider and more varied societal views around climate change or economic growth, the mediation of this appears to be undertaken chiefly through national governments, intergovernmental organisations, and the bodies that shape and lobby for specific policy outcomes, as may be seen by the transitions between the Trump and Biden US administrations. Mediating how the landscape is viewed and responded to by society as a whole is a complex interplay between government, media, and other interest groups. Shaping of the perception of the landscape and policy response to it is an exercise in power, something Geels has previously been criticised for ignoring and subsequently responded to [72,73]. What is clear is that the views of how societies respond to external pressures are strongly mediated through a policy framework and the policy tools that are used to deliver them. The view of retrofit may have shifted from a technical solution to solve large-scale issues to an opportunity space, where leadership in Clean Growth technology and large-scale deployment of retrofit are designed to bring economic and social benefits.

The second layer is the socio-technical system. Rittel and Webber [11] frames this around the sustainable retrofit of the social housing sector, as shown in figure 1.2. This model proposes a series of inter-related domains that map on to specific areas of the socio-technical domain. These systems are made up of three main elements: actors, such as businesses, trade associations, policymakers, and consumers; artefacts, which may range from physical technologies such as batteries or insulation, and rules, such as policy instruments, standards, or market structures. They reflect the hard and soft elements outlined by Ince and Marvin [16]. Table 1.1 provides an outline of the regimes and examples of their relevant actors, rules and artefacts, taking the UK as an example.

When we consider the wholesale retrofit of the domestic energy system, these are the actors, artefacts, and rules that must be connected, mobilised, or adapted. When compared with the early views of the role of retrofit, changes, such as the smart meter rollout and the rise of storage technology, have the potential to shape how we think about retrofit. This must be linked to the connected adaptations that other areas of the regime might make, such as new business models or standards. This is the third layer of Geels' model, one where innovation attempts to enter the socio-technical regime. Innovations can fail, take their place within the existing structure, or disrupt the structure and change the shape of the regime.

The multi-level perspective is not the only way to frame the retrofit problem. Approaches such as assemblage theory have also been applied to explore retrofit at

Table 1.1 Social-technical domain for domestic energy consumption in the UK

Regime	Function	Actors	Rules and artefacts
Government	Development of policy and appropriate instruments to deliver desired policy outcomes.	Business, Innovation and Industrial Strategy, Ministry for Housing Communities, and Local Government, Combined and Local Authorities, Planning, Building Control, Regulators (i.e. Ofgem).	Policy, legislation, tax or subsidy models, standards, regulation, enforcement models.
Infrastructure	Development, operation and maintenance of the gas and electricity networks, supply of energy to homes.	Energy Companies, National Grid, Cadent, Distributed Network and Service Operators (DNO, DSO), energy companies, Data Communications Company (DCC).	Gas network, electricity networks, smart meter infrastructure and customer tools, regulations, policies, consumer advice.
Non-governmental institutions	Development of finance, warranty and insurance services, professional standards, advice and business models.	Insurance companies, mortgage companies, valuation bodies, professional bodies, finance companies, energy advisory services, consumer protection bodies.	Insurance, mortgages, warranties, legal services, standards documents, finance, consumer advice and protection services.
Market users and practices	Market decision-makers at the individual or large scale that may buy products and services or make decisions that impact energy use.	Owner-occupiers, landlords (private and social), residents, housebuilders, energy companies.	Various markets for products and services.
Knowledge regime	Development of knowledge with regard to the operation of the regime.	Universities, Research Technology Organisations (RTOs), consultants, manufacturers, training providers, professional and advisory bodies.	Outcomes of research such as theories, evidence, products and services, test facilities, policy advice.
Production regime	The supply chains responsible for the delivery, operation and maintenance of the domestic energy regime.	Contractors, installers, manufacturers (product and materials), resellers, maintenance workers.	Supply chains, processes for design and delivery, management systems, safety systems.
Technical regime	Physical products and technology that shape the physical aspects of the domestic energy system.	Manufacturers of products and materials.	Fabric, systems (heating, cooling and hot water), renewables and energy storage, appliances, energy management systems, such as Internet of Things devices.

scale [16]. Geels provides a framing device that allows us to see the problem more holistically. The model has been criticised for weakness in terms of shaping and predicting system change and, given the practical need for rapid retrofitting, is highly theoretical. However, this accusation could be levelled at much of the academic thinking around retrofit. While socio-technical systems help us see the boundaries and components of the problem domain, it is possible that the framing of retrofit as a wicked problem gives us a different perspective that shows us the wider limitations of policy responses, even if we are able to conceive of the system in which it operates.

1.5 Policy responses to retrofit

While we have touched on some of the policy initiatives, it may be useful to abstract them in terms of different categories of action and how they relate to the wider socio-technical system. Some are relatively simple, taking the form of grants or direct funding for retrofit projects, while others are more complex, attempting to reflect the complexities of delivering retrofit at scale. The following section is not designed to be an exhaustive consideration of policy but allows us to consider various responses to deliver retrofits in the UK.

Primarily, one of the main difficulties in developing the retrofit sector is consistency [13]. Reshaping an industry to deliver retrofit at scale requires skills, investment in supply chain capability, and, potentially, new products and services. Policy surrounding retrofit in the UK has been, generally, stop-start. We can see cycles of investment across a multi-year cycle and then a gap. This has not been viewed as conducive to the type of shift that is required for retrofit to scale up. The National Audit Office [74] identified that the lack of continuity in policy hampered the development of supply chains, with a potential commensurate increase in costs.

1.5.1 Grants and subsidies

One of the simplest approaches for governments is to engage with market-based solutions. This means providing a grant, subsidy, or low-cost finance to reduce the cost of retrofit technologies. This makes a retrofit cheaper and acts on the market segment of the regime to increase adoption. In the UK, the Boiler Upgrade Scheme does this for air source heat pump technology, providing a direct grant to consumers to reduce the cost of purchase and installation. Some of these programmes differ slightly in their delivery approaches and are outlined below. In addition, the availability of grants also acts on the supply chains as they mobilise to respond to increased market demand.

The *supplier obligation* has been applied across Europe to address energy efficiency by placing an obligation on energy suppliers and distributors to deliver specific amounts of energy savings or reductions in emissions. These types of policies were largely a response to the oil crises of the 70s [37]. The components of the programmes are underpinned by methods to manage the process: targets, technical processes to assess the validity of claims, management and enforcement. In some cases, there may be trading schemes. The UK currently has the ECO4 programme, building on activities of the Carbon Emissions Reduction Target (CERT), which focused on emissions, and the Community Energy Savings

Programme (CESP), focusing on place-based retrofit, both of which operated through the early part of the 2000s [75]. The National Audit Office was generally positive about the achievements of the CERT and CESP programmes in its 2016 review of energy efficiency policy outcomes. The follow-up ECO programmes (Carbon Saving Communities Obligation – CSCO and Carbon Emissions Reduction Obligation – CERO) were more tightly focused on hard-to-treat properties and, therefore, achieved lower carbon savings for similar expenditure.

The *Green Deal* was a complex financial model that was designed to encourage the take-up of retrofits by funding improvements from the energy savings, building on the 'pay as you save' model [60]. The model assessed the property using an extension of the Energy Performance Certificate to generate cost savings under a 'golden rule' where energy savings would underpin the payment model and identify improvements that could be undertaken on a home that would be paid back by these savings. The model was relatively complex and underperformed for several reasons. The cost of the loans, which, during a period of historically low interest rates, were considerably higher than market rates, limiting take-up. Additionally, there was a reliance on the EPC as both an energy modelling and specification tool, neither of which it was designed for. The other main issue was a lack of engagement with customers. People did not have a good understanding of the need for retrofitting and its potential benefits. The policy was withdrawn after only 626 homes were improved under the scheme in its first year. The issues of a failed policy of this nature affect the delivery of retrofits more widely. Individuals and companies who train or develop businesses on the back of the policy lose out on this investment, while a failed policy of this type can also potentially have an impact on public confidence.

The *Feed-in-Tariff (FIT)* for solar panels has been a largely successful policy instrument. It provided three financial incentives: i) A generation tariff, which provided funding for all renewable energy generated, ii) An export tariff, which provided funding for each KW/h exported to the grid, and iii) Reduction in grid electricity used. The scheme was successful, with 1.06GW of solar capacity installed between April 2010 and March 2012, with 88% of these domestic installations. The success of the FIT led to large reductions in solar panel costs, as well as increases in the efficiency of their performance, both in the UK and internationally [76]. This led to a reduction in the level of the FIT before finally tapering off entirely in the UK. The FIT also gave rise to a business model known as 'rent-a-roof', where the capital cost and installation of the array would be undertaken by the third party, who would then claim the FIT, while the occupant took the benefit of the renewable energy. This was common in social housing, as well as residential homes [77]. However, this created legal issues in some cases where the property might be sold and an existing right to the roof needed to be maintained with the original provider of the panels.

Place-based policies focus not on individual homes but communities, as with some of the specifically targeted ECO funding. These policies are often tenure-blind in that they do not focus on social housing, private rentals, or owner-occupiers but deliver retrofits across communities [78]. However, there is still often a focus on social housing-dominated communities, such as work in Germany, Spain, and Portugal [79]. This does have a limitation in the UK, where a social housing tenant's right to buy their property has created a more fragmented distribution of social and private properties within specific areas, giving rise to complexities in project management of retrofit projects. There has been some progress in the UK with more place-based approaches in Wales and Scotland under devolved powers [80]. Recognising that retrofit is location-specific in terms of solutions, skills, and capabilities is an important part of taking a localised approach to retrofit. In terms of delivering retrofit from a technical perspective, the domestic stock is highly variable, something that increases with older stock. In addition, local skills infrastructure, supply chains, and governance structures are all different, creating different capacities to respond to opportunities [81]. Macroire *et al.* [78] highlight the benefits of linking place-based retrofit demand to the supply side in terms of jobs, skills, and economic development.

1.5.2 Regulation and standards

Policy makers can also act on the market by prohibiting products or setting standards, as well as linking in with non-governmental institutions to apply industry standards. While the application of a grant might be viewed as a 'carrot', this approach is very much a 'stick'. This approach can drive adoption of technology or impose quality standards on the delivery of grant-funded retrofits.

Mandates around the phasing out of specific technologies reliant on fossil fuels are one of the major approaches that have been identified. These can be politically difficult and rely on new supply chains and infrastructure. The two key examples in the UK are the phasing out of gas boilers and fossil fuel cars. However, while both of these transitions have been discussed, they remain at the consultation stage, with no formal decision as to exact dates and rules formally announced.

In the UK, rental properties will shortly fall under the Mini*mum Energy Efficiency Standards* (MEES). This means a property with an EPC of F or less is not rentable in England and Wales [82], applicable to less than 5% of all houses in England in 2023 [83]. While the social housing sector generally has higher EPC ratings, the private rented sector remains one of the most poorly performing in terms of energy efficiency [83]. There are potentially two issues with the application of MEES. The first is the reliance on the EPC, which, as stated, can make assumptions about properties that may not reflect their actual performance. Secondly, while large institutional landlords will be likely to respond positively to the standard, it may be difficult to enforce for smaller and 'accidental' landlords. However, there does appear to be a growing move to stronger enforcement, often underpinned by landlord licensing schemes to fund this activity [84].

Another way for policymakers to create improvement is to link grant-funded work to standards. In the UK, all delivery of grant-funded programmes is linked to a number of non-governmental institutional standards. *PAS2035* is a quality standard that was designed to reduce the risk of unintended consequences, such as mould or damp, in retrofit. This creates specific roles and pathways that take responsibility for the retrofit process [59]. *TrustMark* is a government-endorsed accreditation body that provides a quality framework for businesses in the domestic construction sector. *TrustMark* is also a requirement for undertaking ECO. The

Microgeneration Certification Scheme (MCS) operates in a similar way for renewable and low-carbon technology, such as air source heat pumps. MCS-accredited installs of heat pumps account for approximately 60% of all heat pump installations in 2025 [85], largely driven by the Boiler Upgrade Scheme, for which the MCS is mandatory.

These approaches for policymakers, standards or prohibition and conditionality of grants can be powerful tools. They are often a pre-requisite for pushing consumers towards a market-based solution while maintaining quality. However, they do require a change in both how the industry delivers, requiring a shift in skills and supply chains, and the supporting physical and regulatory infrastructure to support this transition.

1.5.3 Skills

Skills for retrofitting are a complex area. It might be viewed that there are no specific retrofit skills as such; rather, they are formed from construction supply chains that deliver new builds, conversions, extensions, and other home improvement activities. These may be interior improvements, such as kitchens and bathrooms, the installation of heating and ventilation systems, or fabric upgrades such as new windows or insulation. There is a clear crossover between new build and retrofit markets; installing a window or heat pump in a new build will require similar skill sets to a retrofit.

Skills can range from labourer to higher-level skills such as designers or surveyors. While higher-level skills, particularly the professions, which may be underpinned by university degrees and institutional memberships, trade skills are often shaped by place-based ecosystems [78], with skills providers and further education institutes responding to local needs of businesses. However, Ravenscroft [86] indicates that there is an unevenness to place-based responses to the retrofit skills issue. This does indicate that place-based approaches to skills must reflect the capacity, appropriate structures, and experience within the local skills infrastructure to generate a meaningful response.

1.5.4 Consumer advice and consumer product

One of the key frictions for retrofit is that it is viewed as poorly understood by the consumer, particularly in the private sector. There is an argument that we might consider retrofitting closely linked with the renovation or refurbishment market, rather than a specific activity separate from these types of home improvements that are carried out regularly and this needs to reflect that homeowners and private landlords, particularly, are not a single homogenous market [23]. Retrofit has been positioned as a technical solution, rather than a consumer product with multiple benefits for the homeowner, and this means that choices such as kitchens and bathrooms will supersede choices around insulation or low-carbon technology.

However, we can see that approaches that frame retrofit in terms of the benefits, such as Octopus Energy's Zero Bills, can have success in the marketplace, although this is an emerging offer in the retrofit market [87], so the details of qualifying homes, links to grants, and costs to the consumer have not yet been fully researched. In terms of renovations, the use of trigger events, where maintenance, either planned or unplanned, and improvements are linked to energy efficiency measures, has the benefit of addressing the disruption barrier that is often highlighted as a major barrier to retrofits as well as reducing costs [88].

Consumer advice and support in defining and procuring works, through models such as one-stop shops, does help remove friction. They have been adopted across Europe [88] and are designed to address barriers to consumers adopting retrofits, such as lack of trust, lack of knowledge, and, in some cases, access to funding. One-stop shops can provide technical expertise to help shape a retrofit design, provide access to trusted supply chains and ongoing support in managing the contract, and offer other supporting services such as financing. They are often supported through the public sector, such as municipalities, and many operate on a not-for-profit basis and are grant-funded, which improves the perceived level of trust but can impact their longevity.

1.5.5 Summary

Many of these policies are focused on the delivery of retrofits. There is often a lack of policy to support the second-order issues that could allow for an increase in delivery. Issues such as skills, for example, are not well linked to the delivery infrastructure. In addition, policy in the UK has been centrally driven to resolve a problem that could be viewed at many levels as place-based. Retrofitting in London does not have the same issues as delivering retrofitting in Greater Manchester, so a 'one-size-fits-all' approach may not be viewed as reflecting the complexities on the ground. This is also compounded by the fact that many of these policies cut across traditional departmental boundaries. The Green Deal shows how complex policy-making can create issues, but potentially what is required is interlinked policy-making that is framed within a wider strategy recognising the systemic nature of the retrofit 'problem'.

1.6 Addressing the retrofit problem

Socio-technical systems identify the systemic nature of the problem of retrofit. It unpacks the components and draws relationships between them, but even this framing does not give us the whole picture. Theories of wicked problems, to some extent, warn us about the hubris of trying to solve national and international policy problems with simple solutions. While academics identify the need for experimentation within living lab environments, we must recognise the limitations of the approach and the wicked problem rule of not being able to afford to 'get it wrong'. Invariably, there will always be a top-down element to this approach [88]. However, there are some elements we can consider that might help steer the UK in a clearer direction.

1.6.1 A UK national strategy for retrofit

Looking specifically at the UK, retrofit has not been guided by a single strategic view or long-term plan. The past 15 years of retrofitting have revealed two potential issues. The first is a lack of a coherent national strategy and supporting infrastructure for retrofit. The second is the lack of a consistent policy, which has meant that flows of grants, prioritisation, and a joined-up approach have created a stop-start sector, making investment decisions difficult. In the EU, there is a requirement for member states to have a National Renovation Strategy under Article 4 of the Energy Performance of Buildings Directive [89].

Much of the policy was designed either to address market failure or stimulate the market. However, the lack of consistency has meant that the desired market development in the UK has not occurred as predicted. The development of a national retrofit strategy may not resolve the multiple issues, but it will provide a space where issues can be mediated and better co-ordinated. This is not to state that a centralised approach will solve issues; indeed, previous centralised policies have been unsuccessful in many cases. What it will do is resolve the policy and strategic uncertainty that has plagued retrofit.

The question of centralised policymaking raises the question of what scale we need to intervene. There are national issues for retrofitting that should be resolved by central government. Issues such as standards for delivery, funding models, measurement tools and standards, valuation standards and taxation all have the power to be levers that may change the economic drivers for the retrofit market.

This would also require the input from professional institutes, standards groups, the financial sector, and warranty providers. However, not all retrofit problems can be resolved at the national level. The impact of policy instruments, such as the supplier obligation, which has driven the CESP, CERT and ECO vehicles, has had some success. Changes to standards for rented accommodation through the Minimum Energy Efficiency Standards also show the influence that can be exerted at the national level. Large-scale programmes such as the Social Housing Decarbonisation Fund can have an impact, although these are often delivered in a local context.

1.6.2 The importance of place

When we consider the UK in a regional context, retrofit as a problem will manifest in different ways in different places. A city such as London will have a different building stock and economic drivers when compared with a Victorian city, such as Manchester, or a rural place such as Cumbria, with a significant heritage retrofit issue. Ultimately, delivery should be focused on place. Place determines individual technical solutions, supply chains, and skills, the elements of which will vary from place to place.

Looking specifically at Greater Manchester, Figure 1.3 outlines the guiding principles and related components for the successful delivery of retrofit as defined by the Greater Manchester Retrofit Accelerator, which evolved into the GM Retrofit Action Plan developed in 2022 [90]. This model, which identified a city-focused approach, recognises some of the complexity we have discussed. The Retrofit Action Plan identifies co-benefits, reflecting the wider Build Back Better

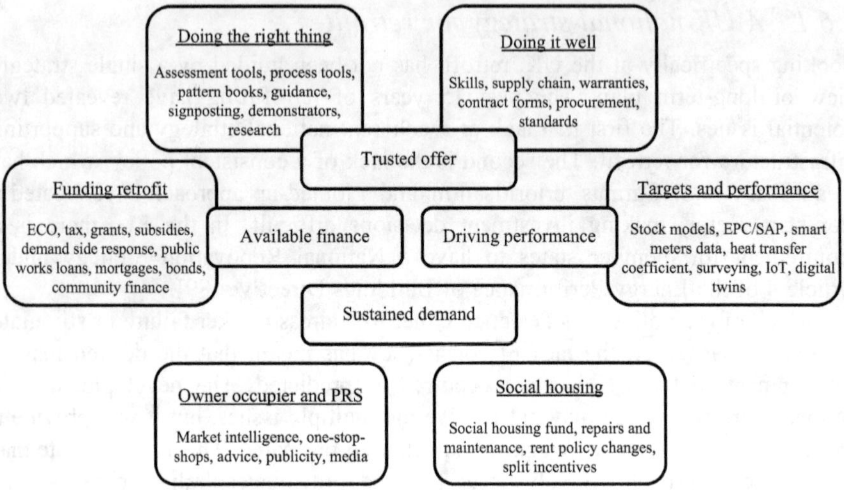

Figure 1.3 Demand and supply factors in retrofit delivery

philosophy. The model also recognises the market failure that is apparent in retrofit, which requires four elements; a trusted accessible offer that achieves its objectives, that has a funding model behind it, and that has demand within each of the sectors.

Figure 1.3 divides the issues between the supply side and the demand side. Retrofit is clearly systemic in nature, so there is a need for the whole system to move forward at the same time. We cannot create demand where there is no availability of supply chain or business models, including finance, to meet that demand. There is a requirement for a single, concerted effort to move from a state where very few retrofits are being undertaken, either through funded programmes or early adopters, to a state where there is a developed market for retrofit. However, what the model tells us, also supported through the socio-technical approach, is that policy needs to act on multiple areas, as well as being supported by wider stakeholders within the system. Here, we consider some of the key elements of this place-based system.

1.6.2.1 Sustained demand

In the domestic setting, there are three main tenures, social housing, the private rented sector and the owner-occupier market. Table 1.2 breaks down the proportion of these households for England [83] and Greater Manchester [90]. This identifies a small, but meaningful, difference in the split by tenure between the national and regional levels, with GM having a smaller proportion of owner-occupiers and larger social and private rented sectors. This indicates that, as a place, Greater Manchester policymakers have a different approach to take in terms of tenure 'markets'.

This high-level perspective does not tell the complete picture. Each of these markets contains a range of sub-sections in terms of willingness and capability to engage with the market, economic drivers, and financial capacity. Considering

Table 1.2 Distribution of homes by tenure for England and Greater Manchester

Tenure	England %	Greater Manchester %
Owner-occupier	65%	57%
Private rented	19%	20%
Social housing	16%	23%

owner-occupiers, a willingness to engage may follow a Rogers adoption curve-type model [91], with retrofit potentially still at the early adopter stage. The persona model developed by Haines and Mitchell [32] identifies different types of consumers within the market for retrofit and energy efficiency, reflecting life stages, financial capacity, and willingness to adopt. Research in Greater Manchester, undertaken prior to the cost-of-living crisis, indicated that 30% of GM households were willing to undertake some form of retrofit [90] but identified knowledge and financial barriers to adoption. Additionally, attitudes to issues such as sustainability or net zero may shape the willingness of households to adopt. It should be noted that this changes over time. For example, the cost-of-living crisis not only reshaped the economic argument for households to undertake retrofit, specifically in the context of energy efficiency, changing demographics and attitudes also play their part. It is important not to assume a static consumer model as the market progresses. This perhaps argues for a regular and more detailed understanding of these market shifts.

1.6.2.2 Financing retrofit

The second major demand-side issue is finance. There are currently a variety of funding programmes that might be viewed as being targeted towards retrofitting. As stated previously, the national level of funded programmes is often delivered through regional or local governmental bodies. However, the scale of the retrofit problem is clearly not going to be solved by the public sector alone. Governments may intervene to support financing, such as low-cost loans, guarantees, or models of repayment through bills [92]. In addition, models such as Energy Supply Companies may finance improvements through service models, such as heat as a service [93]. Brown *et al.* [92] provide a model that identifies the different models of financing residential retrofit, containing six elements for any financial model in delivering retrofit (Table 1.3).

This model provides a useful framing for the development of financial services in retrofit. Financial instruments such as green mortgages, community bonds, heat as a service, or Octopus Energy Zero Bills model all have this underlying structure. There are possible questions around transparency for, and understanding by, the consumer and a level of protection. Issues such as performance guarantees, particularly underpinned by an EPC, rather than measured data, could present a consumer with a situation where the finance costs are not met by savings. Additionally, complex financial issues, such as the ability to provide network services to an electricity

Table 1.3 Model of finance mechanisms in residential retrofit – adapted from [92]

Finance mechanism feature	Description	Examples
Source of capital	Investment may come from multiple sources	Banks, institutional investors, firms, government, community finance
Financial instrument	Debt or equity	Debt to individual households, securitised debt sold in market, equity as ownership
Project performance	Finance may require performance of the property, either EPC or measured performance, applied quality standards	Energy consumption, specific EPC level, kWh, temperatures (for heat as a service)
Point of sale	The customer journey, how the occupant accesses the finance for improvements	Bank, mortgage provider, finance package with retrofit provider
Security and underwriting	Security is a recourse for a finance body to recover non-payment. Underwriting is how credit ratings are used to assess borrowers and identify risk	Security may be a charge on a property or the threat of court proceedings. Underwriting is usually through credit rating agencies
Repayment channel	How funds are repaid to creditor or shareholders	Repayments, energy bills, service charges, rent or mortgage repayments

network operator have a value that might be poorly understood by the consumer [94], creating an opportunity for businesses to exploit this imbalance of understanding.

1.6.2.3 Trusted offer – doing the right thing

Doing the right thing is associated with developing an appropriate technical solution for a property or properties. In the early stages of retrofit, this was very heavily focused on fabric, which still remains a central issue. The UK building stock is one of the oldest in the EU and, as such, has a level of complexity that potentially cannot be solved by a reliance on high-level technical solutions at the archetype level. Baker *et al.* [95] highlight the variability of details, identifying that archetypes rarely display uniform characteristics. This is driven by building vernacular, often at the regional or sub-regional level, and changes to properties, such as additions to the building or the introduction of new services. This can mean retrofit requires adaptability to address issues as they appear on the ground, which can be a challenge in creating standards. However, many issues repeat, and appropriate solutions can be developed and shared [96] to avoid repeated design work.

This problem translates to appropriate standards, guidance, and tools for the delivery of retrofits. The construction industry often operates on heuristics, or rules of thumb. These often determine industry practice and may not have a technical underpinning but can become deeply embedded in practice. Moving the practices of the sector is complex. We need to better understand best practices and ensure these are communicated through appropriate networks. Using builders' merchants and manufacturers, who often provide training, can help.

However, this might be supported with independent research and innovation activity, which can help shape or support practice with guidance, the development of new products or services such as expert systems, or enhanced surveying tools to help us specify and deliver retrofits. This places a requirement on the research and technology community to not only undertake research but also to ensure they are appropriately linked to these networks and share findings not only with the research community but also with industry in a form they can effectively engage with.

1.6.2.4 Trusted offer – doing it well

Doing it well concentrates on delivering the design as intended. This is an issue of quality delivery, and this focuses on the quality of work delivered on-site. Drivers of quality in any construction project are complex and often contextual to the project and project team. For policymakers, the most applicable tools are the use of quality standards and the development of appropriate skills infrastructure, as described previously.

In addition, the skills infrastructure for a specific place will need careful consideration. Construction has an image problem that has persisted for many years [97]. It is often considered a low-skilled sector and has a persistent problem with an ageing workforce, an issue that is repeatedly highlighted in reports [98]. Understanding the problem is not the same as addressing it. In the UK, the construction skills landscape could be considered fragmented and difficult for both learners and employers to understand. There are multiple qualifications that may be applicable and the linkages between work and training may be complex to navigate. This is compounded by competitive funding for skills providers that may not be applicable to specific places and is often evaluated on a numerical rather than a quality basis. While this provides policy accountability, it does not necessarily lead to the skilled people that the industry wants or needs.

In terms of place, skills ecosystems for trade skills are local, while higher-level construction skills, such as surveying or project management, may be regional or national. In terms of trade skills, there may be young people in colleges, people who are re-skilled and up-skilled either from other industries or within construction, or people from groups that may not normally enter construction. While these are addressed differently, they all represent entrants into the retrofit skills supply chain. On the demand side, there are large and small businesses with different capabilities as to how to engage with the skills and education ecosystem. There is potential for places to consider a skills ecosystem model, where there are interrelationships between normally competitive institutions, shared resources, and engagement with industry to, not only provide skilled people, but also to create channels to better understand what those skills should be. Fundamentally, this is about linking trained individuals to employment, so those pathways should be clearly defined.

The image of the construction industry, complex pathways, and demographics all hamper the development, not only of retrofit skills, but also of the wider construction workforce. However, as white-collar jobs are potentially removed by artificial intelligence tools, does the physical nature of construction offer some protection from obsolescence, in what might be a major shift in employment patterns? Would this improve the appeal of the industry?

1.6.2.5 Targets and performance

Largely, the UK has relied on modelled data to understand the performance of retrofits. The EPC has often been the basis for both the definition of policy and the measure of its success. However, unreliability in the process, particularly where an EPC outcome is tied to funding [45] and the potential impact on the collection of the data [99], can make this an unreliable measure of success, whether we are considering emissions savings or fuel poverty.

In the UK, the shift to the smart meter infrastructure and the growth of IoT devices associated with home energy management, create opportunities to greatly reduce the costs associated with the real measurement of fabric performance [100]. This allows us to better understand the performance of our homes, both pre- and post-retrofit, at a lower cost than previously possible. In addition, better understanding and access to data from heating, cooling, and ventilation systems will give us another layer of insight as to their performance. This data also provides the framework for wider services, such as energy flexibility. As stated previously, green finance would be better linked to the actual performance of homes rather than a high-level model, such as the EPC, giving a better understanding of the financial risks associated with funding retrofits for both the funder and consumer. The move from modelled to measured data provides a firmer basis for policy and financial decision-making; it better tracks quality and outcomes and provides much-needed trust in the retrofit market. In addition, these types of data could provide an underpinning for more focused national or local policymaking, allowing a far better understanding of the nature of the housing stock or, potentially, the ability to identify clusters of households in fuel poverty. While we live in a data-rich world, it is clear we are not effectively exploiting it enough to drive energy efficiency in buildings.

1.7 Conclusions

The outlining of retrofit as a 'wicked problem' does help us understand the complexity, uncertainty, and divergence of the problem and recognise isolated actions will not deliver the change required. It is clearly a socio-technical problem, in that it is a problem that cannot be addressed only by having the right technical solution.

How we shape our desired outcomes, how we arrange institutional structures, and the 'rules' that different actors use to deliver and live with retrofit, all have a significant role to play. Looking at the issue from a high-level supply and demand perspective, which is in itself a simplification, shows us the different elements of the retrofit 'system' that need to move at the same time if sustained development of the sector is to be achieved. We cannot stimulate demand without the skills and supply chains ready to meet this demand. Equally, we cannot build supply chains and train without the jobs and projects to fulfil these requirements. Retrofit remains a difficult issue to address without intervention from governments and these interventions must be sustained if the policy objectives around improving the housing stock to achieve net zero or address fuel poverty are to be addressed in the longer term.

This 'wickedness' is the very issue that makes retrofit such a rich vein of research. However, if researchers are not fully engaged and embedded in the problem of delivering retrofits, we will remain in a position where we observe the problem from our own discipline, be it quantitative or qualitative. If ever a problem begged for engagement across the disciplines, retrofit clearly is one. Genuine socio-technical engagement with the problem requires an engagement with each other as researchers, as well as the wider stakeholders in the system. Without this, we run the risk of engaging as an 'objective', disciple-focused researcher, providing another perspective to a problem domain that is already crowded with singular perspectives. The acceptance of our own boundaries, our own uncertainties, and a willingness to engage with the problem across the disciplines reflects the challenge for the sector.

References

[1] Krebel L., Stirling A., Van Lerven F., Arnold S. *Building a Green Stimulus for COVID-19. A Recovery Plan for a Greener, Fairer Future*. London: New Economics Foundation; 2020

[2] Galvin R., Sunnika-Blank M. 'Ten questions concerning sustainable domestic thermal retrofit policy research'. *Building and Environment*. 2017; 118:377–88

[3] HM Government. *Warm Homes, Greener Homes: A Strategy for Household Energy Management*. London: DECC; 2010

[4] HM Government. *Climate Change Act*. London: HM Government; 2008

[5] Business, Energy and Industrial Strategy. *Heat and Buildings Strategy*. London: HM Government; 2021

[6] Green Finance Institute. *Financing Energy Efficient Buildings: The Path to Retrofit at Scale*. London: Green Finance Institute; 2020

[7] Glew, G., Parker, J., Fletcher, M., *et al. Literature Review of Benefits and Risks in Domestic Retrofit Practice and Modelling. Demonstration of Energy Efficiency Potential*. London: BEIS; 2021

[8] EU Commission. *EU Building Data Observatory*. Available from https://building-stock-observatory.energy.ec.europa.eu/database/, [Accessed 12 February 2024]

[9] Crawley J., Biddulph P., Northrop P.J., Wingfield J., Oreszczyn T., Elwell C. 'Quantifying the measurement error on England and Wales EPC ratings'. *Energies*. 2019;**12**(18):3523

[10] Galvin R., Sunikka-Blank M. 'Economic viability in thermal retrofit policies: Learning from ten years of experience in Germany'. *Energy Policy*. 2013;54:343–51

[11] Rittel H.W., Webber M.M. 'Dilemmas in a general theory of planning'. *Policy Sciences*. 1973;**4**(2):155–69

[12] Chiu L.F., Lowe R.J. 'Eliciting stakeholders' requirements for future energy systems: A case study of heat decarbonisation in the UK'. *Energies*. 2022; **15**(19):7248

[13] Bergman N., Foxon T.J. 'Reframing policy for the energy efficiency challenge: Insights from housing retrofits in the United Kingdom'. *Energy Research & Social Science*. 2020;63:101386

[14] Lowe R., Chiu L.F., Oreszczyn T. 'Socio-technical case study method in building performance evaluation'. *Building Research & Information*. 2018; **46**(5):469–84

[15] Swan W. Retrofit innovation in the UK social housing sector. In W. Swan, P. Brown (eds.), *Retrofitting the Built Environment*. 2013. pp. 36–52

[16] Ince R., Marvin S. 'Constructing domestic retrofit as a new urban infrastructure: Experimentation, equitability and contested priorities'. *Local Environment*. 2019;**24**(9):825–42

[17] Gupta R., Gregg M. 'Do deep low carbon domestic retrofits actually work?'. *Energy and Buildings*. 2016;129:330–43

[18] Saffari M., Beagon P. 'Home energy retrofit: Reviewing its depth, scale of delivery, and sustainability'. *Energy and Buildings*. 2022;269:112253

[19] Rydin Y., Devine-Wright P., Goodier C., *et al. Powering Our Lives: Foresight Sustainable Energy Management and the Built Environment Project: Final Project Report*. London: Government Office of Science; 2008

[20] Boardman B. *Fixing Fuel Poverty: Challenges and Solutions*. London: Earthscan; 2010

[21] Baruah P.J., Eyre N., Qadrdan M., *et al.* 'Energy system impacts from heat and transport electrification'. *Proceedings of the Institution of Civil Engineers-Energy*. 2014;**167**(3):139–51

[22] Rinaldi A., Yilmaz S., Patel M.K., Parra D. 'What adds more flexibility? An energy system analysis of storage, demand-side response, heating electrification, and distribution reinforcement'. *Renewable and Sustainable Energy Reviews*. 2022;167:112696

[23] Swan W., Ruddock L., Smith L., Fitton R. 'Adoption of sustainable retrofit in UK social housing'. *Structural Survey*. 2013;**31**(3):181–93

[24] Hong S.H., Gilbertson J., Oreszczyn T., Green G., Ridley I., Warm Front Study Group. 'A field study of thermal comfort in low-income dwellings in England before and after energy efficient refurbishment'. *Building and Environment*. 2009;**44**(6):1228–36

[25] Jenkins D.P. 'The value of retrofitting carbon-saving measures into fuel poor social housing'. *Energy Policy*. 2010;38:832–9

[26] Reeves A., Taylor S., Fleming P. 'Deep carbon emission reductions in existing UK social housing: Are they achievable and how can they be funded'. 9th European Council for an Energy Efficient Europe: Act! Innovate! Deliver! Reducing Energy Demand Sustainably, La Calle Sur Loup, France, Jun 2009. Stockholm: ECEEE; 2009, pp. 919–30

[27] Bell M., Lowe R. 'Energy efficient modernisation of housing: A UK case study'. *Energy and Buildings*. 2000;**32**(3):267–80

[28] Kelly M.J. 'Retrofitting the existing UK building stock'. *Building Research and Information*. 2009;**37**(2):196–200

[29] Loveday D.L., Vadodaria K., Haines V., *et al.* 'Refurbishing the UK's 'Hard to Treat' dwelling Stock: Understanding Challenges and Constraints'. Presented at CIBSE Technical Symposium; Leicester, UK, 2011

[30] Gupta R., Gregg M., Passmore S., Stevens G. 'Intent and outcomes from the Retrofit for the Future programme: Key lessons'. *Building Research & Information*. 2015;**43**(4):435–51

[31] Husiev O., Arrien O.U., Enciso-Santocildes M. 'What does Horizon 2020 contribute to? Analysing and visualising the community practices of Europe's largest research and innovation programme'. *Energy Research & Social Science*. 2023;95:102879

[32] Haines V., Mitchell V. 'A persona-based approach to domestic energy retrofit'. *Building Research & Information*. 2014;**42**(4):462–76

[33] Brown P., Swan W., Chahal S. 'Retrofitting social housing: Reflections by tenants on adopting and living with retrofit technology'. *Energy Efficiency*. 2014;7:641–53

[34] Den X.L., Riviere M., Lessmann F., Nesbit M., Paquel K., Illes A. 'Ex-post evaluation of ERDF and CF support to energy efficiency interventions: Findings in public and residential buildings across EU member states'. *European Structural & Investment Funds Journal*. 2017;**5**(2):134–46

[35] Sovacool B.K. 'Fuel poverty, affordability, and energy justice in England: Policy insights from the Warm Front Program'. *Energy*. 2015;93:361–71

[36] Sebi C., Nadel S., Schlomann B., Steinbach J. 'Policy strategies for achieving large long-term savings from retrofitting existing buildings'. *Energy Efficiency*. 2019;**12**:89–105

[37] Bertoldi P., Rezessy S., Lees E., Baudry P., Jeandel A., Labanca N. 'Energy supplier obligations and white certificate schemes: Comparative analysis of experiences in the European Union'. *Energy Policy*. 2010;**38**(3):1455–69

[38] Glew D., Smith M.B., Miles-Shenton D., Gorse C. 'Assessing the quality of retrofits in solid wall dwellings'. *International Journal of Building Pathology and Adaptation*. 2017;**35**(5):501–18

[39] Swan W., Fitton R., Smith L., Abbott C., Smith L. 'Adoption of sustainable retrofit in UK social housing 2010-2015'. *International Journal of Building Pathology and Adaptation*. 2017;**35**(5):456–69

[40] Recart C., Dossick C.S. 'Hygrothermal behavior of post-retrofit housing: A review of the impacts of the energy efficiency upgrade strategies'. *Energy and Buildings*. 2022;262:112001

[41] Gori V., Marincioni V., Altamirano-Medina H. 'Retrofitting traditional buildings: A risk-management framework integrating energy and moisture'. *Buildings and Cities*. 2021;**2**(1):411–24

[42] Khoury J., Alameddine Z., Hollmuller P. 'Understanding and bridging the energy performance gap in building retrofit'. *Energy Procedia*. 2017;122:217–22

[43] Calì D., Osterhage T., Streblow R., Müller D. 'Energy performance gap in refurbished German dwellings: Lesson learned from a field test'. *Energy and Buildings*. 2016;127:1146–58

[44] Galvin R. *The Rebound Effect in Home Heating: A Guide for Policymakers and Practitioners*. London: Routledge; 2015

[45] Gledhill T., Swan W., Fitton R. 'A practitioner study into the variability of UK domestic energy assessments'. *International Journal of Building Pathology and Adaptation*. 2023;**43**(5):1161–78

[46] Jenkins D., Simpson S., Peacock A. 'Investigating the consistency and quality of EPC ratings and assessments'. *Energy*. 2017;138:480–9

[47] Mathisen O.V., Sørbye M.E., Rao M., Tamm G., Stantchev V. 'Smart energy in smart cities: Insights from the smart meter rollout in the United Kingdom'. In Visvizi, A. Lytras, M. (eds.) *Smart Cities: Issues and Challenges*. Amsterdam: Elsevier; 2019, pp. 283–307

[48] Zhou S., Brown M.A. 'Smart meter deployment in Europe: A comparative case study on the impacts of national policy schemes'. *Journal of Cleaner Production*. 2017;144:22–32

[49] Darby S., Liddell C., Hills D., Drabble D. *Smart Metering Early Learning Project: Synthesis Report*. London: DECC; 2015

[50] Barai G.R., Krishnan S., Venkatesh B. 'Smart metering and functionalities of smart meters in smart grid-a review'. Proceedings of the Electrical Power and Energy Conference; London, Canada, Oct 2015. New York: IEEE; 2015:138–45

[51] Bergman N., Foxon T.J. 'Drivers and effects of digitalization on energy demand in low-carbon scenarios'. *Climate Policy*. 2023;**23**(3):329–42

[52] Zhou B., Li W., Chan K.W., Cao Y., Kuang Y., Liu X. *et al.* 'Smart home energy management systems: Concept, configurations, and scheduling strategies'. *Renewable and Sustainable Energy Reviews*. 2016;61:30–40

[53] Brown D., Kivimaa P., Sorrell S. 'An energy leap? Business model innovation and intermediation in the 'Energiesprong' retrofit initiative'. *Energy Research & Social Science*. 2019;58:101253

[54] Eyre N., Fawcett T., Topouzi M., Killip G., Oreszczyn T., Jenkinson K. *et al.* 'Fabric first: Is it still the right approach?' *Buildings and Cities*. 2023;**4**(1):965–72

[55] Lamb N., Elmes D. 'Increasing heat pump adoption: Analysing multiple perspectives on preparing homes for heat pumps in the UK'. *Carbon Neutrality*. 2024;**3**(1):10

[56] Rye C. 'Are traditional buildings really carbon villains? Why it matters when insulating solid walls'. *Journal of Building Survey, Appraisal & Valuation*. 2015;**4**(2):119–26
[57] Hamilton I., Milner J., Chalabi Z., *et al.* 'Health effects of home energy efficiency interventions in England: A modelling study'. *BMJ Open*. 2015;**5** (4):e007298
[58] Burns P., Coxon J. *Boiler on Prescription Trial Closing Report*. Sunderland: Gentoo Group; 2016
[59] Fylan F., Glew D. 'Barriers to domestic retrofit quality: Are failures in retrofit standards a failure of retrofit standards?' *Indoor and Built Environment*. 2022;**31**(3):710–18
[60] Rosenow J., Eyre N. 'A post mortem of the Green Deal: Austerity, energy efficiency, and failure in British energy policy'. *Energy Research & Social Science*. 2016;**21**:141–44
[61] National Audit Office. *Green Homes Grant Voucher Scheme*. London: Business, Energy and Industrial Strategy; 2021
[62] Fitton R., Swan W. 'Retrofit assessment: Getting it right from the start'. *Journal of Building Survey, Appraisal & Valuation*. 2023;**12**(3):208–17
[63] Koppenjan J.F., Klijn E.H. *Managing Uncertainties in Networks: A Network Approach to Problem Solving and Decision Making*. London: Routledge; 2004
[64] Head B.W. 'Wicked problems in public policy'. *Public Policy*. 2008;**3** (2):101–18
[65] Geels F.W. *Technical Transitions and Systems Innovations: A Co-Evolutionary and Socio-Technical Analysis*. Cheltenham: Edward Elgar Publishing; 2005
[66] Genus A., Coles A.M. 'Rethinking the multi-level perspective of technological transitions'. *Research Policy*.2008;**37**(9):1436–45
[67] Geels F.W., Sovacool B.K., Schwanen T., Sorrell S. 'The socio-technical dynamics of low-carbon transitions'. *Joule*. 2017;**1**(3):463–79
[68] Verbong G., Geels F. 'The ongoing energy transition: Lessons from a socio-technical, multi-level analysis of the Dutch electricity system (1960–2004)'. *Energy Policy*. 2007;**35**(2):1025–37
[69] Geels F.W., Kern F., Fuchs G., *et al.* 'The enactment of socio-technical transition pathways: A reformulated typology and a comparative multi-level analysis of the German and UK low-carbon electricity transitions (1990–2014)'. *Research Policy*. 2016;**45**(4):896–913
[70] BEIS. *The Clean Growth Strategy: Leading the Way to a Low Carbon Future*. London: Crown Copyright; 2017
[71] Department of Health and Social Care. *Build back better: Our plan for health and social care* [online] Available from https://www.gov.uk/government/publications/build-back-better-our-plan-for-health-and-social-care [Accessed 30 June 2025]
[72] Meadowcroft J. 'Engaging with the politics of sustainability transitions'. *Environmental Innovation and Societal Transitions*. 2011;**1**(1):70–5

[73] Geels F.W. 'Regime resistance against low-carbon transitions: Introducing politics and power into the multi-level perspective'. *Theory, Culture & Society*. 2014;**31**(5):21–40
[74] National Audit Office. *Green Deal and Energy Company Obligation*. London: National Audit Office; 2016
[75] Rosenow J., Eyre N. 'The green deal and the energy company obligation'. *Proceedings of the Institution of Civil Engineers-Energy*. 2013;**166**(3): 127–36
[76] Cherrington R., Goodship V., Longfield A., Kirwan K. 'The feed-in tariff in the UK: A case study focus on domestic photovoltaic systems'. *Renewable Energy*. 2013;**50**:421–6
[77] Elnokaly A., Martin A. 'The impact of the UK governments spending review into the Feed in Tariff (FIT) on the installation on photovoltaic panels in the residential sector'. *World Journal of Science, Technology and Sustainable Development*. 2014;**11**(2):102–16
[78] Macrorie R., Arbabi H., Eadson W., *et al.* 'Support place-based and inclusive supply chain, employment and skills strategies for housing-energy retrofit' in Crowther A., Foulds C., Gladkykh G. (eds.) *Strengthening European Energy Policy: Governance Recommendations From Innovative Interdisciplinary Collaborations*. Cham: Springer Nature Switzerland; 2024. pp. 73–85
[79] Desvallées L. 'Low-carbon retrofits in social housing: Energy efficiency, multidimensional energy poverty, and domestic comfort strategies in southern Europe'. *Energy Research & Social Science*. 2022;**85**:102413
[80] Killip G., Fawcett T., Jofeh C., Owen A.M., Topouzi M., Wade F. *Building on our strengths: A market transformation approach to energy retrofit in UK homes* [online]. London: CREDS; 2021. Available from Building-on-our-Strengths.pdf [Accessed 12 July 2026]
[81] Bedford T., Catney P., Robinson Z. 'Going down the local: The challenges of place-based net zero governance'. *Journal of the British Academy*. 2023;**11** (s4):125–56
[82] Sayce S.L., Hossain S.M. 'The initial impacts of minimum energy efficiency standards (MEES) in England'. *Journal of Property Investment & Finance*. 2020;**38**(5):435–447
[83] Ministry of Housing, Communities and Local Government. *English Housing Survey*. London: MHCLG; 2024
[84] Cowan D., Marsh A., Harris J. 'Local authority intervention in private renting: From compliance to hardline enforcement'. *Journal of Law and Society*. 2024;**51**(3):390–412
[85] Microgeneration Certification Scheme *MCS Data Dashboard* [online] 2025. Available from https://mcscertified.com/about-the-mcs-data-dashboard/ [Accessed 21 August 2025]
[86] Ravenscroft C. *Ready for retrofit?: An analysis of Local Skills Improvement Plans in England*. London: The Gatsby Charitable Foundation; 2024
[87] Octopus Energy. *Octopus Energy trials "Zero Bills" upgrade to existing homes*. [online] 2025. Available from https://octopus.energy/press/octopus-energy-trials-zero-bills-upgrade-on-existing-homes/ [Accessed 21 August 2025]

[88] Claude S., Ginestet S., Bonhomme M., Moulène N., Escadeillas G. 'The Living Lab methodology for complex environments: Insights from the thermal refurbishment of a historical district in the city of Cahors, France'. *Energy Research & Social Science*. 2017;**32**:121–30

[89] Broc J., Castellazzi L., Economidou M., Labanca N., Paci D., Panev S., *et al. Assessment of second long-term renovation strategies under the Energy Efficiency Directive*. Luxembourg: Publications Office of the European Union; 2019

[90] Greater Manchester Combined Authority. *retrofitGM: Accelerating the Renovation of Greater Manchester's Buildings* [online]. Available from https://www.greatermanchester-ca.gov.uk/media/5954/gmca-retrofit-action-plan.pdf [Accessed 21 June 2025]

[91] Rogers E.M., Medina U.E., Rivera M.A., Wiley C.J. 'Complex adaptive systems and the diffusion of innovations'. *The Innovation Journal: The Public Sector Innovation Journal*. 2005;**10**(3):1–26

[92] Brown D., Sorrell S., Kivimaa P. 'Worth the risk? An evaluation of alternative finance mechanisms for residential retrofit'. *Energy Policy*. 2019; 128:418–30

[93] Britton J., Minas A.M., Marques A.C., Pourmirza Z. 'Exploring the potential of heat as a service in decarbonization: Evidence needs and research gaps'. *Energy Sources, Part B: Economics, Planning, and Policy*. 2021;**16**(11–12):999–1015

[94] Carmichael R., Gross R., Hanna R., Rhodes A., Green T. 'The Demand Response Technology Cluster: Accelerating UK residential consumer engagement with time-of-use tariffs, electric vehicles and smart meters via digital comparison tools'. *Renewable and Sustainable Energy Reviews*. 2021;**139**:110701

[95] Baker C., Smith L., and Swan W. 'Make no assumptions: The selection of domestic retrofit improvements' in Swan, W., Brown, P. (eds.) *Retrofitting the Built Environment*. Chichester: John Wiley & Sons; 2013:97–112

[96] Red Coop *Retrofit pattern book* [online]. 2025. Available from https://retrofit.support [Accessed 23 June 2025]

[97] Haupt T., Harinarain N. 'The image of the construction industry and its employment attractiveness'. *Acta Structilia*. 2016;**23**(2):79–108

[98] Chartered Institute of Building. *Exploring the Impact of an Aging Population of the Workforce and Built Environment*. London: CIOB; 2015

[99] Hardy A., Glew D. 'An analysis of errors in the Energy Performance certificate database'. *Energy Policy*. 2019;**129**:1168–78

[100] Allinson D., Crawley J., Elwell C., *et al.* Technical Evaluation of SMETER Technologies Project, Phase 2 Data: Heat Transfer Coefficients, In-Use Temperature and Energy Data, Local Weather, Dwelling Information, and Occupancy Surveys From 30 English Dwellings, 2019–2021. Colchester, UK: Data Service; 2024

[88] [illegible] S., [illegible] S., Bo[illegible] M., [illegible] N., [illegible] [illegible] methodology for [illegible] insights from the [illegible] of a [illegible] in the city [illegible]. [illegible] 2017;[illegible]:1–30.

[89] [illegible] J., [illegible], [illegible] M., [illegible] N., [illegible] O. [illegible] *[illegible] strategies [illegible]* [illegible] Publications Office of the [illegible] 2019.

[90] [illegible] [illegible] combined [illegible] [illegible]. Available from: https://[illegible] [accessed [illegible] June 2020].

[91] [illegible]

[illegible] 2021;[illegible]

[illegible] [accessed [illegible] June 2020].

[illegible] [illegible] [illegible]

[illegible] 2016;[illegible].

[illegible] Technical [illegible] [illegible] Data [illegible] Information and [illegible] 2019–2021 [illegible].

Chapter 2

Urban energy models for retrofit decision-making

Pamela Fennell[1], Shyam Amrith[1] and Paul Russevelt[1]

2.1 Introduction

The World Green Building Council estimates that buildings are responsible for 39% of global energy-related carbon emissions: 28% from operational emissions, from the energy needed to heat, cool, and power them, and the remaining 11% from materials and construction [1]. The C40 Cities global network of mayors highlights the critical role of cities in responding to the climate crisis. Although urban areas only occupy 2% of the world's land surface, they consume over 67% of the world's energy [2]. For cities, retrofitting buildings to reduce energy consumption is critical to achieving carbon reduction targets. However, retrofit decision-making is a multi-faceted problem that, when addressed, also has the potential to reduce social harm and its associated costs, in particular:

- Reducing heat loss from homes will reduce the number of people exposed to a range of serious and life-limiting health conditions as a result of being unable to heat their homes to adequate levels [3].
- Access to clean energy sources for cooking, heating, and electricity generation will have a significant impact on the 3.8 million premature deaths each year due to household air pollution [4].

As Hong *et al.* [5] highlight, urban energy analysis is a complex, multi-scale, multi-sector challenge that demands a new breed of tools to support the rapid pace of decision-making, which is vitally needed.

In this chapter, we explore how Urban Building Energy Models can be used to support retrofit decision-making by municipal authorities. We begin by defining Urban Building Energy Models (UBEMs) and setting out their potential uses. We then explore their input data requirements and possible sources of input data. We set out a typical workflow for a UBEM for retrofit decision-making and then examine three case studies where UBEMs have been used to inform energy

[1]Energy Institute, Bartlett School of the Environment, Energy and Resources, University College London, United Kingdom

planning by local decision-makers. We conclude with a survey of emerging developments in urban building energy modelling.

Urban building energy models (UBEMs) are numerical simulations of the energy and environmental performance of groups of buildings. UBEMs aim to assess the aggregated dynamics of the group of buildings and, to differing extents, to take account of the interaction between each building and its surroundings. In this chapter, we follow Langevin *et al.* [6] and define a UBEM as a model which:

- represents multiple buildings, which are often geographically co-located but may also represent a stock of buildings of a particular use type (e.g., a study of supermarket buildings in Japan [7]).
- produces energy use metrics as an output but may produce other metrics (e.g., internal temperatures) as well.
- can generate out-of-sample predictions.

Langevin *et al.* [6] classify UBEMs according to whether they are top-down or bottom-up and the extent to which calculations are based on underlying physical models or statistical models (white-box vs. black-box models). These different types of models have different roles to play in supporting decision-making by municipal authorities.

Top-down black-box models, which estimate aggregate building energy use from sector-wide socio-economic and/or technical variables, can be used to assess the historical impact of system-level changes but are unable to represent disruptive change due to their reliance on historical data.

While bottom-up black-box models are also reliant on historical data and, therefore, cannot explicitly represent underlying dynamics, they have an important role to play in understanding how measures might be taken up across a city.

Top-down white-box models can represent the complexity of building stock energy use, and its components, at the aggregate level, making them useful for prediction of long-term change, but they cannot link aggregate building energy use to building-level operations and require expert knowledge to fully represent system components and causal flows.

Bottom-up white-box models simulate the physical relationships of processes at the building, or energy end-use level, and explicitly represent key dynamics influencing building energy end uses, building stock diversity, and the aggregate energy effects of changes to operations at the individual building level, making them particularly important for city-level retrofit decision-making. This is the form of model on which we will focus for the rest of this chapter.

2.1.1 Different types of bottom-up UBEMs

Bottom-up UBEMs are the most computationally intensive form of UBEM since they aim to simulate the energy and environmental performance for each building in the stock. For large building stocks, this might require vast quantities of input data and can only be possible by distributing calculations across large numbers of computing cores. Some UBEMs avoid the need for this by defining archetype buildings that represent specific sections of the stock and creating simulations for these buildings alone, which are then scaled to give a representation of the whole stock. The US National Laboratory of the Rockies, formerly National Renewable Energy Laboratory, ComStock model [8] is an example of this approach, where 330,000 archetypes are used to represent the non-domestic building stock of the whole country. While still requiring a very large number of simulations, it is still only a small fraction of the total number of non-domestic buildings in the US.

Research by Mathur *et al.* [9] identified 85 UBEMs in the literature but highlighted an uneven pattern of use. UBEMs are predominantly used in the developed cities of the global North and generally applied to temperate climates, as can be seen in Figure 2.1, in which the radius of the point represents the number of UBEMs and the colour shows the corresponding climate zone.

UBEMs can be used to support retrofit decision-making in many ways:

- **Identification**: UBEMs can be used to explore the energy performance of a building stock and identify which segments of the building stock are performing well and which are performing poorly. This type of information allows city authorities to examine in more detail the underlying causes of better or worse performance.
- **Scenario analysis**: A key use of models in general, and UBEMs in particular, is to assess the effects of interventions. While individual building models allow the impact of interventions, such as increasing insulation or inclusion of renewable technologies, to be quantified, they cannot explain the impact across a whole neighbourhood or city. UBEMs are particularly useful for evaluating different options for implementing change. For example, the effect of bringing all buildings with an Energy Performance Certificate rating of D up to C.
- **Optimisation**: Advanced optimisation of UBEMs enables the evaluation and comparison of numerous solutions to identify the best interventions.
- **Broader impacts**: UBEMs are often able to produce additional outputs, beyond energy, such as internal temperatures, which can be used to assess exposure to overheating risks. UBEMs can also be combined with other forms of data to understand impacts on different groups of people, linking socio-economic data allows fuel poverty to be explored, for example. This type of analysis can be combined with optimisation to develop packages of measures which target specific criteria, such as minimising energy spend for specific households.
- **Visualisation and communication**: Post-intervention, UBEMs can play a part in tracking energy consumption. While a UBEM itself is a simulation model, the data underlying it can act as a digital twin of a city's energy consumption characteristics and can provide an important resource for visualising and communicating model results.
- **Resilience to future climate**: Weather data for future climate scenarios have been created by researchers. These can be used to test the impacts on the stock and its inhabitants, as well as strategies to mitigate unwanted effects, such as overheating.

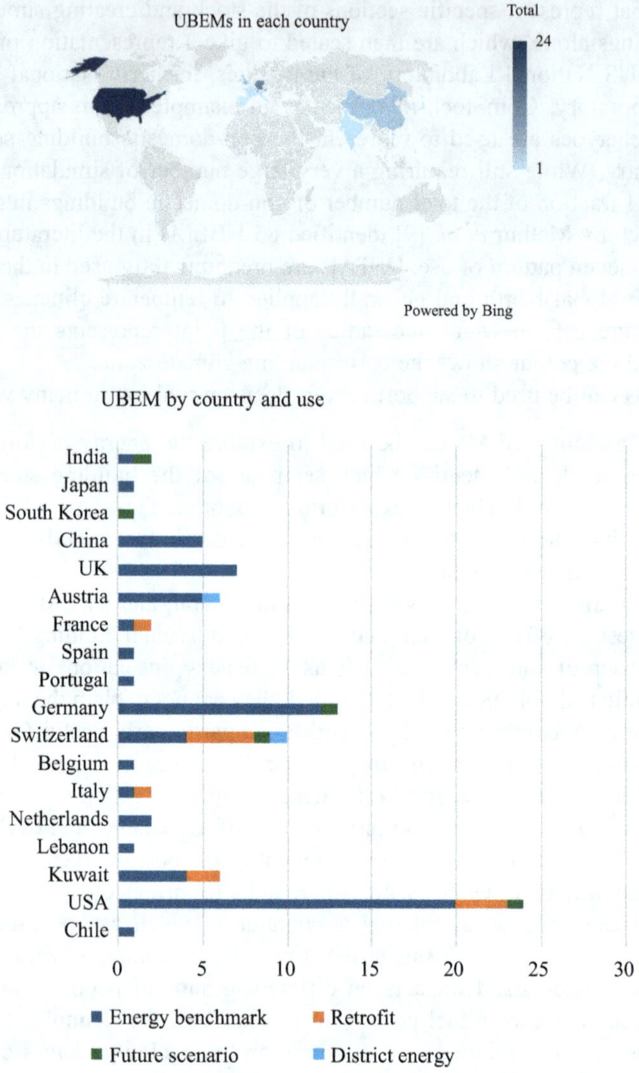

Figure 2.1 Location and number of UBEMs

- **Renewables and integration**: With detailed geometry and surrounding context, the potential for installing solar photovoltaic (PV) panels on rooftops can be estimated. Incorporating roof angles, shading, and local weather into the model, the power output can be modelled with time. This can also be modelled within a grid network to assess their integration.
- **Demand analysis**: UBEMs can be simulated with a high temporal resolution, allowing detailed predictions and analysis of peak loads. Opportunities for load shifting and demand-side response can therefore be identified and quantified.

- **District-level systems**: District heating and/or cooling systems can be modelled alongside the buildings on the network. This makes it possible to predict system performance and size the system according to predicted peak demands.

2.2 Input data and data sources

All bottom-up, building physics-based UBEMs have similar core data requirements, but the way in which that information is represented varies considerably from UBEM to UBEM. The input data must characterise the requirements for heating, cooling, power usage of building occupants, and how they interact with buildings in order to determine energy demand. Building geometry and fabric need to be characterised to understand energy flows between the building and the environment, while building systems must be detailed so the efficiency of heating and cooling can be considered. The interactions between buildings and the environment need to be considered, including the impact of the local microclimate and the effect of surrounding structures. Figure 2.2 groups these types of input into three levels: user, building, and environmental. In this section, we describe the

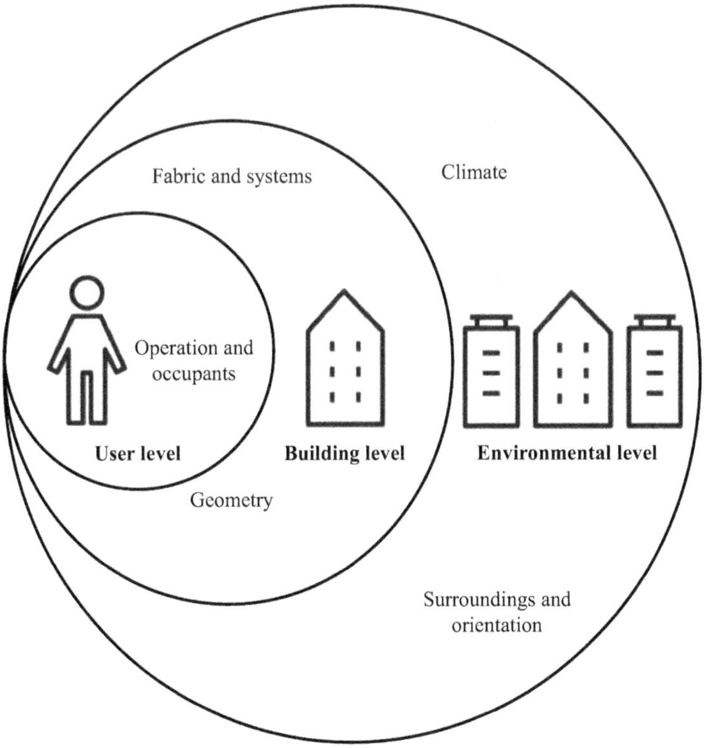

Figure 2.2 Input data requirements

types of input data that are required, the different modelling approaches and the possible sources of data for each.

2.2.1 User level

Occupants interact with buildings to operate equipment, lighting, and other systems, in addition to being sources of heat themselves, which means time-series data is required concerning the presence of occupants and their actions. Models vary in how they treat occupant interactions with buildings and equipment in two principal dimensions: whether they are deterministic or stochastic and whether they are static or dynamic. Static, deterministic inputs are fixed and fully known in advance, for example, a schedule of occupant presence or lighting usage that is fully specified and is not affected by feedback from other parts of the model. Static-stochastic approaches introduce an element of uncertainty, where schedules are still unaffected by feedback but where a probabilistic approach is used to generate a range of different schedules. Dynamic approaches introduce feedback from other parts of the model; the lighting schedule might be varied depending on the availability of natural daylight. If the same lux level has the same rule-based effect on the lighting schedule, then the approach would be classed as dynamic-deterministic, but if the effect of the interaction is not fully specified in advance, then this would be a dynamic-stochastic approach. Agent-based occupancy models, which use probability distributions to determine the result of an interaction between an occupant and system, are a key example of this. Barbour *et al.* [10] proposed tracking the 3.5 million inhabitants of Boston from mobile phone usage using a statistical Time-Geo Framework. With this approach, they observed that occupancy patterns are 5 times lower than suggested with a standard deterministic approach. A difference of 15–21% was noted in the energy consumption due to this. The computational burden of dynamic-stochastic approaches to occupant modelling is considerable, as is the challenge of accessing reliable data on which to base models, which means that these approaches are still applied relatively infrequently.

2.2.2 Building level

2.2.2.1 Geometry

The geometry of a building determines the volumes of space that must be heated or cooled, the surface areas through which heat transfer takes place, and the thermal properties of its envelope. This geometry can be made available in the form of Semantic City 3D models. Traditionally, 3D city models can be developed with levels of detail ranging from 0 to 4 [11]. Findings from the literature show that increasing levels of detail do result in increasing accuracy, but the additional computational burden is significant.

- 2D geometry (CityGML LoD0) - Buildings are represented as footprints or roof edge polygons; in the absence of building geometry, arbitrary box models are developed.

- 2.5D extrusion (CityGML LoD1) - 2D geometry of the building footprint is extruded to their respective height. This is the most widely used approach for UBEMs. Cerezo Davila *et al.* [12] combined a 2D GIS database with cadastral data for the number of floors to extrude the 2.5D model. They observed that the differences in conditioned volumes to the modelled volume resulted in some errors when not combined with other semantic details.
- 3D geometry (CityGML LoD2) - Represents the actual 3D geometry of the buildings, accounting for different shapes of the roofs as opposed to the prismatic flat roof LoD1 geometry. Monien *et al.* [13] used LiDAR data and building reconstruction methods to develop the same for Essen, Germany. They found an increase in accuracy of 10% compared with LoD1. Nouvel *et al.* [14] also observed an increase of 15–20% in the result accuracy for buildings with pitched roofs and attics while modelling a 3D geometry as compared to 2.5D extrusion.
- 3D with external features (CityGML LoD3) - 3D models with detailed wall and roof structures including doors, windows, and other external features. Saretta *et al.* [15] found that while greater precision in the definition of glazing ratios and shading elements improved the accuracy in estimating the potential of BIPV in an urban district in Switzerland, it increased computational time by a factor of 4.

Official maps, such as the Ordnance Survey in the UK, provide basic geometric information in many contexts. This is typically 2D data, which can be supplemented with height data from LiDAR surveys to provide a basic 3D representation. In some cities, full 3D models have been built, for example, Helsinki, Singapore, and New York. In many contexts, geometric information may only be available with approval from the relevant local authorities, and in others, data may not be accessible at all. In these contexts, satellite images may be the only available data set, with researchers using image processing and machine learning to extract details of buildings.

2.2.2.2 Building fabric and systems

How quickly heat is lost or gained through building fabric is a key determinant of energy consumption in a building stock, but even an individual building can be a complex amalgamation of fabrics, and detailed data is difficult to find. To reduce the data collection burden for UBEMs, models often use standard sets of data for different types of buildings. The approach to defining the group of standard specifications that describe a building stock is generally based on the age and function of individual buildings in the stock. In jurisdictions where building regulations have been implemented and enforced, the age of a building is a good guide to its fabric. Where such regimes do not exist, modellers may use expert judgement to define the segments [16]. Building fabric is often significantly influenced by the use of the building; highly glazed façades are much more likely in commercial office buildings than in residential, making data on use a key characteristic. Building use is also a key determinant of the building systems, with domestic properties having relatively homogeneous systems in most countries. Non-domestic buildings are much more diverse, but nonetheless, building function will drive system demands, such as, dictating the need for cooling or heating. The way in which building systems are modelled in UBEMs depends on both the scale and purpose of the model. It is common for models that cover large areas to represent building systems by their efficiencies rather than modelling them in detail. This contrasts with models that focus on district energy systems, which require a detailed representation of the proposed systems and their interactions with the buildings.

The Energy Performance Certificate regimes enacted in EU, and former EU countries, have created rich datasets which provide details of building fabric and some information on building systems. However, these datasets are not comprehensive, and other approaches will also be needed to fill in gaps. The Tabula project [17], for example, aimed to define standard archetypes which can then be applied based on age, built form, and building use. Some models rely completely on reference buildings, such as the US Department of Energy's commercial reference building project [18]. Detailed models of the reference buildings are scaled up to the whole stock.

2.2.3 Environmental level

Local weather conditions are a critical influence on energy consumption. Most UBEMs use weather files which are based on historic measurements, with data available for around 30,000 locations worldwide. Weather stations are usually located in open areas to avoid obstructions, often at airports, and therefore do not include the effects of urban morphology on microclimate (Urban Heat Island) effects. These effects can be included in models, either by measuring local conditions directly or by using tools such as the Urban Weather Generator [19], to account for the effects of local urban morphology.

Effects of local shading, shared walls between buildings, radiation from the sky, and neighbouring buildings, and wind patterns can be included in UBEMs. However, the extent to which surroundings are included varies by model and depends on the data available. Models based on scaling results from reference buildings or archetypes may not include any contextual information beyond local weather data [20]. Including mutual shading and adjacencies requires a 3D model of the surroundings, even where those surrounding buildings may not be included in the simulation, necessarily increasing simulation time. More advanced approaches include long-wave radiation exchange between buildings and microclimate effects, with additional impacts on simulation time, in addition to requiring greater detail on surface properties [21].

2.2.4 Model resolution

When developing a UBEM, it is necessary to decide on a suitable model resolution, both in the spatial and temporal sense. This choice depends on a range of factors including the purpose of the model, the availability of data, and the balance between detail and computational expense.

Temporal resolution refers to the frequency at which the model is evaluated over a certain modelled time frame. A decision must be made as to whether the model is to be simulated once per day, once per hour, many times per hour, and so on. Note that the temporal resolution of the model does not necessarily correlate with the reporting frequency of model outputs. Certain dynamics occur over relatively short timescales and, hence, require a high temporal resolution to be fully captured. This can apply to UBEMs which include power systems integration, load shifting and demand-side response, integrating renewables, matching supply with demand, and sizing systems. Since the model is evaluated a higher number of times for the same run period, a high temporal resolution increases the computational burden. By contrast, a lower temporal resolution is sufficient for applications where it is not necessary to capture these fast fluctuations but rather the effects of interest occur over a longer timeline, such as, the quantification of greenhouse gas emissions or calculating payback periods for retrofit measures.

Spatial resolution refers to the level of detail used to represent the geometry of buildings, covered in section 2.2.2.1. While not strictly defined as such, it is useful to also consider archetypes under this overarching concept, since the reason for employing these typically stems from the same considerations.

A high spatial resolution is needed when specifics about geometry, building orientation, and/or surrounding context have a significant impact on the outputs of interest. Examples include evaluating the potential of widespread solar PV, identifying buildings at risk of overheating, and peak load analysis, where diversity of the stock is important. A lower spatial resolution, including the use of archetypes, is well suited to any analysis where the specific characteristics of individual buildings are less important, such as, modelling at a national level or mapping decarbonisation pathways.

2.3 UBEM workflows for retrofit decision-making

Retrofit decision-making using UBEMs requires two key stages as shown in Figure 2.3: firstly, development of a baseline model and, secondly, construction and simulation of retrofit scenarios, either selectively or more systematically, as part of an optimisation process. These aspects are explained in more detail in the following sections.

2.3.1 Baseline modelling

1. City data modelling - Sourcing and processing necessary data and filling gaps. In many locations, this will be the most time-consuming step, particularly in places where accessible data is limited. The target information will often need to be extracted from a range of urban data sets collected for diverse purposes, such as taxation, which may include information on the use and floor area, certification, providing information on building fabric, plant and equipment, and flood risk management information on geometry from LiDAR mapping, etc.

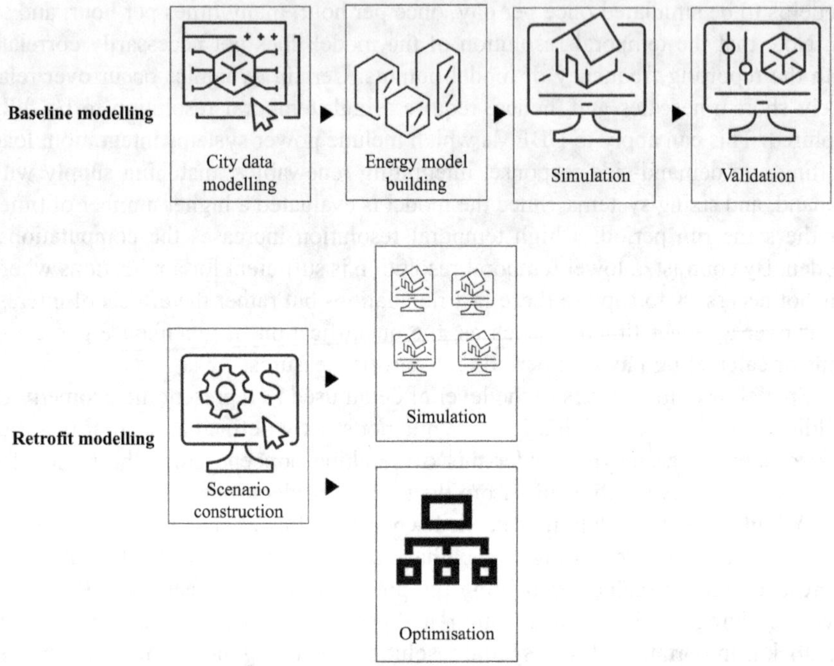

Figure 2.3 Typical UBEM workflow for retrofit decision-making

2. Energy model building - Compiling data and urban geometry model into a 'simulation ready' state. While pre-processed CityGML models are available for some cities, these are in the minority and developing a UBEM for most contexts will require consolidating and processing of input data. These steps often involve combining data sets collected for a range of different purposes. Linking data is a key challenge, as data collected for different purposes may not have a single consistent unit of reference. The volume of data involved in a UBEM brings a need for automatic processing of the large volumes of data needed to characterise the geometry, building fabric, occupancy, and loads in many buildings.
3. Simulation - Generate load profiles, indoor temperature, and energy use.
4. Validation - Harmonising simulation output with measured data. This step is often the most challenging for urban scale models, as typically only aggregate data is available due to privacy concerns regarding household data. Where aggregate data is available, this is often only at an annual level, which means results at finer temporal resolutions are unvalidated. Recent research efforts have focussed on developing inter-model comparison cases, which can be used to assess model capabilities, or researchers may address this by taking detailed measurements of a few cases and using these to calibrate individual buildings, updating similar buildings in the stock following this step.

2.3.2 Retrofit modelling

The way in which retrofits are modelled depends on the desired output. If the output of interest is the final energy consumption or GHG emission levels, then a single, post-intervention model may be run, allowing comparison of pre- and post-intervention scenarios. However, as can be seen in the case studies that follow, municipal authorities are often interested in the way in which intervention programmes can be structured. In which case, it is necessary to define the take up rates for the interventions as well. Take up rates are generally defined exogenously, based either on historical evidence of take up or practical considerations of capacity. In some cases, it may be defined endogenously, by selecting properties which meet a particular threshold from the baseline model to apply retrofit to. For example, cooling measures might be applied to properties which demonstrate overheating in the baseline model, for example.

2.3.3 Optimisation

The ability to predict the effects of interventions means that different potential options or scenarios can be compared to one another. This is the basis of scenario analysis, which allows us to answer 'what if' questions. Optimisation takes scenario analysis a step further by allowing a huge number of options to be tested and ultimately aims to identify the 'best' performing strategies with respect to some pre-defined criteria. The efficiency of this search is greatly enhanced via the use of an optimisation algorithm which aims to arrive at near optimal solutions without needing to test every possible option.

When planning a stock-level retrofit, there are many retrofit options available to choose from. These can be applied in different combinations and their combined effect depends on how they interact. Moreover, this is the case for every building in the stock, and retrofitting one building can have impacts on others in the surrounding area. The purpose of any retrofit is to yield some improvements over the previous state of the buildings. The factors of interest often fall under environmental, energy-related, social, and financial criteria. The decision-maker would need to be able to assess whether the criteria are improved by a given intervention and quantify these effects in order to make informed decisions.

When modelling, this kind of problem can be parameterised; each retrofit type e.g., window replacement, is represented by a parameter with a set of possible values e.g., single glazing, double glazing, or triple glazing. These different options can be fed one-by-one into the model input for simulation, and the effects on model outputs can be observed with each change. In theory, every combination of parameter values can be tested and the solutions that best satisfy the criteria can be chosen. However, in practice, this would require a tremendous number of simulations, which would more than likely be infeasible in terms of the time and computational resources required. In many cases, this problem can be circumvented via the application of an optimisation algorithm, which aims to get as near to the optimal solutions as possible while avoiding the need to test every possible option.

Optimisation has been widely applied at the building level [22]. By contrast, very few studies have attempted to directly apply optimisation to UBEMs. The associated challenges are significant, which exist on top of the aforementioned challenges faced when modelling the building stock. For one, optimisation requires a model to be evaluated many thousands of times, which, for an already computationally expensive UBEM, can render the task impossible to complete in any reasonable time frame. Secondly, by considering many buildings at once, it is very likely that the resulting parameter space will be extremely large. Referred to as the 'curse of dimensionality', this is a challenge for all optimisation algorithms and ultimately leads to sub-optimal results. Finally, there is a lack of methods with which to set up, perform, and analyse stock-level optimisation, and specifically to deal with the added layer of complexity which arises at this scale.

2.4 Case studies

While many UBEMs were originally developed as academic exercises, they are increasingly finding a role in retrofit decision-making by municipal authorities. In this section, we explore three cases, in the UK, USA, and Belgium, where municipal authorities have employed a UBEM to either guide capital investment in their own building stock or to help define the policies to incentivise and guide private investment in decarbonisation.

2.4.1 Comparing pathways to net-zero in Islington

Islington is an inner London borough with over 200,000 inhabitants who occupy a diverse building stock, with buildings dating from the eighteenth century, interspersed with flats and maisonettes from the post-war period, and more recent developments in a mix of high-, medium- and low-rise blocks. A significant proportion of the building stock in the borough is owned by the local authority, Islington Borough Council (IBC), which is responsible for 33,300 dwellings in 4,500 buildings. IBC has a statutory duty to ensure that public funds deliver value for money and sees modelling as one of the most important ways to achieve this. Decisions about investments in their building stock need to consider not only techno-economic factors but also factor in social and environmental impact, as well as being driven by manifesto commitments made by the political leadership of the council. IBC has some in-house modelling capacity and typically undertakes high-level feasibility studies itself before outsourcing more detailed, larger-scale modelling work once a business case has been established.

In 2021, IBC commissioned a team from the University College London Energy Institute to develop a model that could be used to inform capital investment in the Council's building stock. The aim was to develop a strategic plan for the housing stock that would underpin decision-making across a variety of different departments. In particular, the project aimed to answer the following questions:

- Is decarbonisation of a whole borough's social housing stock by 2030 possible?
- Which technologies make the biggest contribution to achieving decarbonisation and how much do they cost?
- What are the most effective ways of planning retrofit programmes?

The UBEM tool used on this project, Pathways [23], is a bottom-up tool that models each building explicitly and applies a hybrid approach, combining building details and historical information to quantify the impact of different retrofit options. The tool is based on a static snapshot of the building stock at a given point in time but considers the implementation of packages of measures and the ways in which they are rolled out. In total, six groups of retrofit measures were considered:

1. do nothing,
2. envelope only,
3. envelope plus Air Source Heat Pumps (ASHP),
4. envelope plus ASHP plus Ground Source Heat Pumps (GSHP) for blocks of flats where the site has sufficient free area for boreholes,
5. envelope plus ASHP plus GSHP plus solar PV,
6. all in, all measures in group 5 plus roof and floor insulation.

In this study, each package of measures was applied to all the relevant buildings but the order in which buildings were retrofitted varied based on whether the priority was to improve the lowest-performing properties first, treat households in fuel poverty first, treat buildings with the highest emissions first, or order by cost-effectiveness of measures.

The results highlighted the relatively small impact on decarbonisation that results from envelope measures and the need to prioritise electrification of heating within the context of a decarbonised electricity supply.

Funding for local government projects is often released on an ad hoc basis and, before the pathways model for Islington was developed, IBC invested significant internal resources in developing feasibility studies as part of applying for funds. A key objective of the project was to reduce the need for repeated small-scale modelling work and establish a more consistent evidence base for funding applications and decision-making across the whole organisation. In the future, IBC sees the model being developed to assist communication about strategic priorities across the council, acting as an important visualisation tool.

2.4.2 Integrating energy and sociodemographic data in Santa Barbara

Santa Barbara County is situated in Southern California and has a population of over 400,000. The County, which comprises a diverse range of urban and rural communities, is facing increasing impacts from climate change, in particular from wildfires. In response, the County is developing a Climate Action Plan (CAP) for 2030, which sets out a roadmap for achieving a 50% reduction in greenhouse gas emissions from 2018 levels by 2030. In addition to quantifying current and projected future Greenhouse Gas (GHG) emissions and determining emissions targets, the CAP will specify evidence-based measures for achieving those targets.

An important part of the CAP is the inclusion of equity 'guardrails'. These require:

- the benefits of the CAP are shared equitably with marginalised communities and those most affected by climate change.
- equitable access to affordable funding and job opportunities for all communities.
- support and incentives minimise the burdens and impacts associated with adopting new technologies or changing behaviours.
- promote housing affordability and avoid displacement.

Modelling underpins the delivery of all of these objectives, enabling the County to assess the impact of proposed programmes and vary assumptions. The UCLA Energy Atlas [24] combines building energy consumption data from utility providers, building characteristics, and sociodemographic data in a spatial model. One output of the model is a map of energy consumption across counties and neighbourhoods, as well as by building type, age, type of energy, and GHG emissions. The UCLA Energy Atlas model is a bottom-up, statistics-based model.

The UCLA Energy Atlas model combines utility data on energy consumption with local government data on zoning and building stocks at a building-by-building level. This includes the age, floor area, and use of buildings, which is combined with building codes applicable when each building was constructed to understand the building fabric. This allows the most energy-intensive parts of the building stock to be identified and the impact of specific policies to reduce demand to be evaluated.

By combining energy and buildings data with census data, the impact of policies on marginalised communities can be explored, enabling policy design to be tested for compliance with the CAP equity guardrails before being released for public consultation.

The inclusion of utility data in the model provides important insights at a stock level but has significant implications for the level of granularity that can be displayed. In order to protect the privacy of customers, data can only be shown at an aggregate level, typically at the zip code level. If the County were able to access data down to the neighbourhood level, insights could be significantly enhanced, particularly as 15-minute electricity consumption data becomes increasingly available.

2.4.3 Developing neighbourhood-specific strategies in Ghent

Ghent is Belgium's second-largest municipality with over 260,000 inhabitants. The oldest of Ghent's buildings dates from 942 and 50% of the building stock was built before 1946 [25]. A specific challenge for the municipal authority is addressing the energy performance of historic buildings in low-income neighbourhoods where inhabitants lack the means for significant capital investment.

The city of Ghent has developed a data-driven approach that uses models of current and potential energy performance for two key purposes: strategic planning and policy advocacy. The city's building stock is affected by, at times, conflicting policies at the city, regional (Flanders), and national (Belgium) levels. Relative prices of gas and electricity are a key example of this. Belgium has some of the highest household electricity prices in the EU, with taxes and levies accounting for around 35% of the cost to consumers, while gas prices are almost 4 times lower, with taxes and levies accounting for 15% of the cost [26]. As a result, in addition to developing policy at the local level, to enable delivery of its climate change commitments [27], the city also attempts to influence policy at regional and national levels.

The city uses the Feature Manipulation Engine (FME) [28] to link data from cadastral, energy consumption, business, and social need databases to provide a core data set which is used for energy. Spatial analysis of the linked data is used to identify sources of waste heat and levels of heat demand to suggest possible locations for district heating networks, while household energy demand under a range of retrofit scenarios is modelled using a physics-based, bottom-up model developed by researchers at the University of Ghent.

A key priority for the city has been to understand how to calibrate more precisely levels of retrofit to be undertaken, taking into account income levels in different neighbourhoods. The strategies developed aim to take into account the fact that deep retrofits of older buildings are unaffordable for many households and to prioritise these areas for heat networks. The model allows different combinations of heating system and retrofit levels to be tested across the city, with proposals tailored for each neighbourhood. The model also facilitates consultative processes, which aim to provide communities with data on as many different options as possible to allow informed choices to be made.

2.5 Future developments

Urban building energy modelling is a fast-developing field, and future developments are likely to focus on coupling multi-domain simulation models, which include buildings, district energy systems, urban microclimate, transportation, and electricity transmission networks, to effectively model complex urban systems. Automatic integration of city data is another key development strand [29].

References

[1] World Green Building Council. '*Bringing Embodied Carbon Upfront: Coordinated Action for the Building and Construction Sector to Tackle Embodied Carbon*'. London, UK: World Green Building Council; 2019

[2] C40 Cities. '*The multiple benefits of deep retrofits - A toolkit for cities*'. London: C40 Cities; 2020

[3] Clair A., Baker E. 'Cold homes and mental health harm: Evidence from the UK Household Longitudinal Study'. *Social Science & Medicine*. 2022; 314:115461
[4] World Health Organisation. '*Global Health Observatory: Household air pollution attributable deaths*' [online]. 2024. Available from https://www.who.int/data/gho/data/indicators/indicator-details/GHO/household-air-pollution-attributable-deaths [Accessed 22 January 2024]
[5] Hong T., Chen Y., Luo X., Luo N., Lee S.H. 'Ten questions on urban building energy modeling'. *Building and Environment*. 2020;168:106508
[6] Langevin J., Reyna J.L., Ebrahimigharehbaghi S., *et al.*, 'Developing a common approach for classifying building stock energy models'. *Renewable and Sustainable Energy Reviews*. 2020;133:110276
[7] Yamaguchi Y., Suzuki Y, Choudary R., Booth A., Shimoda Y. 'Urban-scale energy modelling of food supermarket considering uncertainty' in: *Proceedings of BS2013. 13th Conference of International Building Performance Simulation Association*. Chambéry, France. 2013, pp. 1382–88
[8] Parker A., Horsey H., Dahlhausen M., *et al.* '*Comstock reference documentation: Version 1*'. NREL/NREL/TP-5500-83819. Golden, CO: National Renewable Energy Laboratory; 2023
[9] Mathur A., Fennell P., Rawal R., Korolija I. 'Assessing a fit-for-purpose urban building energy modelling framework with reference to Ahmedabad'. *Science and Technology for the Built Environment*. 2021;**27**(8):1075–1103
[10] Barbour E., Davila C.C., Gupta S., Reinhart C., Kaur J., González M.C. 'Planning for sustainable cities by estimating building occupancy with mobile phones'. *Nature Communications*. 2019;**10**(1):3736
[11] Biljecki F., Stoter J., Ledoux H., Zlatanova S., Çöltekin A. 'Applications of 3D city models: State of the art review'. *ISPRS International Journal of Geo-Information*. 2015;**B**(4): 2842–89
[12] Cerezo Davila C., Reinhart C.F., Bemis J.L., 'Modeling Boston: A workflow for the efficient generation and maintenance of urban building energy models from existing geospatial datasets'. *Energy*. 2016;117:237–50
[13] Monien D., Strzalka A., Koukofikis A., Coors V., Eicker U. 'Comparison of building modelling assumptions and methods for urban scale heat demand forecasting'. *Future Cities and Environment*. 2017;3:1–3
[14] Nouvel R., Zirak M., Coors V., Eicker U. 'The influence of data quality on urban heating demand modeling using 3D city models'. *Computers, Environment and Urban Systems*. 2017;64:68–80
[15] Saretta E., Bonomo P., Frontini F. 'A calculation method for the BIPV potential of Swiss façades at LOD2.5 in urban areas: A case from Ticino region'. *Solar Energy*. 2020;195:150–65
[16] Oraiopoulos A., Wieser M., Verdiere M., Fennell P., Ruyssevelt P. 'Thermal comfort in low-income housing in informal settlements in Lima, Peru Towards a localised adaptive comfort standard' *18th Conference of IBPSA. Presented at the Building Simulation 2023*. Shanghai, China, 2023

[17] Ballarini I., Corgnati S.P., Corrado V. 'Use of reference buildings to assess the energy saving potentials of the residential building stock: The experience of TABULA project'. *Energy Policy*. 2014;68:273–84

[18] Deru M., Field K., Studer D., *et al.* (2011) 'U.S. department of energy commercial reference building models of the national building stock'. Golden, CO: National Renewable Energy Laboratory; 2011 NREL/TP-5500-46861, 1009264

[19] Bueno B., Norford L., Hidalgo J., Pigeon G. 'The urban weather generator'. *Journal of Building Performance Simulation*. 2013;**6**(4):269–81

[20] Berthou T., Duplessis B., Stabat P., Rivière P., Marchio D. 'Urban energy models validation in data scarcity context: Case of the electricity consumption in the French residential sector'. *Building Simulation*. 2019; 16:3140–7

[21] Hong T., Ferrando M., Lou X., Causone F. 'Modelling and analysis of heat emissions from buildings to ambient air'. *Applied Energy*. 2020;277:115566

[22] Amrith S., Korolija I., Fennell P. Rovas D., Ruyssevelt P. 'Optimising building stock retrofits for urban-scale action planning: Improving the feasibility without losing information'. *Journal of Building Performance Simulation*. 2025:1–16

[23] Evans S., Godoy-Shimizu D., Steadman P., Amrith S., Humphrey D., Ruyssevelt P. 'Getting to net zero: Islington's social housing stock'. *Buildings and Cities*. 2023;**4**(1):524–44

[24] Pincetl S., Gustafson H., Federico F., Fournier E.D., Cudd R., Porse E. 'Building an energy atlas'. in S. Pincetl *et al.* (eds.) *Energy Use in Cities: A Roadmap for Urban Transitions*. Cham: Springer International Publishing; 2020, pp. 45–64

[25] Cihan Kayacetin N., Versele A. 'A circular and bio-based renovation strategy for low-income neighbourhoods'. *IOP Conference Series: Earth and Environmental Science*. 2022;**1078**(1):12080

[26] Eurostat. '*Natural gas price statistics*' [online]. 2024. Available from https://ec.europa.eu/eurostat/statistics-explained/index.php?title=Natural_gas_price_statistics [Accessed 4 April 2025]

[27] Heyse T. *City of Ghent Climate Plan 2020-2025*. Ghent: City of Ghent; 2020

[28] Sekon J.B.H. 'A review of feature manipulation engine'. *Jurnal Kejuruteraan*. 2018;**1**(1)

[29] Hong T., Chen Y., Luo X., Luo N., Lee S. H. 'Ten questions on urban building energy modeling'. *Building and Environment*. 2025;168:106508

Chapter 3

Retrofit as a service: a human-centred design approach

Val Mitchell[1], *Victoria Haines*[1] *and Stuart Cockbill*[1]

3.1 Introduction

To enable effective retrofit of homes at scale, new strategies are needed that encourage timely and meaningful engagement between householders and relevant stakeholders. In this chapter, we explore how Service Design, a human-centred and systemic approach to innovation, can be applied to domestic energy retrofits to achieve beneficial outcomes for householders and stakeholders. Service Design is a relatively new design discipline [1] that orchestrates the human, communication, infrastructure, policy, and material components of a service to create value for service users and 'stakeholders' (adapted from an interview with Birgit Mager [2]). Stickdorn *et al.* [3] define Service Design as a 'human-centred, collaborative, interdisciplinary, iterative approach which uses research, prototyping, and a set of easily understood activities and visualisation tools to orchestrate experiences that meet the needs of the business, the user, and other stakeholders', with stakeholders defined as those that have a role in delivery of, or benefit from, a service, directly or indirectly.

In this chapter, we draw on over 20 years of energy sector research in household behaviour and user-centred design to propose how Service Design can be used to tackle many of the complex systemic barriers and help accelerate retrofit at scale. We refer to the human-centred research by ourselves and colleagues in the School of Design and Creative Arts at Loughborough University that has, over time, led us to conceptualise 'retrofit as a service' and to embrace Service Design as a holistic approach to working collaboratively with disciplines beyond design and wider stakeholders in industry and government. We first draw parallels between retrofit and the core principles of Service Design, to describe the evolution of our approach and thinking and conclude with our reflections on the role of Service Design for the future of domestic energy efficiency retrofit as a contribution towards tackling climate change.

[1]School of Design and Creative Arts, Loughborough University, United Kingdom

Retrofit is an essential part of the decarbonisation of the UK housing stock. In 2022, domestic energy consumption accounted for about a third of total energy use in the UK [4], primarily through use of energy to heat living space and hot water. Heat is lost through the building fabric, especially where there is little or no insulation to walls, roofs, and floors, and through gaps where warmed air can escape. Forty-eight per cent of homes in England have no cavity or solid wall insulation and 62% have less than 200 mm of loft insulation [5]. To meet the carbon reduction target set by the UK Government [6] and with no prospect of the housing stock being demolished and rebuilt in the near to mid-future, homes must be retrofitted with energy efficiency measures to reduce energy demand. This includes the addition of wall, floor and roof insulation, replacement of windows with double or triple glazing, replacement of heating systems from condensing gas boilers to electric heat pumps and district heat networks, and inclusion of ventilation systems to prevent overheating. There is the added pressure of scale, with 28 million homes in the UK [7] and only 12% of homes in England and Wales with an EPC rating of B or above [8]. Almost all homes will need some sort of retrofit measure (walls, floors, roofs, heating system, glazing) [9] and with the lifespan of some measures, such as glazing and heat pumps, of less than 20 years, many will require repeated refurbishment before 2050. To meet our energy demand targets, a retrofit of the domestic mass market is essential.

Whilst a wide variety of energy efficiency technologies already exist and are effective, the process of retrofitting them into homes is complex. The housing stock is not homogenous, nor is the occupant population. Beyond types of tenure (owner-occupier, private rental, or social housing), human factors such as individual aspirations, expertise, interest, and budget affect a householder's willingness to engage in the retrofit of their home. For the retrofit process to be effective, others in the supply chain must be equally considered, including planners, suppliers, installers, and those who train the workforce. The process of retrofitting even a single home can be a drawn-out process that can extend over many years, subject to financial or other life pressures [10] and may involve multiple occupants as homes are bought and sold. Multiple service providers may be involved including skilled craftspeople, retrofit specialists, general trades, such as plumbers and electricians, and householders undertaking work themselves, often referred to as 'Do-It-Yourself' or DIY. Multiple financial stakeholders might also need to be navigated by the householder to explore options and secure funding for work. Retrofitting needs to be effective at an individual level, to ensure a positive experience for the householder, as well as at scale, to provide viable and adaptable business models to achieve the carbon reduction targets set locally, nationally, and globally in a timely way. As a potential solution, human-centred design offers a focus on users and associated stakeholders to consider the perspectives, goals, and motivations of people, often as part of complex systems.

Services are the 'soft infrastructure of society' [2]; they are everywhere, often invisible yet essential to getting things done. In the simplest terms, a service is 'something that helps someone do something' [11]. A service exists to enable people to achieve a goal; this might be relatively straightforward, for example, buying a cup of coffee, or more complex and drawn out, for example, moving house. Although the parts of a service might be provided by several different organisations, Downe [11] asserts that 'from the point of view of the user, a service is a set of continuous actions towards that goal regardless of who is providing it'. However, services do not exist solely to meet user needs; they also exist to deliver the goals or intended outcomes of the organisation(s) that provide them. In the public sector, the outcomes intended for the service provider are likely to be less about profit and more about the delivery of policy intent, for example, meeting the UK's targets for decarbonisation of existing housing stock.

Service Design is concerned with systematically applying design methodology and principles to the design of services [12,13]. The practice of Service Design can be applied to a single service, or a combination of services, that enable a person to achieve a goal. In any case, it begins with a focus on the customer and brings user and stakeholder needs together through a consideration of the end-to-end service journey. This builds a bridge between currently disconnected services and the multiple stakeholders within a supply chain. Service Design draws attention to both the tangible and intangible elements of a service, as well as considering the interwoven digital and physical interactions that enable service delivery. Service Design makes frequent use of visual tools to facilitate decision-making and communicate, for example, through the creation of a 'service blueprint', how a service is currently delivered or might be delivered in the future. This allows people from different backgrounds and perspectives to build a shared understanding of how a service works and co-ordinate activities and decision-making processes, strategically and operationally. Whilst this chapter does not attempt to provide a comprehensive overview of Service Design, we draw on its core principles and use examples from our work, and that of others, to propose where it can add value within energy demand reduction and, in particular, domestic retrofit.

3.2 The core principles of Service Design

Service Design is a human-centred design approach that seeks to innovate new, or improve existing, services to make them more useful, desirable, and effective for both customers and the organisations delivering them [14]. Service Design approaches are increasingly being adopted to tackle complex and interrelated issues by understanding the desired outcomes a customer wishes to achieve, their end-to-end experience of the service that is set up to deliver these outcomes, and how this service is designed and delivered by the service provider(s). Service Design as a holistic, collaborative, and user-centric approach to design is underpinned by six key principles [3]:

1. Service Design is human-centred
2. Service Design is collaborative
3. Service Design is iterative

4. Service Design is sequential
5. Service Design is real
6. Service Design is holistic

Each of these is discussed in relation to domestic energy efficiency retrofit, drawing on our experience from past research projects, including CALEBRE[1], REFIT[2], LEEDR[3], Challenge 100[4] and others.

3.2.1 Service Design is human-centred

Service Design begins with identifying the current and future needs of all service users and others directly or indirectly impacted by the service. These needs are identified with particular attention paid to the problems or frustrations users currently face and what they will need from the service being improved, or newly designed, to meet their goals. For retrofit solutions, they need to be tailored to meet the householders' specific requirements and preferences. With over 60% of homes in England being owner-occupied [5], and these householders independently making decisions on whether or not to retrofit, it is particularly important to engage them in the retrofit process in a positive way, ensuring their varying needs and concerns are addressed.

A key aspect of human-centred Service Design is the defining of services from 'the outside in' [15]. This means starting with an understanding of what the service users are trying to achieve and then re-defining everything that is involved in delivering and using the service from the point of view of how the users of the service will conceive it. This goes beyond defining the steps in the 'user journey' to include how all aspects of the service are represented and structured including what operational and frontline teams do and how, how support models and processes work, and how data are collected and shared.

The One-Stop Shop (OSS) is becoming a popular model within Europe, including the UK, for driving forward retrofit at scale, and the widespread adoption of this model provides evidence of human-centred design approaches beginning to influence how policymakers are tackling the retrofit problem. OSS models offer homeowners all the services needed to complete a home energy retrofit, from a home energy assessment, applying for grants and financing, commissioning contractors, carrying out work, and providing quality assurance [16]. OSS often chooses consumer-friendly names that highlight the experiential benefits of retrofit alongside economic and environmental benefits, for example, 'Cosy Homes Oxfordshire' [17]. By providing householders with a single point of contact, OSS

[1]CALEBRE project - Consumer Appealing Low Energy Technologies for Building Retrofitting, Grant Reference EP/G000387/1

[2]REFIT project - Personalised Retrofit Decision Support Tools for UK Homes using Smart Home Technology, Grant Reference EP/K002457/1

[3]LEEDR project - Low Effort Energy Demand Reduction, Grant Reference EP/I000267/1

[4]Challenge 100 – Tackling fuel poverty for 100 families in 100 homes, in 100 days. E.ON UK plc initiative

aims to guide homeowners through the whole retrofit process including simplifying and supporting decision-making throughout the renovation journey [18].

A challenge facing the OSS model is how to understand and support the diverse needs of householders through one service offering. Householders are likely to have different needs and motivations regarding the outcomes they want from retrofit and, therefore, a 'one size fits all' service offering is unlikely to support retrofit at scale. Whilst it is impractical to understand and meet every household's bespoke requirements, the design tool of 'personas' can help to group types of people and provide a summary of their goals and behaviours, preferences, and pain points. These representations of archetypal people are widely used within Experience Design to support the design of digital products and services [19,20] but have also been used to inform policy design, including design for social good, [21] and systems design, including energy transitions [22]. Personas, based on in-depth qualitative research, allow designers to consider the needs of a diverse range of users whilst keeping a focus on the context of use. They differ from market segments, which are usually based on large sets of quantitative market research data and used primarily to identify groups of consumers who will be receptive to a similar product, service, or marketing campaign. Segments are usually defined by a quantitative breakdown of specific socio-demographic variables such as age, gender, income, and location, whereas personas focus on rich qualitative data from smaller samples, with the primary purpose to aid design decision-making [23].

When used within a Service Design approach, persona sets can help tailor retrofit service offerings to different groups of householders so that they are more easily woven into established home improvement behaviours. Within the CALEBRE project, we created a set of personas with a home improvement focus a decade ago [24], which have since informed further persona sets including the early adopter retrofit personas created by Carbon Co-op as part of the People Powered Retrofit project [25]. The use of personas has allowed us to direct attention towards the emotional, as well as rational, aspects of retrofit decision-making, which, as confirmed by the work of Bobrova *et al.* [26], is often overlooked within energy retrofit projects. They argue that retrofit research and policies typically focus on the cognitive aspects of retrofit decision-making, including financial benefits such as savings on energy bills and whether improvements will lead to a return on investment. From the perspective of homeowners, energy retrofit is a home improvement process, and, therefore, from the outset of our human-centred research, we have proposed that the emotional and rational barriers to undertaking retrofit are best understood by researching how households currently go about making improvements to their homes [27]. We used creative storytelling methods, including a bespoke timeline tool [28] on the CALEBRE project, to capture the home improvement experiences over time of households living in hard-to-treat homes; those with solid walls, without a cavity that could be filled with insulation. This rich qualitative information was used to create a set of personas that represented archetypal owner-occupier households living in solid-wall UK homes [24], including:

- The idealist restorer - where the property is a project, the householders are keen to regain or preserve the aesthetic and period features of their home and will take the time to research retrofit alternatives.
- The affluent service seeker - where the property is a pleasure, and the householders seek quality in their retrofit but also value for money from the specialist professionals they employ to undertake the work.
- The property ladder climbers - where the property is purchased to get into the property market, and the householders make improvements with the intention of selling on quickly.
- The functional pragmatists - where the property is a home and a place to live, but home improvements are a hassle rather than a hobby. These householders are only likely to make improvements when time and finances permit or repairs become urgent.

We also identified personas who were stalled in their home improvement journey through life pressures or a lack of finances.

Informed by the CALEBRE persona approach, Carbon Co-op created a set of retrofit personas for their People Powered Retrofit service to represent householders likely to access a One-Stop Shop service as early adopters but differing from each other with respect to their motivations for retrofit [29]. These were:

- Civic-minded retirees - motivated to contribute to society and looking to spend their retirement in a comfortable home.
- Climate pragmatists - motivated to reduce their emissions, willing to invest time and money in doing so and seeking an expert to help them do this.
- Climate idealists - highly motivated to achieve the highest degree of emission reductions, irrespective of costs.
- 'Techno-optimists' - motivated by wanting to try new technologies and identify as eco-minded/wealthy innovators. This group was also likely to be attracted to premium-branded products from aspirational brands such as Tesla.

The Loughborough personas have also been used to inform research and development of retrofit services, for example, modelling and order of retrofit [30]. The Carbon Co-op personas reflect a more recent trend towards using personas or more generic archetypes to inform the design of customer engagement strategies, particularly within the context of OSS [31].

3.2.2 Service Design is collaborative

Fundamental to the Service Design approach is creating a context where users and stakeholders from different backgrounds can be actively engaged in the generation and evaluation of service concepts. This is particularly important when new groups of users or stakeholders are involved. Alongside those who will use a service, Service Design considers the needs and goals of all who have a stake in providing, delivering, and maintaining a service and the relationships that exist, or need to exist, between them. This is critical in the successful delivery of domestic energy efficiency retrofits, given the complexity of the wide range of stakeholders involved, including local authorities, planners, energy companies, construction professionals, etc., and the complex interactions necessary. By applying a Service Design approach, the roles and goals of all these stakeholders can be considered and services designed to meet these through appropriate collaboration.

Our work on the REFIT project used co-design with householders, researchers, designers, and building energy technologists as a collaborative approach to developing new retrofit services [32]. A series of co-design workshops, as part of the wider project, were carried out to generate a set of 12 future energy-related services that would benefit householders (Figure 3.1). Co-design focuses on designing 'with', not 'for', users and views people as 'experts of their own experiences' [33]. This ensures a deeper understanding of the barriers and enablers from the users' perspectives, leading directly into design, with the benefit of greater market success. The co-design activities in REFIT included first sensitising a set of ten households to the retrofit context using their own personalised smart meter and smart home data before ideas for future services were constructed based on insights captured. The householders explored what types of information they would find useful and through which touchpoints it would be most meaningfully delivered to them. A round-table discussion involving technologists and human-centred designers followed, with the aim of developing the feasibility and viability of the future service propositions. Householders then returned in a closing workshop to explore whether the services presented were meaningful, useful, and desirable,

Figure 3.1 Co-design of future energy services with REFIT project participants [32]

noting any further improvements and challenges to their introduction. This collaborative approach increased the breadth and number of potential future service propositions identified and allowed a critical assessment of the services.

Service Design uses the theatre metaphor of frontstage (facing the audience) and backstage (behind the scenes) to conceptually distinguish between the side of the service visible to the customer and the support processes in operation that deliver the customer journey and experience. The frontstage and backstage components are separated by a conceptual boundary referred to as 'the line of visibility' [34]; in retrofit, this helps to differentiate between householder-facing interactions and supporting processes 'behind the scenes'. Examples of frontstage activities in domestic retrofit might include a householder exploring early options through a home energy audit and interacting with a retrofit co-ordinator and trade professionals, firstly to scope the work and then again when the work is carried out. Backstage activities include planning and approval processes, procurement of materials and scheduling of work by contractors, invoicing and customer support, as well as more fundamental activities, such as training of skilled workers at college. By including stakeholders from both frontstage and backstage in co-design of a retrofit service, there is an obvious opportunity to develop a deep and detailed understanding of a service from all perspectives and make sure frontstage and backstage activities align to provide a timely and convenient service to householders.

Carbon Co-op extended the use of personas within their Service Design approach to include the development of frontstage 'contractor personas' to demonstrate that contractors, like householders, have varied motivations for engaging in retrofit and different trade professionals face different barriers when seeking to provide retrofit services. The project recognised that, in order to provide the best possible service to householders, the needs of the RMI (Refurbishment, Maintenance and Improvement) contractors also need to be understood [25].

3.2.3 Service Design is iterative

Service Design recognises that complex problems are more likely to be successfully addressed when the design process follows cycles of exploring needs and evaluating potential solutions with users and stakeholders through rapid and iterative prototyping. Iteration allows Service Designers to incrementally explore and test interventions at different scales, from singular touchpoints to complete end-to-end service experiences.

Our work on the REFIT project explored a highly focused prototype touchpoint for a future energy advice service through iteratively sensitising householders to their own energy data in the context of their own home; essentially, an improved version of a home energy audit [35]. The householder engagements showed the benefits of iterative sensemaking around energy use – the data needs to be personalised to have any meaning, and householders need help understanding it – although immediate impacts on retrofit decisions were limited [36]. Nevertheless, the value of iterative cycles of testing with householders using their own real energy data was demonstrated and increased tailoring of design outcomes in response to user feedback was shown to uncover further user needs.

Living Labs provide a real-world context for iterative user-centred innovation and have been applied to design energy retrofit services. Living Labs use co-design approaches to focus on users' needs and values and, through fostering communication and collaboration between householders and stakeholders, support a participatory decision-making process [37]. Voytenko *et al.* [38] assert that Living Labs must be local or small-scale, in some way experimental, participatory, and provide a platform for direct user involvement in decision-making. Sovacool *et al.* [39] describe an Energy Systems Catapult trial that used the Living Lab model to evaluate smart heating controls in 100 homes spread across the four locations in the UK: Birmingham, Bridgend, Manchester, and Newcastle. The trial provided insight into the real-life heating practices of the participating households and provided recommendations for heating system designers, energy retailers, and policymakers. Interestingly, the smart heating preferences and practices of householders were found to co-evolve over time, with users coming to 'learn' their own needs and shift behaviours to meet them. Therefore, as found in the REFIT study, deeper insight into user needs was achieved than one would expect from attitudinal user research by providing participants with the opportunity to make sense of new technology over time in the context of their everyday lives.

3.2.4 Service Design is sequential

Service Design visualises services as a sequence of interrelated steps and interactions that unfold over time. It does this using customer journey maps and service blueprints, which document the whole process of delivering a service as a sequence of steps and interactions over time (depicted horizontally), detailing the customer journey and experience, with layers of supporting activities and processes (depicted vertically) that build out a picture of the complexity of the backstage. In a service blueprint, the frontstage and backstage actors are detailed with interactions and dependencies using directional arrows to mark 'flows' of data, information, or processes. Visual representations of customer journeys and how one or more services deliver these can be co-created with users and stakeholders to create a shared understanding of their needs and to generate, prototype, and evaluate service concepts. The service blueprint includes details of how a service is experienced by users through 'touchpoints' in the frontstage and delivered through dependencies and processes in the backstage.

In the delivery of retrofit as a service, these touchpoints refer to the points of contact between the householders and the service. For the OSS delivery model, Service Design considers the design of all touchpoint interactions within the retrofit customer journey with the goal of making sure households receive the right information at the right time in the most appropriate form. This includes designing interactions with frontstage staff, website content and navigation, and the design of energy advice reports that might be read on screen or as a printed report. The Retrofit Playbook [31] provides a detailed customer journey that defines the stages of an OSS from the point of view of the householder, together with the myriad of touchpoints that could be used to deliver this. The detail of this customer journey was created by Otley Energy, a Community Benefit Society that has championed the use of Service Design approaches as a means to accelerate domestic retrofit in the UK. Otley Energy proposes a retrofit customer journey broken down into the following stages and sub-stages: Awareness; Delivery of a whole house plan (Consideration, Acquisition, Delivery); Delivery of retrofit measures (Further consideration, Acquisition, Delivery); and Follow Up (Figure 3.2). When combined with a persona-led approach, Service Design enables retrofit stakeholders to consider how service touchpoints for each stage of the service journey can be tailored to meet the needs of different persona groups, taking into account their motivations and the channels of communication they are likely to already be engaging with. For example, local estate agents could be used to raise awareness of an OSS retrofit service in a way that motivates 'Property Ladder Climbers' to include energy retrofit as part of their plans to renovate and add value to the property price prior to re-sale [40].

The experience of the retrofit process can be 'overwhelmingly difficult' [41] and can impact the satisfaction of the retrofit outcome for the customer or householder. A step-by-step approach is therefore usually favoured [42]. Customer journeys can extend beyond the immediate retrofit process to advocacy to others about their retrofit experience, and this has been a powerful tool in schemes, such as SuperHomes [43], a community of energy-conscious individuals supporting the retrofit of UK homes, where householders have shared their experiences with others earlier in the journey. SuperHomes is a charity supported by the National Energy Foundation, set up to help achieve net zero by facilitating the delivery of home retrofits. Beyond the UK context, De Wilde and Spaargaren [44] have also used customer journey maps to gain feedback on three designs for a retrofit service, which revealed the need for retrofit advice services to be tailored to the diversity of homeowners. Their work suggests that retrofitting can be enabled through customer-centric Service Design, highlighting the importance and value of understanding customer journeys and experiences at different levels.

As a retrofit can be a complex process, often over an extended period of time, it is also important to plan the retrofit process as a series of steps that homeowners can understand and manage, from initial assessment to the final installation, maintenance, and beyond. Our research on the CALEBRE project into home improvements by people in older, hard-to-treat homes identified how retrofitting measures can take place over very long periods of time and often stall through the pressures of life or lack of finance [45]. In many cases, it is not possible to completely retrofit a whole house as a single package of measures, as the occupants cannot afford the capital costs, cannot move out to avoid the disruption, nor want to. As a result, a retrofit can become a long, drawn-out, step-by-step process [42]. Therefore, it is important to consider the whole range of planned improvements from the outset to ensure that refurbishment measures are introduced in an appropriate order [30]. For example, sizing a boiler or heat pump should be based on the required heating load of the property, influenced by the level of insulation and

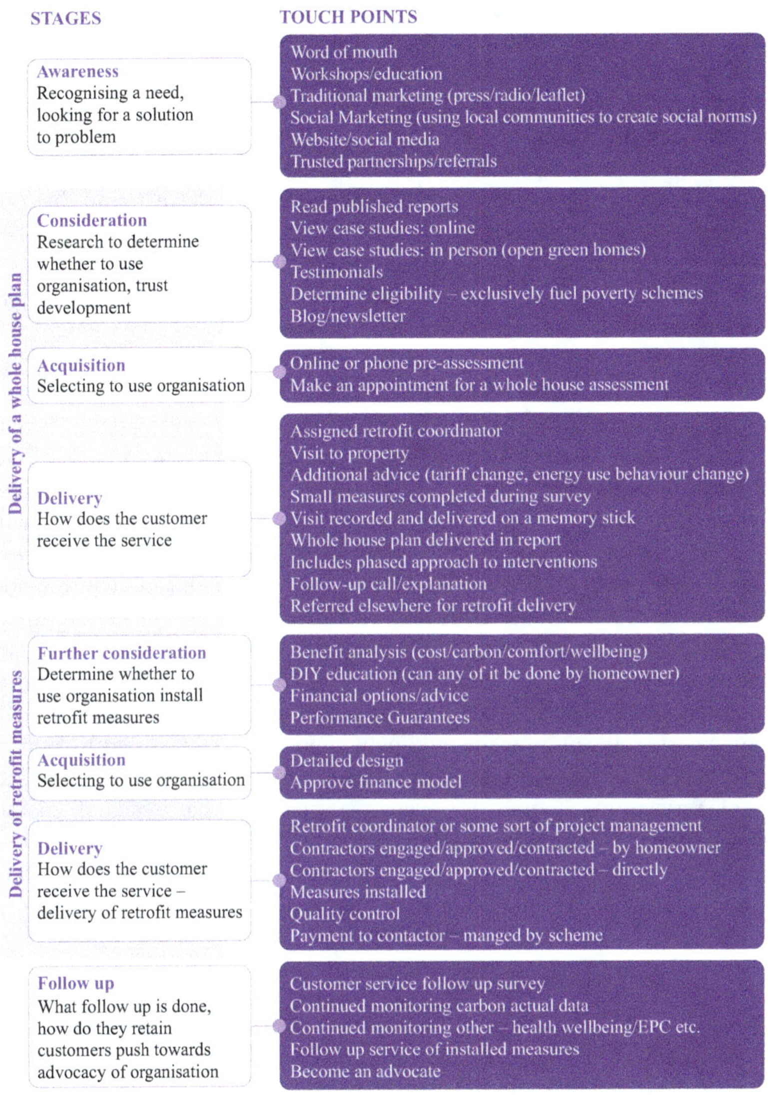

Figure 3.2 A retrofit customer journey for an OSS (from [31], credit Otley Energy)

airtightness. However, if the boiler is replaced before the other fabric measures are completed, it is very likely that it will be oversized for the eventual demand. Another consideration for the sequence of retrofits relates to thermal storage. Many homes now have a combination boiler, which heats hot water on demand, rather than the supply coming from a hot water storage tank. This meets the occupants' needs by providing a constant supply of hot water and by releasing and repurposing additional space in their home, resulting from the removal of the storage tank. However, with the move to heat pumps, which require a thermal store, homes will need to find space that may no longer be available.

Despite policy interventions favouring whole-house retrofitting, financial constraints mean that many householders are likely to opt to retrofit their homes in a piecemeal fashion, when sufficient resources are available. Resources also include people's appetite for the refurbishments, and we have found that people stall in their retrofit journey for all sorts of reasons, including changes to health, work, family, and interests. As a result, any retrofit service needs to be able to cope with this, providing a realistic overview of the work that could be carried out, the stakeholders who would need to be involved, and the financial commitment at each stage. It is also important to manage expectations of the retrofit process from all perspectives. The emphasis on including stakeholders in the service, not just focusing on end users, gives the opportunity to design and develop success metrics for all actors early on so that possible retrofit measures can be tested against meaningful criteria that evidence value. This might include, building in sensors in a home at the time of the retrofit to monitor the performance of a measure or intervention over a longer period of time, rather than retrospectively trying to understand smart meter data without wider context – all of which can help to evidence the added value of retrofitting, especially to those who 'stall'. Once a retrofit service is conceptualised as a customer journey, with supporting frontstage interactions and backstage processes, different personas can be used to play out different use scenarios. For example, predictions of where customers are likely to stall on their retrofit journey can be made, and new opportunities identified to better evidence the benefits of retrofitting to them. Co-design can then be used to refine the design of service touchpoints to overcome predicted barriers and encourage engagement with the service, as illustrated in 'Toastie', a conceptual retrofit Service Design by Kinnings [46] (Figure 3.3), that co-designs tailored retrofit plans with householders given their personal and contextual circumstances.

3.2.5 Service Design is real

Service Design pays attention to the context of use and emphasises that user and stakeholder needs should be researched and prototyped in real-world contexts. This focus on the reality of life is critical to ensuring services, like retrofitting, meet the real needs of users and stakeholders, not just those assumed. Our research has often looked at these real-world contexts to understand the messy reality of everyday life, using tools such as timelines or tours to help situate the participants within a particular context, enabling them to tell us about their real experiences, grounding our findings in the reality of what people actually do in their homes, and why. In the CALEBRE project, we used a timeline tool, developed for the project, to encourage householders to share their stories about their own home improvement activities [45]. This allowed us to collect data in a structured format but as part of a relaxed, storytelling approach. As householders volunteered information, the timeline was drawn up in front of them, using a pen on magnetic boards, supported by a bespoke set of magnetic cards. This

Figure 3.3 Toastie – a service concept designed to offer a seamless and efficient retrofit-planning experience [46]

allowed participants to reflect on the reality of their experiences of retrofitting their homes through decoration, window and heating system replacement, and extending and modifying their homes, including both highs and lows.

Within the LEEDR Project, video home tours were undertaken, where householders spoke about energy use in the context of their own homes. An initial interactive floorplan activity, as a generative task, allowed householders to talk about how they made use of the space in their homes as part of their family routines and dynamics [47]. As well as a trust-building activity, it allowed the research team to understand the realities of people's lives and for the family members to challenge each other on the truthfulness of their average day. Home video tours followed, to explore everyday practices in context, where researchers accompanied household members around their homes to gain a deep insight into the use of the home from an energy demand perspective. This allowed the research team to capture real experiences of people's lives to create meaningful stories to inform the design of concepts that aligned with everyday life [47–49].

In an award-winning scheme, retrofit at a community scale was tackled by the energy utility company E.ON through Challenge 100 [50], supported by expertise from Loughborough University. E.ON carried out deep retrofits to 100 homes in fuel poverty by installing tailored energy efficiency measures, including new heating systems, draught exclusion, wall and loft insulation, and providing support and financial advice to householders. This project enabled homes to be more energy efficient with comfortable living conditions and highlighted many of the realities of retrofitting homes at a moderate scale, including the individual nature of the retrofits to match the property and its household's requirements and the importance of home visits to build trust and explain the details of the renovations. The project found examples of associated work that was needed, such as clearing or replacing guttering, that were not covered by the grant funding available.

Service blueprints and customer journey maps used in this context can help retrospectively understand the householders' complete retrofit process, including the pleasure and pain points, and identify further opportunities to elevate both their experience with the service and the outcomes they can achieve by retrofitting. As a way of capturing the learnings from Challenge 100 and other pilot schemes, mapping the customer experience and supporting service makes tangible the value of the service as delivered and informs the development of future schemes.

Research into the willingness of householders to adopt retrofitted improvement measures has shown that people might not be willing to accept all measures even when potential savings are modelled with their own data [51]. A bespoke set of measures is, therefore, needed that fits with the context of the householders' everyday lives, which can then be modelled to provide a more realistic pathway to retrofit. Scaling this up would allow a model of the decarbonisation of the housing stock to be created with more accuracy.

3.2.6 Service Design is holistic

Whilst the retrofit process is seldom carried out as a single and complete package, it is important to consider it holistically. Service Design takes a holistic approach to understanding problems and proposing and evaluating potential solutions by considering the complete service ecosystem. This includes a careful consideration of the end-to-end customer experience, with multiple touchpoints, both tangible and intangible, multiple stakeholders, and in some cases, multiple services, and all the interactions between them. Service Design also allows an interdisciplinary approach, using tools from a range of origins, to support the understanding and development of the system as a whole. This enables the design and development of the retrofit experience from multiple stakeholder perspectives and the development of a system that connects all these stakeholders and services together. By repeatedly zooming in on the detail (a customer-specific service experience) and then back out to the big picture (the wider retrofit ecosystem), a Service Design approach is able to consider the wider impacts of any actions, interactions, or changes. The power of Service Design is its ability to create shared understanding at different scales concurrently, which is of critical importance where a service or system, like retrofit, is complex, fragmented, and disjointed.

For those who require empirical justification for allocation of resources, such as policymakers and grant providers, this holistic perspective provides an insight into the very real barriers to change at the household level that prevent retrofitting taking place, whilst simultaneously understanding how current policy and service provision are failing. For example, householder concerns about disruption to décor and the hassle of the work being undertaken are not always allayed by installers, where time and budgets drive quality and training provision and motivation to do a 'good job' are varied [27]. Understanding these factors needs deep-dive qualitative research with households to gather a holistic perspective of retrofitting that uncovers underlying concerns and motivations. Although people are driven by a desire to improve the energy efficiency of their homes, in particular reducing energy bills, many are also keen to improve comfort and appearance of the property. This broader relationship with the home is encapsulated by several schemes, for example, the 'Better Homes Yorkshire' [52] campaign does not mention retrofit, focusing rather on health, warmth, and energy bills; 'Home efficiency upgrades' are offered instead. The added value of a holistic view is that the insight generated from zooming in can be better actioned and leveraged by the Service Designer whose role is to join up the dots and to help organisations, service providers, and policymakers improve what they are doing. This can be as simple as improving training provision for skilled workers and installers, through to reframing the retrofit problem and messaging around it.

3.3 Conclusions

Retrofitting the UK housing stock to achieve net zero targets is a complex problem that will require intersecting policy and socio-technical innovations to address. Service Design does not simplify this complexity, but by taking a whole systems approach and using visual tools, such as journey maps and blueprints as 'boundary objects' [53], it can help build a holistic picture of this complexity and provide the context for the multi-stakeholder engagement needed to accelerate retrofit at scale.

Service Designers, and their tools, enable key stakeholders to zoom out and 'see the system' [54], whilst simultaneously enabling zooming in to understand, from the customer perspective, the many real and nuanced contextual factors that householders face when engaging with energy retrofit. The visual and collaborative tools of Service Design, including journey maps and service blueprints, when used within a Service Design process, provide policymakers and retrofit service providers with the means to make complex intangible processes tangible. The possibilities that arise from this are far-reaching, from observing and interpreting needs and behaviours at the customer level and expressing and evaluating the quality of existing service designs, through to envisaging solutions that do not yet exist and choreographing how these can be brought to life [55].

Service Design, and Service Designers, help understand the end-to-end retrofit sequence, both from the perspective of the householder and other key stakeholders involved in the supply chain. This enables offerings – at the customer level – to be tailored to the needs of different customer groups and local contexts, whilst at the same time ensuring that these outcomes are feasible and viable in terms of delivery by service providers. Service Design, by making explicit processes and linkages between different retrofit stakeholders, brings to the foreground of decision-making how the process is undertaken, step-by-step and over time, where the touchpoints are, and what interactions are most critical. As retrofit sequences are understood,

Service Design helps us to iterate existing services and prospective service concepts. Through cycles of exploring user and stakeholder needs and evaluating potential solutions through rapid and iterative prototyping, greater affinity between customer outcomes and service delivery can be achieved. In retrofit as a service, meaningful value is co-created across the end-to-end retrofit customer experience; each touchpoint is optimised to ensure the sustained engagement of householders with domestic retrofits, action is prompted and taken, confidence is instilled, and longer-term support is provided. Missed opportunities can be capitalised on, such as at the time of home purchase, through this nuanced understanding of retrofit personas and wider home improvement journeys.

For service providers, Service Design provides greater visibility and understanding of existing and new business goals by making visible the value flows, or lack thereof, within the retrofit ecosystem. New connections can be made and new value provided through connecting previously unconnected stakeholders, all of whom can keep decision-making focused on customer needs. Service Design and Service Designers become the 'glue' that facilitates multi-stakeholder engagement and delivery of retrofit as a service. For policy, Service Design enables more strategic targeting and usage of government funding. Through 'seeing the system', weak links and unintended consequences, such as misuse of funds, can be mitigated and corrected, and proper usage of funds enabled. Policy makers can see and understand the complexities of the retrofit experience at multiple levels and, importantly, how service providers respond to policy through service interventions. Policies can then be targeted to encourage service provision of the *right* type to enable the change we all need.

Recent years have seen an increase in funding towards the establishment of one-stop shops (OSS) for retrofitting homes in the UK, with an emerging role for cities as the catalysts for accelerating energy retrofit in the domestic housing sector [31]. The initial target market for these initiatives has been early adopters with the required finance and enthusiasm to embark on retrofitting and social housing providers who can be supported to retrofit multiple properties in a locality. But to engage the wider population within the UK, an inclusive approach is needed. This must focus on understanding the needs of householders and tenants, landlords and suppliers, beyond the technological solutions. Service Design is used successfully in a range of other applications, such as healthcare and government digital services, but is in its infancy for domestic energy efficiency retrofits. The principles discussed in this chapter demonstrate how Service Design lends itself to the retrofitting of UK homes, dominated by owner-occupiers who each have a bespoke set of circumstances, against a backdrop of decarbonising the housing stock. As the UK moves towards summertime overheating [56], future retrofitting to keep homes cool in the summer will be needed alongside the more familiar wintertime home improvements. Service Design offers a credible way forward to consider new and evolving climate impacts on the UK housing stock and its occupants and what this means for achieving our net zero goals in the longer term.

References

[1] Holmlid S. 'Interaction design and service design: Expanding a comparison of design disciplines'. Presented at Nordes 2007: Design Inquiries; Stockholm, Sweden, 2007

[2] Penin L. *An Introduction to Service Design: Designing the Invisible*. London: Bloomsbury Publishing; 2018

[3] Stickdorn M., Hormess M.E., Lawrence A., Schneider J. *This is Service Design Doing: Applying Service Design Thinking in the Real World*. Sebastopol, CA: O'Reilly Media; 2018

[4] Department for Energy Security and Net Zero. *Energy Consumption in the UK* [online]. 2023. Available from https://www.gov.uk/government/statistics/energy-consumption-in-the-uk [Accessed 26 June 2024]

[5] Department for Levelling Up, Housing and Communities. *English Housing Survey 2022-2023: Headline report* [online]. 2023. Available from https://www.gov.uk/government/collections/english-housing-survey-2022-to-2023-headline-report [Accessed 26 June 2024]

[6] Wetherill M., Swan W., Abbott C. 'The influence of UK energy policy on low carbon retrofit in UK housing' in: *Presented at Retrofit 2012*. Salford, UK, January 2012

[7] National Audit Office. *Decarbonising Home Heating. HC 581*. London: National Audit Office; 2024

[8] Department for Levelling Up, Housing and Communities. '*Live tables on energy performance of buildings certificates*' [online]. 2024. Available from https://www.gov.uk/government/statistical-data-sets/live-tables-on-energy-performance-of-buildings-certificates [Accessed 26 June 2024]

[9] Butt B., Jones R.V., Fuertes A. 'Opportunities and barriers to business engagement in the UK domestic retrofit sector: An industry perspective'. *Building Services Engineering Research and Technology*. 2021;**42**(3):293–305

[10] Simpson K. '*Energy efficiency refurbishment of UK owner-occupied homes: The householders' perspective*'. PhD Thesis, Loughborough University; 2017 [online]. Available from https://repository.lboro.ac.uk/articles/thesis/energy_efficiency_refurbishment_of_uk_owner-occupied_homes_the_householders_perspective/9453971 [Accessed 26 June 2024]

[11] Downe L. *Good Services: How to Design Services that Work*. Amsterdam: BIS Publishers; 2020

[12] Bruce M., Bessant J. *Design in Business: Strategic Innovation through Design*. Harlow: Pearson Education; 2002

[13] Holmlid S., Evenson S. 'Bringing design to services' in: Presented at IBM Service Sciences, Management and Engineering Summit: Education for the 21st Century; New York, USA, 2006

[14] Moritz S. '*Service design – Practical access to an evolving field*' [online]. 2005. Koln International School of Design. Available from https://issuu.com/st_moritz/docs/pa2servicedesign/4 [Accessed 12 June 2024]

[15] Tarling K. *The Service Organization: How to Deliver and Lead Successful Services, Sustainably*. London: London Publishing Partnership; 2023
[16] McGinley O., Moran P., Goggins J. 'Key considerations in the design of a One-Stop-Shop retrofit model' Presented at Civil Engineering Research in Ireland Conference; Cork, Ireland, 2020, pp. 354–9
[17] Cosy Homes Oxfordshire. '*Cosy Homes Oxfordshire: About us*' [online]. 2024. Available from https://cosyhomesoxfordshire.org/about/ [Accessed 18 June 2024]
[18] Bjørneboe M.G., Svendsen S., Heller A. 'Evaluation of the renovation of a Danish single-family house based on measurements'. *Energy and Buildings*. 2017;150:189–99
[19] McKay E. *UI is Communication: How to Design Intuitive, User-Centered Interfaces by Focusing on Effective Communication*. San Francisco, CA: Morgan Kaufmann; 2013
[20] Mulder S., Yaar Z. *The User is Always Right: A Practical Guide to Creating and Using Personas for the Web (Voices That Matter)*. Berkeley, CA: New Riders; 2006
[21] Salminen J., Wenyun Guan K., Jung S.G., Jansen B. 'Use cases for design personas: A systematic review and new frontiers'. *Proceedings of the 2022 CHI Conference on Human Factors in Computing Systems*; New Orleans, USA. New York: ACM; 2022, pp. 1–21
[22] Cherry C., Thomas G., Groves C., *et al*. 'A personas-based approach to deliberating local decarbonisation scenarios: Findings and methodological insights'. *Energy Research & Social Science*. 2022;87:102455
[23] Pruitt J., Adlin T. *The Persona Lifecycle: Keeping People in Mind Throughout Product Design*. San Francisco, CA: Morgan Kaufmann Publishers; 2006
[24] Haines V., Mitchell V. 'A persona-based approach to domestic energy retrofit'. *Building Research & Information*. 2014;**42**(4):462–76
[25] Atkinson J., Heaslip M., Grimshaw O., *et al*. *People Powered Retrofit: A Community-Led Model for Owner-Occupier Retrofit*. Manchester: Carbon-Coop and URBED; 2019
[26] Bobrova Y., Papachristos G., Chiu L.F., Tikhomirova S., Coon T.M. 'Home for the Common Future (HCF): The use of home-meanings to promote domestic energy retrofit'. *Energy Research & Social Science*. 2024;107:103358
[27] Mallaband B., Haines V., Mitchell V. 'Barriers to domestic retrofit: Learning from past home improvement experiences' in Swan W., Brown P. (eds.) *Retrofitting the Built Environment*. Chichester: Wiley Blackwell; 2013. pp. 184–99
[28] Haines V.J., Mitchell V.A., Mallaband B. 'Using a practice-orientated approach to inform the design of energy efficiency measures for older homes'. Presented at ERSCP-EMSU Conference Proceedings. Delft & Cape Town; Delft, Netherlands, 2010
[29] Oxenaar S. '*People powered retrofit – Carbon co-op business model overview*' [online]. 2021. Available from https://www.rescoop.eu/toolbox/business-models-for-citizen-led-renovation [Accessed 26 June 2024]

[30] Banfill P., Simpson S., Haines V., Mallaband B. 'Energy-led retrofitting of solid wall dwellings: Technical and user perspectives on airtightness'. *Structural Survey*. 2012;**30**(3):267–79

[31] Wheeler J., Alker J., Box P. '*The retrofit playbook - Driving retrofit of existing homes – A resource for local and combined authorities*' [online]. 2021. Available from https://ukgbc.org/wp-content/uploads/2021/02/Retrofit-Playbook.pdf [Accessed 26 June 2024]

[32] Cockbill S.A., Mitchell V., May A.J. 'Householders as designers? Generating future energy services with United Kingdom home occupiers'. *Energy Research & Social Science*. 2020;69:101615

[33] Sleeswijk Visser F., Stappers P.J., van der Lugt R., Sanders E.B.N. 'Context mapping: Experiences from practice'. *CoDesign*. 2005;**1**(2):119–49

[34] Bitner M.J., Ostrom A.L., Morgan F.N. 'Service blueprinting: A practical technique for service innovation'. *California Management Review*. 2008; **50**(3):66–94

[35] Cockbill S.A., May A., Mitchell V. 'The assessment of meaningful outcomes from co-design: A case study from the energy sector'. *She Ji: The Journal of Design, Economics, and Innovation*. 2019;**5**(3):188–208

[36] Cockbill S.A. '*The role of co-design in supporting energy-related retrofit by householders*'. Doctoral dissertation, Loughborough University; 2017 [online]. Available from https://repository.lboro.ac.uk/articles/thesis/The_role_of_co-design_in_supporting_energy-related_retrofit_by_householders/9360428 [Accessed 26 June 2024]

[37] Karvonen A. 'Towards systemic domestic retrofit: A social practices approach'. *Building Research & Information*. 2013;**41**(5):563–74

[38] Voytenko Y., McCormick K., Evans J., Schliwa G. 'Urban living labs for sustainability and low carbon cities in Europe: Towards a research agenda'. *Journal of Cleaner Production*. 2016;123:45–54

[39] Sovacool B.K., Osborn J., Martiskainen M., Anaam A., Lipson M. 'Humanizing heat as a service: Cost, creature comforts and the diversity of smart heating practices in the United Kingdom'. *Energy and Climate Change*. 2020;1:100012

[40] Rightmove. '*Greener homes report*' [online]. 2023. Available from https://www.rightmove.co.uk/guides/content/uploads/2023/07/Rightmove_Greener_Homes_Report_2023.pdf [Accessed 3 July 2024]

[41] Bobrova Y., Papachristos G., Chiu L.F. 'Homeowner low carbon retrofits: Implications for future UK policy'. *Energy Policy*. 2021;155:112344

[42] Fawcett T. 'Exploring the time dimension of low carbon retrofit: Owner-occupied housing'. *Building Research & Information*. 2014;**42**(4):477–488

[43] SuperHomes. '*SuperHomes*' [online]. 2022. Available from https://superhomes.org.uk/ [Accessed 2 July 2024]

[44] de Wilde M., Spaargaren G. 'Designing trust: How strategic intermediaries choreograph homeowners' low-carbon retrofit experience'. *Building Research & Information*. 2019;**47**(4):362–74

[45] Haines V., Mitchell V., Mallaband B. 'Merging a practice-orientated approach with an engineering-driven product development: A case study on home improvement'. *Journal of Design Research*. 2012;**10**(1–2):28–49
[46] Kinnings A. '*Toastie – Where sustainability just got affordable*'. Loughborough University [online]. 2024. Available from https://sdcashow2024.lboro.ac.uk/student/albert-kinnings/ [Accessed 4 July 2024]
[47] Mitchell V., Mackley K.L., Pink S., Escobar-Tello C., Wilson G.T., Bhamra T. 'Situating digital interventions: Mixed methods for HCI research in the home'. *Interacting with Computers*. 2015;**27**(1):3–12
[48] Pink S., Ardèvol E., Lanzeni D. (eds.). *Digital Materialities: Design and Anthropology*. London: Bloomsbury Publishing; 2016
[49] Pink S., Mackley K.L., Morosanu R., Mitchell V., Bhamra T. *Making Homes: Ethnography and Design*. Abingdon: Routledge; 2020
[50] E. ON. *Challenge 100: Tackling Fuel Poverty for 100 Families in 100 Homes, in 100 Days*. Coventry: E.ON UK plc; 2010
[51] Cosar-Jorda P., Buswell R.A., Mitchell V.A. 'Quantifying energy demand reduction considering householder willingness to apply measures'. *Building Research & Information*. 2019;**47**(7):850–64
[52] YES Energy Solutions. '*Better Homes Yorkshire*' [online]. 2024. Available from https://www.yesenergysolutions.co.uk/schemes/better-homes-yorkshire [Accessed 4 June 2024]
[53] del Olmo M.V., Morelli N. 'Service journeys as boundary objects in participatory processes for multi-stakeholder engagement: The case of the easy-rights journeys'. Presented at Design Research Society Conference 2022; Bilbao, Spain, 2022
[54] Lee S.H., Yang M., de Weck O.L., *et al*. 'Service design in action: Transformation, consideration, and system thinking'. *Proceedings of the Design Society*. 2023;3:3145–54
[55] Service Design Network. '*Service design network*' [online]. 2005. Available from https://www.service-design-network.org/ [Accessed 5 July 2024]
[56] Taylor J., McLeod R., Petrou G., *et al*. 'Ten questions concerning residential overheating in Central and Northern Europe'. *Building and Environment*. 2023;234:110154

Chapter 4
Quality retrofit assurance

Peter Rickaby[1]

4.1 Introduction

In Glasgow, Scotland, on a stormy night in 2014, four tonnes of external wall insulation (EWI) fell from the top storeys of a residential tower block. The following day, during the same storm, at a primary school in Edinburgh, the brick outer leaf of an insulated gable collapsed. Nobody was hurt in either case, but these dangerous incidents were not isolated examples of retrofit failure.

A year earlier, in the Fishwick district of Preston in north-west England, nearly three hundred owner-occupied terraced houses were insulated with external wall insulation. The work was recommended by the local authority as a measure to alleviate fuel poverty and was funded via the UK Government's Community Energy Saving Programme (CESP). Unfortunately, no proper assessments were made of the existing homes, no designs were prepared, untrained and incompetent installers bodged the job, and there was no quality control. Within a short time, rain had penetrated behind the external insulation layer and, with nowhere else to go, it appeared internally as damp walls on which mould began to grow. Eight years later, twenty-two houses had been repaired, but over two hundred other households were still waiting for remediation of cold, damp, and unhealthy homes that they could not sell. At the time of writing this chapter, many of those households are still waiting.

There are other, similar examples of poor or failed EWI in north and south Wales and several thousand individual cases of failures where cavity wall insulation (CWI) was installed where it should not have been, mostly in locations with high exposure to driving rain. The culmination of this catalogue of failures occurred in west London on the night of 14 July 2017, when a cladding fire at Grenfell Tower, a 24-storey block of apartments, destroyed the building, killed 72 residents, injured 70 more, and left over two hundred people homeless. The block had recently been retrofitted with EWI and over-cladding that proved to be highly combustible and give off toxic fumes when it burnt [1]. Any domestic retrofit failure is a disaster for the people who live in the building, but all of these failures could have been avoided.

[1]UK Centre for Moisture in Buildings, University College London, London, UK

4.1.1 Retrofit for the Future

A few years earlier, between 2009 and 2014, the UK's Technology Strategy Board, now known as Innovate UK, funded a programme called Retrofit for the Future, in which 115 occupied houses owned by 86 housing associations were subject to retrofit. The aim was to learn about deep retrofitting of domestic buildings, and most of the projects were designed and overseen by small teams of experienced retrofit professionals. The challenging target was to reduce carbon dioxide emissions associated with energy use in each home to not more than 17 $kgCO_2/m^2/yr$. All of the projects involved improvements to the building fabric, the building services, and the installation of renewable energy systems. The projects were subjected to independent post-construction review, post-occupancy evaluation, two years of detailed instrumented monitoring of performance, and academic evaluation. An expert technical panel spent two years reviewing, summarising, and analysing all the data, and reporting their conclusions [2].

Only a few of the Retrofit for the Future projects achieved the post-retrofit performance target, but many came close. More importantly, many lessons were learned about the technical risks associated with retrofit, how to manage them, and what good practice looks like. These lessons subsequently became the foundations of the UK's domestic retrofit quality assurance system.

Four key lessons stood out. First, the places where retrofitting goes wrong are corners, junctions, edges, and interfaces. That is, places where building elements, i.e., floors, walls, windows and doors, and roofs, connect with each other and where, without careful detailing, thermal bridging, thermal by-pass, and air infiltration and air leakage will occur; and interfaces between the building fabric, the building services, and people, where matched capacities and effective controls are important.

Second, the critical role of ventilation in retrofits, in securing healthy indoor air quality when air infiltration rates are reduced, is not well understood by consultants, contractors, installers, landlords, or householders. In Retrofit for the Future, it is fair to say that ventilation systems were poorly chosen, poorly designed and specified, poorly installed and commissioned, poorly handed over, poorly maintained, and poorly operated by users. Some projects did not adopt coherent ventilation strategies, and some ventilation systems were disabled by users because of noise, draughts, or perceived cost.

Third, retrofitting involves technical risk, which can be assessed and managed. Most retrofit risks are associated in some way with moisture, so moisture risk management is critical. Rainwater penetration into retrofit construction, condensation, damp, and mould (CDM) inside retrofitted homes because of thermal bridging, poor airtightness or inadequate ventilation, timber rot and decay, and spalling of insulated masonry walls are all moisture-related problems.

Fourth, the 'performance gap' in retrofit, defined as the difference between predicted and actual savings in fuel use, fuel costs, and emissions, is better characterised as an 'attention to detail deficit'. Paying attention to the details of the issues described in the last three paragraphs goes a long way towards closing the performance gap.

As one member of the expert technical panel put it, 'Retrofit is not rocket science – but it is complicated'. Consequently, the valuable lessons learned from Retrofit for the Future took several years to assimilate into a retrofit quality assurance system.

4.2 Addressing the lessons learned

Identifying gaps in industry knowledge, highlighted by Retrofit for the Future, was identified as a major issue for the UK. As such, a number of initiatives and follow-up projects were established to raise awareness of the underpinning knowledge to address this gap in understanding.

4.2.1 CoRE, the STBA and the UKCMB

Learning from Retrofit for the Future was consolidated by the EU-funded Centre of Refurbishment Excellence (CoRE), where it was embedded in a training programme for the new role of 'Retrofit Coordinator'. The Sustainable Traditional Buildings Alliance (STBA) was set up and received Government funding to research and promote responsible retrofitting of traditionally constructed buildings and the UK Centre for Moisture in Buildings (UKCMB), based at University College London, was established to research moisture risks in buildings and disseminate guidance. The Code of Practice for managing moisture in buildings (BS 5250) was extensively updated. Domestic retrofit guidance was also published by the Construction Products Association [3] and by the Institute for Sustainability [4], amongst others. In all this work, the importance of managing technical risks to buildings and serious health risks to occupants was emphasised.

4.2.2 The Green Deal, ECO and PAS 2030

At the same time that retrofit professionals were learning from Retrofit for the Future and from follow-up programmes, such as Scaling Up Retrofit, the UK Government was removing professional input from its domestic retrofit programmes. In 2010, the new Government started developing its flagship retrofit policies, the Green Deal and the Energy Company Obligation (ECO). The Green Deal subsequently failed, but the ECO, in which improvement measures are paid for by the six largest domestic energy suppliers, remains in place in 2024 and has so far delivered thousands of energy efficiency measures in UK homes. In neither scheme was there a role for professionals. Surveyors, architects, and engineers were seen as not only unnecessary but also too expensive for retrofit at scale, and both programmes were placed in the hands of installers of energy efficiency measures, a group that has a limited understanding of building physics, moisture management, or other technical risks.

However, Ministers were correctly paranoid about 'cowboy builders' damaging people's homes and health and thus damaging confidence in their retrofit programmes through poor-quality retrofit work. Ministers, therefore, commissioned the British Standards Institution (BSI) to develop a Publicly Available Specification (PAS – essentially a preliminary national standard) for the installation of domestic retrofit measures. This was PAS 2030 [5], against which installers were required to be certified annually by Certification Bodies (CBs), themselves accredited by the national accreditation body, the UK Accreditation Service (UKAS). There was already a certification industry to quality-assure the work of Energy Performance Certificate (EPC) assessors and gas-fired central heating installers, amongst others, but the introduction of PAS 2030 catalysed a huge expansion, eventually encompassing over a hundred CBs.

Unfortunately, the first edition of PAS 2030 was widely condemned as not fit for purpose, and despite updates in 2014 and 2017, neither PAS 2030 nor the CBs seem to have had any discernible effect on retrofit outcomes, although they have certainly added to costs. This was the inadequate quality assurance regime that presided over many of the issues described at the beginning of this chapter.

4.2.3 Each Home Counts

In the face of mounting retrofit problems, in July 2015 the Secretaries of State at the Departments of Energy and Climate Change (DECC) and Communities and Local Government (DCLG) initiated an industry-led review under the chairmanship of Peter Bonfield, then Chief Executive of the Building Research Establishment (BRE), the UK's largest building research and technology organisation (RTO) for the built environment. The Bonfield Review, as it was initially referred to, was later re-named the Each Home Counts review. Officially, it was 'An independent review of consumer advice, protection, standards, and enforcement for energy efficiency and renewable energy'. The aim was to restore the confidence of Government, funding bodies, and the public in the domestic retrofit industry as a prerequisite for further policy initiatives or investment. Hundreds of members of the industry gave up their time to contribute to several work streams, and the report of the review was eventually published in December 2016 [6]. The report included 27 recommendations, and an Implementation Board was established with the support of the Department for Business, Energy and Industrial Strategy (BEIS) to put them into effect.

Three key recommendations of the Each Home Counts review are worthy of consideration here. First, domestic retrofit should move away from a measures-based approach, in which separately funded energy efficiency measures are installed in homes individually, as in ECO, towards a 'whole dwelling' approach. This was seen during the review as the best way of ensuring that appropriate attention is given to corners, junctions, edges and interfaces and that maxims such as 'no insulation without ventilation' and 'build tight, ventilate right' are embedded in the technical culture of the industry. Subsequently, the whole dwelling approach has also been recognised as the best way to move the housing stock towards the UK's statutory 'net zero carbon' target by 2050, medium-term improvement plans being established for each home, even if immediate installation of all the necessary measures cannot be afforded.

Second, a 'Quality Mark' should be established to identify retrofit work that complies with appropriate standards and differentiates it from other work. The Quality Mark should be supported by a Customer Charter, setting out what consumers should, and should not, expect from the retrofit industry, and by a Code of Conduct, setting out how the industry should behave, especially in matters of customer care, the handling of complaints, remediation, and redress. The role of the Quality Mark was subsequently entrusted to TrustMark, a pre-existing Government-endorsed consumer standards body, which adopted the Each Home Counts Customer Charter and Code of Conduct. The Government subsequently made the delivery of publicly funded or publicly endorsed retrofit programmes, such as ECO, subject to TrustMark approval.

Third, a Code of Practice for retrofits, in the form of a framework of technical standards, should be established. The development of the Code of Practice was entrusted to the technical standards workstream of the Each Home Counts review, the BSI Retrofit Standards Task Group (RSTG), which continues to advise BSI on strategy for retrofit standards and oversees their development. The RSTG is a group of invited technical experts, which is now complemented by an industry-representative retrofit technical committee known as CB 401; the two groups work together in complementary roles, with overlapping membership.

A fourth key recommendation of the Each Home Counts review was not implemented. This was the establishment of an online retrofit advice hub to provide impartial advice to householders, landlords, and members of the industry. The reluctance of BEIS to fund the hub, which in principle cannot be commercially sponsored as well as impartial, significantly impeded the overall implementation of Each Home Counts, leaving a hole in the intended system of support for consumers and industry and of quality assurance for domestic retrofit.

4.3 The BSI Retrofit Standards Framework and PAS 2035

The BSI Retrofit Standards Task Group commissioned research into domestic retrofit standards, identifying approximately five hundred technical standards covering all aspects of retrofit. Evaluation of those standards revealed that some were more technically robust and more readily adopted than others, although there were some gaps. The BSI Retrofit Standards Framework (see Figure 4.1) was subsequently established to bring the best standards together and fill the gaps.

At the heart of the Retrofit Standards Framework is the Publicly Available Specification, PAS 2035 [7]. This is the UK's overarching comprehensive domestic retrofit standard, published in June 2019, covering the assessment of dwellings for retrofit, improvement option evaluation, the development of medium-term improvement plans, the design and specification of retrofit, and the evaluation of the outcomes of retrofit projects. Alongside PAS 2035, the installation standard PAS 2030 was updated [8] and the two standards were 'locked together', essentially forming one standard. Retrofit designs compliant with PAS 2035 must be installed in accordance

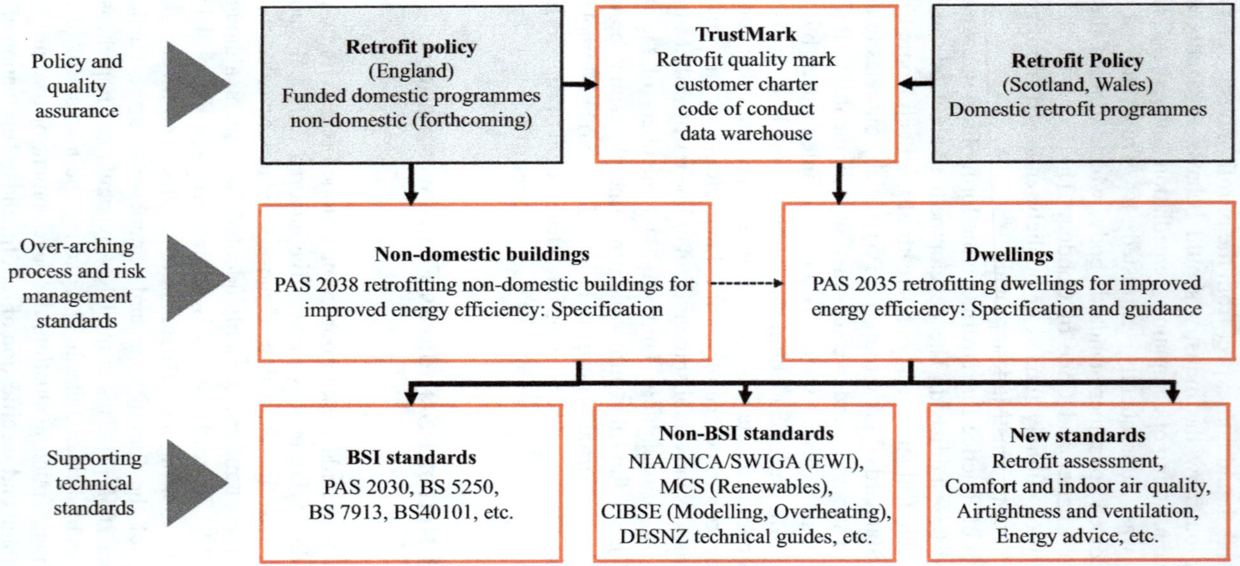

Figure 4.1 The BSI Retrofit Standards Framework

with the requirements of PAS 2030, and PAS 2030-compliant installations must be based on PAS 2035-compliant designs. The only exception to this interlocking of PAS 2035 and PAS 2030 is renewable energy systems, covered by the Microgeneration Certification Scheme. (MCS). Renewable technologies, such as solar panels, air-source heat pumps, or battery storage, must be installed in accordance with the requirements of the MCS standards, rather than those of PAS 2030.

The original edition of PAS 2035 was risk-based: projects were required to follow one of three paths through the retrofit process, identified as Paths A, B, and C, according to the outcome of an initial risk assessment based on five criteria: the number of dwellings to be improved; the average number of measures per dwelling; the technical risk associated with the proposed measures; the extent of interaction between the proposed measures; and the construction and built form of the dwellings. Figure 4.2 provides a simplified diagram of the original PAS 2035 retrofit process. PAS 2035 was updated in 2023 [9] and the risk paths were removed in order to simplify the process.

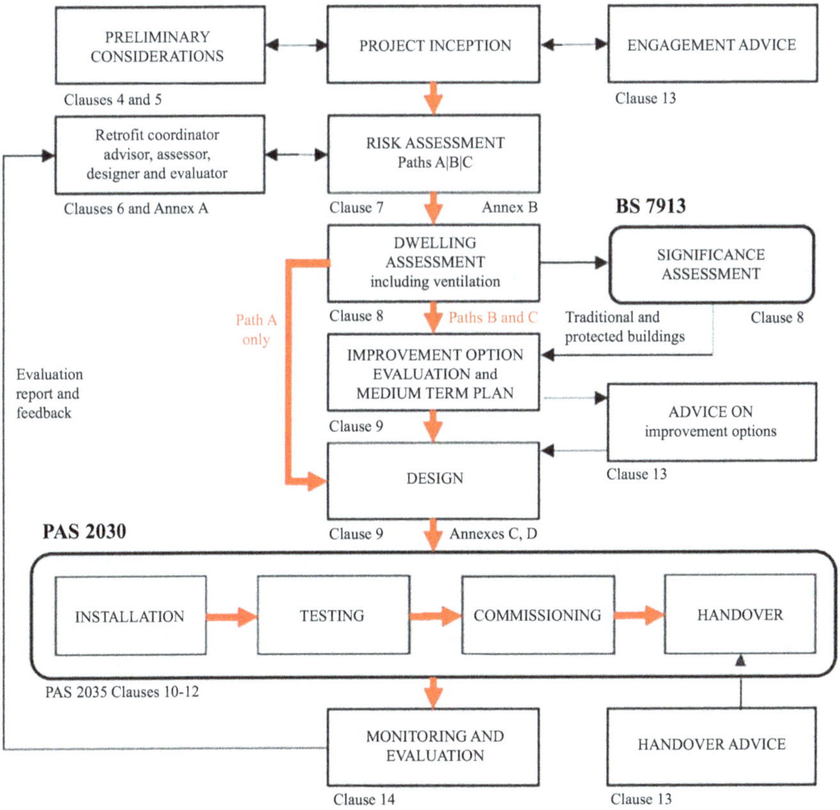

Figure 4.2 A simplified diagram of the PAS 2035:2019 retrofit process. Note that the risk Paths A–C were removed from PAS 2035:2023.

Retrofit designs prepared in compliance with PAS 2035 must be based on whole-dwelling assessments covering the context, condition, occupancy, energy performance, ventilation, and historical significance of the buildings. Retrofit designs must include construction details for key corners, junctions, and edges. Compatible specifications for elements that interact, even if they do not connect physically, should also be considered. Ventilation upgrades will need to be addressed if insulation or airtightness measures are included. Finally, there needs to be consideration of measures to minimise both overheating risk and careful management of moisture risks.

An important feature of PAS 2035 is that it requires compliance with many of the other standards that are part of the BSI Retrofit Standards Framework, thus ensuring that all current best practice standards are brought to bear. Standards with which PAS 2035 requires compliance include BS 5250 [10], BS 7913 [11], and the MCS product and installation standards [12]. The BSI Retrofit Standards Task Group continues to promote and support the development of new standards to expand and complete the framework, including standards for retrofit assessment, for retrofit advice, for airtightness and ventilation, for thermal comfort and indoor air quality (IAQ), and for building performance evaluation. Thus, the developing framework is a mechanism for continuous improvement of retrofit standards.

PAS 2035 also specifies minimum qualifications for people who fill various retrofit roles identified in the PAS, including advisors, assessors, designers, and evaluators. The key retrofit role is the Retrofit Coordinator, which every project must have. The Retrofit Coordinator's role is to coordinate the project from inception to evaluation, to protect the client's interest and the public interest, and to claim compliance with PAS 2035 by submitting evidence to whoever requires it, usually TrustMark. The introduction of the role of Retrofit Coordinator is an initial step towards 're-professionalising' retrofit, bringing in knowledge of building physics and risk management, especially moisture risk management, that assessors, designers and installers often do not have.

4.3.1 Quality assurance

TrustMark and the BSI Retrofit Standards Framework, including PAS 2030 and 2035, constitute the UK's domestic retrofit quality assurance system, as illustrated in Figure 4.3. This system has been applied to all Government-funded or endorsed domestic retrofit programmes since July 2021. It may also be used, on a voluntary basis, by any other organisation engaged in domestic retrofit, such as the devolved administrations, local authorities, housing organisations, funding bodies, or even individual householders. Implementation is supported by the training of a large number of Retrofit Coordinators, Retrofit Assessors, and Retrofit Designers and by the development of software tools and guidance to support the PAS 2035 process.

Similar arrangements are also being developed for non-domestic retrofit in the UK, following publication of PAS 2038 [13] in early 2021. The development of the overall retrofit quality assurance system has been described in detail in a BSI 'White Paper' [14].

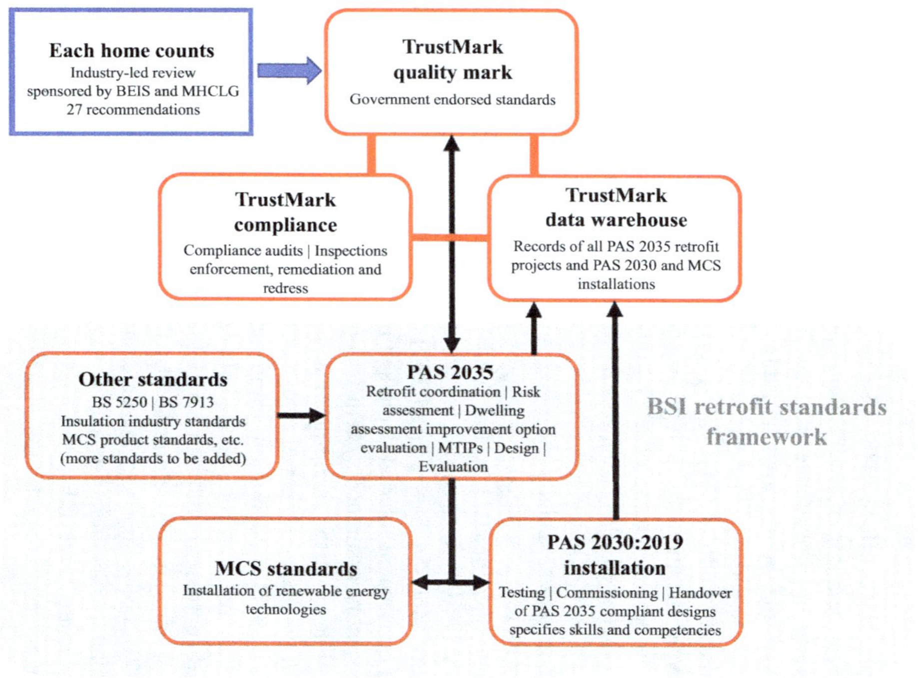

Figure 4.3 The UK's domestic retrofit quality assurance system

The transition from earlier, unsatisfactory, ways of working to TrustMark approval and compliance with PAS 2035, over a two-year period, proved painful for the energy efficiency industry. New qualifications had to be gained; new ways of working had to be developed and adopted; new requirements, such as for design and ventilation, had to be met and new costs had to be assimilated into prices and financial structures. However, almost everyone engaged in the Each Home Counts review and its implementation is confident that the costs associated with the new quality assurance regime are far outweighed by the cost of continuing to get retrofitting wrong.

Two examples of the effects of the quality assurance system may be seen in the improvement of the detailing of EWI. Before the system was put in place, EWI was often terminated at roof lines (eaves and verges) with metal or plastic trims that were poorly fixed and poorly jointed, leading to water penetration behind the insulation layer. This was the main cause of the problems at Fishwick in Preston. Consequently, the 2019 editions of both PAS 2035 and PAS 2030 banned the use of such trims, leaving the industry to rely instead on costly eaves extensions and 'gable ladders' to protect the top of EWI from weather. The consequence was a safer retrofit but increased EWI costs, which prompted the industry to develop new trims, now known as 'roofline closures'. These are better fixed and sealed to walls, properly sealed at joints, include purpose-made corner pieces, and ensure continuity of EWI with roof insulation to minimise thermal bridging. The new roofline closures were approved in 2023 by the Solid Wall Insulation Guarantee Agency (SWIGA) and by TrustMark as complying with the requirements of PAS 2035 and PAS 2030. Their adoption has reduced EWI costs and reduced risks of water penetration and consequent internal damp and mould.

Before the quality assurance system was put in place, EWI was usually terminated at a low-level at the damp-proof course (DPC) just below ground floor level, leaving an uninsulated thermal bridge at the floor-wall junction, which was often a cause of internal condensation, damp and mould. The 2019 editions of PAS 2035 and PAS 2030 required compliance with best practice guidance from SWIGA that EWI should continue to 400 mm below the DPC, or at least to ground-level, using waterproof insulation to eliminate the thermal bridge. This prompted the publication by BEIS of technical guidance on appropriate methods of compliance, including acceptable arrangements and a method of calculating the moisture-related risks associated with other configurations. Adoption of this guidance has reduced risks of internal condensation, damp and mould, and reduced heat losses from homes.

4.4 Conclusion

The UK's retrofit quality assurance system is based on knowledge of technical risks gained from the Retrofit for the Future programme and subsequently consolidated by work at the Centre of Refurbishment Excellence (CoRE). Application of this knowledge was promoted by the Each Home Counts review, following a series of retrofit failures, and was embedded in the BSI Retrofit Standards Framework in the form of PAS 2035, PAS 2038, and the other standards with which they require compliance. The main purpose of the quality assurance system is to protect homes and their occupants against the unintended consequences of poor retrofits caused by inappropriate or ill-considered designs and specifications, and careless installation of improvement measures.

Currently, it is too early to say whether the new quality assurance system for domestic retrofit in the UK has improved outcomes, but initial results are encouraging – there have been no more serious large-scale failures. However, there will always be organisations that 'fly under the radar' and avoid complying with standards. Nevertheless, delivering 27 million domestic retrofits by 2050 demands that we do better retrofits and that we get them right the first time, at least most of the time. The cost of retrofitting existing buildings in the UK is likely to be of the order of £1 trillion, and it must be done well – there is neither the time nor the money to get it wrong and do it all again.

References

[1] Apps P. *Show Me the Bodies: How We Let Grenfell Happen*. London: One-world; 2022

[2] Technology Strategy Board. *Retrofit for the Future: Reducing Energy Use in Existing Homes – A Guide to Making Retrofit Work*. Swindon: Technology Strategy Board; 2014

[3] Rickaby P. A., Willoughby J., Warren E., McLaren Webb, C. *An Introduction to Low Carbon Domestic Refurbishment*, 2nd ed. London: Construction Products Association and RIBA Publishing; 2014

[4] Rickaby P. A. (ed.). 'Low Carbon Domestic Retrofit Guides' in: *A Suite of Twelve Web-Based Guides*. London: Institute for Sustainability; 2011

[5] British Standards Institution. 'Specification for the installation of energy efficiency measures (EEMs) in existing buildings'. London: BSI; 2011

[6] Bonfield P. *Each Home Counts: An Independent Review of Consumer Advice, Protection, Standards and Enforcement for Energy Efficiency and Renewable Energy*. London: Department for Business, Energy and Industrial Strategy and Department for Communities and Local Government; 2016

[7] British Standards Institution. '*Retrofitting dwellings for improved energy efficiency: Specification and guidance'. PAS 2035:2019*. London: BSI; 2019

[8] British Standards Institution. *Specification for the installation of energy efficiency measures in existing dwellings, and insulation in residential park homes. PAS 2030:2019*. London: BSI; 2019

[9] British Standards Institution. *Retrofitting dwellings for improved energy efficiency: Specification and guidance. PAS 2035:2023*, London: BSI; 2023

[10] British Standards Institution. *Management of moisture in buildings. Code of Practice. BS 5250:2021*. London: BSI; 2021

[11] British Standards Institution. *Guide to the Conservation of Historic Buildings*. BS 7913:2013. London
[12] Microgeneration Certification Scheme (MCS). '*Product standards*' [online]. Available from https://www.microgenerationcertification.org/mcs-standards/product-standards, and Microgeneration Certification Scheme (MCS) Installer Standards at www.microgenerationcertification.org/mcs-standards/installer-standards [Accessed 9 July 2026]
[13] British Standards Institution. *Retrofitting non-domestic buildings for improved energy efficiency: Specification. PAS 2038:2021*. London: BSI; 2023
[14] Rickaby P.A. *The Importance of Standards for Safe Energy Retrofit. BSI White Paper*, London: British Standards Institution; 2023

Chapter 5

Supply chains: who is going to deliver retrofit?

Alice Owen[1] *and Gavin Killip*[2]

5.1 Introduction

This chapter makes the case that the existing construction industry has an essential role in delivering retrofits at the scale needed to transform the existing building stock. They carry out repair, maintenance, and improvement (RMI) projects and have influence over customer decisions. Existing construction firms are the only group of people operating at the scale needed across all geographical areas. Once that is accepted, then the task of retrofit delivery can be addressed in terms of how the industry needs to change, starting from a good understanding of current industry practices. Construction involves physical changes to the environment through the use of materials and technologies, achieved through the collective efforts of many different people. It is a *socio-technical* system with many sub-systems: people, institutions, and technologies interacting and influencing each other, changing over time. A socio-technical perspective means that systems of technology and social actors are seen as two facets of a complex whole, with people and technologies intertwined and interdependent. Socio-technical thinking and analysis require us to grapple with complexity if we are to develop solutions that will actually make a difference.

This emphasis on the socio-technical is important because of how we think about retrofit. For this analysis, retrofit is more than technology rollout; retrofit is a process of integrated, high-quality work to upgrade homes. Retrofit involves making physical changes to buildings and their technical systems. From an energy perspective, there are changes that reduce heat loss, such as insulation, minimal thermal bridging, improved airtightness and controlled ventilation. Additionally, low-carbon heating systems such as heat pumps, as well as efficient lighting and appliances may be installed. The process may also include installing building-integrated renewable energy technologies, such as solar panels. These physical changes can reduce a household's energy consumption and carbon dioxide emissions, while also delivering other benefits in terms of health, well-being, comfort and aesthetics.

[1]Sustainability Research Institute, University of Leeds, UK
[2]Environmental Change Institute, University of Oxford, UK

The technical understanding of buildings as structures has advanced tremendously in recent decades. The design professions – in architecture, structural design, and M&E design – have similarly taken significant steps forward to change their professional norms. Getting from design and specification to real buildings that use less energy, with lower carbon impacts, is not an automatic step. In the same way that a technology-led approach is insufficient to meet the entire challenge of designing changes to existing buildings, we also need a socio-technical approach to examining the practicalities of delivering retrofit. Changing technology means changing the social systems that are part of those technologies. Those social systems include the building occupants but also the people brought in to carry out retrofit projects. Project teams are made up of designers, project managers, general builders, specialist installers, and consultants. Teams typically vary from one project to the next, as do the contractual relationships between them. By understanding the social organisation of projects, it is possible to understand how teams have agency, the capability to do things, and make decisions, but also how that agency is often limited. Limits to agency might be self-imposed, with construction workers being unwilling to adopt change, or they might come from industry norms, such as the division of tasks among professions and craft trades, or they might come from regulatory controls. Thus, we need to attend to the governance of the construction industry and how the incumbency of the existing industry helps or hinders change [1,2].

The agency of construction workers operates between the 'top-down' requirements of policy and the 'bottom-up' demands of market consumers. The industry does not simply do whatever it is asked to do without question. Instead, the people in the industry draw on their experience and expertise to try and influence decisions in various ways, from lobbying governments through trade associations to advising individual clients on the specification of projects. A 'middle out' perspective [3,4] captures these dynamic influences, which can be exerted upstream to policy, downstream to clients and, importantly, sideways among peers in the industry. An example of the upstream influence that construction firms can exercise is that practitioner experience might influence product regulation, e.g. by demonstrating that particular materials are usable in specific contexts. In practice, this upstream influence is rarely seen, and this creates a problem in terms of the practical applicability of regulations, often undermining the credibility of those regulations with practitioners [5]. Downstream influence is when a builder or installer influences customer specification or choice of a technology or material for an RMI project. The builders' previous negative experience, or lack of familiarity with a technology, might mean they advise a customer away from an innovative approach as a way of reducing the perceived risk associated with a project [4]. Sideways influences, between peers, are very important because small firms and micro-enterprises often collaborate and form virtual organisations, which are rooted in one town or region and are very stable over time [6]. Sideways influences are seen in what local contractors choose to use as materials, based on what their peers recommend, or their local supplier suggests [3].

As well as thinking about how to transform construction across many aspects, we also need to reflect upon how to create an industry that will continue to adapt and deliver as technological possibilities and regulatory and environmental demands also change. Buildings need regular maintenance, and the needs of occupants change over time, leading to a steady flow of projects for repairs and improvement. Retrofit should not be thought of as a single event; retrofit processes will need to be revisited, updated, and repeated as circumstances change and the need for maintenance re-appears.

Maybe it seems self-evident that retrofit requires the construction industry; after all, who else can deliver it? But starting with this assumption leads policy-makers, and many large firms, to look at the nature of the industry as it exists today and simply despair at the seeming impossibility of delivering a wholesale transformation of building stock using the resources currently deployed on repair, maintenance, and improvement of buildings. Even in the sphere of new builds, there seems to be little capacity available and deploying massive innovation with quality assurance is difficult. This view of the construction industry leads us to a technology-centred view of construction itself. Too difficult to access skilled labour? Then remove labour from sites and rely on off-site and modular construction instead, emulating the tight specifications and quality control of other forms of manufacture. However, construction projects are fundamentally different from manufactured goods; construction projects are products of a 'complex systems industry' [7], where a small change in one area leads to changes in many other areas of a product, and users – both building users and the installation team – are actively involved in problem-solving and finding acceptable solutions. There needs to be innovations in industry practices and processes, not just in products.

'Market transformation' is a concept that helps us think through how the existing construction industry could deliver retrofits at scale, where the commercial context of the firms in the industry is completely altered. The term 'market transformation' refers to a policy-driven programme of change over time, whereby regulators work with existing market actors to bring about change. The key principles of market transformation are: working 'with the grain' of the dominant market; long-term industry engagement; new standards developed with technical and market intelligence; sufficient time allowed for market actors to learn and invest and policies applied in combination over time, not piecemeal [8].

5.2 What are the opportunities and trigger points for making retrofitting a mainstream construction industry?

There have been advances in delivering energy retrofits through large programmes, often under the umbrella of social housing providers. However, this progress is not readily translated into the owner-occupier and private rented markets for home improvements, although huge potential to achieve energy demand reduction could be made through ongoing repair, maintenance, and improvement (RMI) work. While there is some understandable reluctance from policymakers because of the messy complexity of construction work in millions of individually adapted homes, the opportunity offered by transforming the existing construction industry is critical to universal retrofit. It is the existing repair, maintenance, and improvement (RMI) industry that installs heating systems, carries out repairs to roofs and floors, builds extensions, and creates the structures around energy-using appliances. Thus, it is the existing construction industry that is also best placed to carry out work which will reduce energy demand, or carbon emissions, associated with each of these activities. Builders and tradespeople routinely offer advice and influence design and specification decisions, based on their experience and what they have learnt from trusted colleagues and suppliers [9,10]. It is also clear that the mainstream RMI construction industry is a long way from acting in a way that mainstreams a responsibility for energy demand reduction into its work.

Realising the opportunity to reduce energy demand means taking advantage of 'trigger points' in the existing RMI market. 'Trigger points' are opportunities in the RMI market where energy-saving measures can be fitted at the same time as other works [11]. It might be replacing windows, repairing a roof, converting an attic or basement, changing a home layout to accommodate different people, or updating a kitchen or bathroom. The core idea is that, as the homeowner has already agreed to some level of disruption and cost, there is an opportunity to deliver on energy efficiency improvements at the same time. Equally, if the opportunity to integrate energy retrofit works is not taken when other home improvements are carried out, then it may be several years before another opportunity comes about. For example, there is a large annual market for new kitchens, which involves removing old furniture, fittings, and appliances, leaving an empty shell. This is an ideal opportunity to insulate floors and walls to save energy and to fit heating and ventilation systems. Conversely, once the new kitchen has been installed, it is unrealistic to rip it out and do the energy-related works separately.

There have been attempts to embed the 'trigger points' approach through the policy of 'consequential improvements'. 'Consequential improvements' were proposed as part of Building Regulations in the UK, requiring additional works to improve energy efficiency, but the idea was rejected in 2013 because of industry fears that it would lead to a drop in demand for RMI work. Energy retrofit work includes activities such as the installation of heat pumps or microgeneration, or the improvement of existing building fabric, ideally as part of a whole-house plan (an idea developed below). But the other side of the coin, carrying out repair, maintenance, and improvement in altered ways, requires a slightly different mode of thinking. General RMI work is rarely commissioned with the primary aim of reducing energy demand; it is commissioned for reasons of modernisation, comfort, changing how space can be used, or aesthetics [12] and energy demand reduction, or carbon emissions reduction, must be achieved as a co-benefit of the primary purpose.

5.2.1 Co-benefits

Following on from recognising that there are trigger points when reducing energy demand becomes more possible, the issue of co-benefits is worth digging into a little more deeply in order to understand why retrofit and mainstream construction are not automatically coupled. Co-benefits are usually conceptualised as other things that accompany improvements in energy efficiency. Cost savings to the user through reduced energy bills is the first co-benefit cited, used as a primary justification to undertake energy retrofit work at all, although there are limits to this, not least because households in fuel poverty are likely to take increases in comfort over reductions in cost, and energy use can rise in these conditions [13]. Our suggested approach to mainstreaming retrofitting into RMI work on existing buildings inverts this. Why not start with the desired outcome that the homeowner already desires – a new open-plan living space to make family life easier; warmth throughout the day; the ability to dry clothes on an airer without needing a tumble drier; better air quality or reduced damp to help with asthma or breathing difficulties – and then consider how energy efficiency can be delivered as a co-benefit of that desired outcome?

A homeowner's or tenant's reluctance to add retrofit activity into the scope of works, on the basis of perceived additional cost or unnecessary work and disruption to their home, will be reduced if a benefit that they want is prioritised. This is particularly important given the reluctance of many households to allow other people into their homes for potentially invasive and disruptive work. Builders have strong views on what their clients want, based on their experience of undertaking work so that remedial works are minimised and bills are paid on time, an essential part of running a viable small building firm [14], and this is closely linked to the builder's focus on creating and maintaining a positive local reputation for high-quality work [15]. Builders believe that their clients prioritise work that is straightforward to commission and operate and which fits with the aesthetic of their home. A pleasing appearance is often used as a proxy for quality [16].

Trying to persuade people to invest in energy efficiency as a standalone set of activities does not work; delivering energy retrofit outcomes at scale requires a fundamentally different approach. You do not have to start from the position of persuading the householder that energy efficiency is needed; you can start from the position of offering them energy efficiency as a by-product of home improvements that they already want or maintenance work that is needed anyway.

5.3 What capabilities does the construction industry need to deliver effective retrofit?

Energy-focussed retrofits are widely accepted to need new project coordination roles, termed 'integrators' or 'retrofit co-ordinators' [17,18]. In the UK, this role is codified into the standards which define what quality of work is required: PAS2035. The need for these roles is translated into the prospect of increased and higher-skilled employment [19]. PAS 2035 is a requirement for government-funded programmes in the social housing sector but is not mandated for privately owned housing. However, for the small firms that undertake conventional RMI projects, the reality is that using the PAS2035 guidance brings additional people and costs to a project. Existing practitioners could extend and develop their skills to be able to take on the new integrator roles, but there is little incentive to do that at present. There will certainly need to be new aspects of work across all roles, including a need for greater understanding of the carbon implications of work undertaken; a kind of 'carbon literacy' in construction skills, not only in design standards [20]. Tradespeople with a deep, but narrow, understanding of their own domain have to develop a greater appreciation of how their work connects with other domains. An electrician cutting a hole in airtight insulation to run cables will damage the integrity of the building's thermal envelope. This example might seem a little clichéd, but it illustrates neatly the kind of inter-trade connections that need to be understood, and that understanding then needs to change practice – not just within individual trades and professions, but across all members of project teams and the industry more widely. The 'skills gap' in construction is about much more than technical knowledge – important though that is. Better technical understanding needs to be complemented with skills that translate understanding into changes in practice – specifically, communication skills and a commitment to taking and sharing responsibility for end results. These are important steps towards a more collaborative, less adversarial industry culture [21,22].

5.3.1 What are the characteristics of the existing repair-maintenance-improvement (RMI) industry?

Before further developing the list of changes needed, it is worth examining the attributes of the existing RMI construction industry in order to understand the starting point for the practicalities of making retrofitting a mainstream activity.

The RMI construction industry is large, multi-skilled, busy, and organisationally diverse. Over 1.3 million individuals work in construction in the UK [23], typically in firms that are either very large or very small. In the UK, around one in ten construction workers work in firms employing more than 1200 people, while nearly four times as many work in firms of seven people or less, including sole traders [23]. The dominance of small firms means that, at the firm level, there tends to be a focus on one trade or on one type of construction. However, the construction industry is project-based and has long and complex supply chains (or 'value networks') that provide more services than just product sales [24]. For example, builders' merchants often extend lines of credit to small building firms, which helps them with cash flow in their business, where labour needs paying rapidly, but the customer may not pay their invoice for several weeks after work is complete. Information and advice flow across the system by formal and informal channels, and there are often inconsistencies. For example, an architect may specify a material or component to achieve a desired technical function. But, if the specified product is not routinely stocked by the local builders' merchant habitually used by the lead contractor, then there will be resistance to using that product. Special orders for materials can lead to delays, with serious consequences for the smooth running of projects. To avoid project delays, one product might be substituted for another, which is superficially similar but is more readily available and may be cheaper. This kind of decision-making is not necessarily a resistance to innovation, but it shows that there can be many constraints to innovation in the system. Here, it is the result of a clash between the trusted advice and support from the merchant and the detailing provided by the architect; the architect is not part of the mutually dependent supply chain for materials, and so a substitute product gets chosen for reasons that are perfectly logical. An unwillingness to stock an item may be justified by the merchant in terms of price and lack of local demand, but this is circular; there is a lack of local demand because there is a lack of local stock that can be used on a just-in-time basis that characterises materials flow for many small domestic RMI projects.

5.3.2 The place-specific nature of retrofit

Thinking about the networks of skills and firms needed to deliver retrofit leads to considering retrofit as an area-based programme of projects drawing on a pool of local labour and knowledge. If retrofit is to become a mainstream activity for these existing construction networks, then we need locally relevant skills and knowledge, which can reflect local geography, economy, policies, housing markets, type of housing stock and building materials, and social norms [1,25]. 'Place-based' approaches recognise that different locations need different action, using an area's assets and also responding to the specifics of that area's challenges. Place-based approaches to retrofit may use a similar process to determine what needs to be done, how much it will cost, and how it will be funded, but any two areas will answer these questions differently and, therefore, arrive at a different programme of work, carried out by different firms and organisations, although all adhere to the same standards in terms of the quality of work undertaken and energy performance achieved. The industry-led development of a 'National Retrofit Strategy' concluded that the only way in which a huge national programme can be delivered is through a co-ordinated programme of locally focussed schemes tied together with a rigorous quality assurance process [19].

The current RMI market is dominated by small firms whose primary motivations are to achieve a decent standard of living and be their 'own boss'. Understanding of individual motivations, which, for sole traders and small firms, are the same as organisational motivations, is important so that change initiatives can start from the realities of what are important for practitioners; strong motivations are being able to take pride in completed projects, with the quality of work leading to 'word of mouth' recommendations rather than formal advertising; the opportunity to interact with people while carrying out projects; and running a viable business while still maintaining control and flexibility over how and when work is done [14]. Unlike large corporations, small businesses do not have a focus on maximising profit; many are concerned with a consistent and sufficient income, rather than profit or growth as ends in themselves [15]. These diverse motivations mean that RMI practitioners are resistant to policies that offer only financial incentives for retrofit. Retrofit projects will not be delivered just because they could earn firms a bit more money. Compared with retrofitting, the existing RMI market is a better-known, less risky option with steady demand and mature supply chains.

Retrofit is certainly an opportunity for construction firms, but it is currently an opportunity that is easier to ignore than seize. We will only be able to deliver retrofits in existing buildings if we take actual motivations as a starting point and work with them, rather than trying to displace existing motivations – of installers or households – with new ones of energy saving or responding to climate change.

The diversity of RMI projects, with teams varying from project to project, means there is no way to fully escape the fragmentary nature of this part of the construction industry, at least in the market-led economy of the UK, with two-thirds of homes being owner-occupied. RMI projects on individual homes are always bespoke. Such projects are assemblages of heterogeneous tasks: renewable energy specification and installation (and integration into dwelling patterns of use), work on building fabric, interactions between fabric, use and demand, appliance selection, and behaviour change support. There is also an emerging set of activities in smart systems integration – in individual homes or in neighbourhoods – which may feature in future waves of retrofit. This means that standardisation of products has limited use in increasing the scale of retrofit activity and impact, but the standardisation of processes and practices is vital. Change is possible; the challenges of a fragmented market working on diverse projects can be overcome by coordination and learning. When these various attributes of retrofit projects are considered together, it becomes clear that for technology to be installed and high-quality outcomes to be achieved, a market transformation approach is needed.

5.3.3 What standards are needed to underpin retrofit?

For construction RMI work to include energy considerations routinely, then there are both a set of technical norms and standards, which are negotiated between regulators and technology providers, and a set of occupational standards. The socio-technical nature of retrofit means that both technical and occupational (or skills) standards are needed and they need to be co-ordinated. Technical standards are the focus of building regulations and while Part L of England's building regulations is concerned with conservation of fuel and power, low carbon retrofit also relies on Part F on ventilation and Part G on water (the technically approved documents are different in Scotland and Northern Ireland). The consistency between different Parts of building regulations is not trusted by many small building firms [4]. Technical standards for energy generation are also relevant to whole-house retrofits, in particular the Microgeneration Certification Scheme (MCS), which has different standards for each energy source. The standards which specify the process of assuring quality energy retrofit and who is responsible for ensuring standards are met are laid out in PAS2030 and PAS2035. Such standards should give the RMI client some confidence that the work they are commissioning will be fit for the future, and it is true that ensuring compliance with standards is one way to decrease the risk that large financial investors perceive in retrofit. However, these standards are voluntary, other than for state-funded work, and only a tiny fraction of the RMI workforce has been trained to meet those standards thus far. Further, in the domestic RMI market, most clients are investing in their homes and increasing their own asset base, using funds that are limited. The RMI client may need to prioritise between different project outcomes which appear to be in conflict, and without the driver of regulation, many clients will choose aesthetics and comfort – and cost – over energy efficiency [25]. The fragmentation of the standards picture makes it time- and expertise-intensive for a small firm to deal with and can, therefore, increase resistance to applying technical standards as part of scheme design. It is all too easy to suggest that a particular technical standard adds cost without adding much value, and standards may be ignored or only partially observed on-site, especially where compliance checks are rare [26]. Technical standards need to be well designed, widely understood, and properly policed across the industry; quality controls need to be embedded in practice, not left to people 'making a judgement' on-site about which rules to follow based on, sometimes, poor understanding of the consequences.

Policy aims with regard to safety, or carbon, or climate change resilience are similarly translated into guidance and regulation. It is vital that – for retrofit – three policy domains: planning, building regulations, and business/industrial strategy are moving in lock step. There is also a need to ensure that policy focuses on outcomes, rather than inputs. Building regulations that are performance-based without being prescriptive about how those outcomes are achieved, used in an industry with the capacity to monitor installed performance, would help to alleviate a considerable demotivation to improve energy efficiency reported by RMI practitioners – a lack of trust in the expertise underpinning standards [4]. The current need to add bathroom ventilation fans in houses with mechanical ventilation (MVHR) control systems is one example of how specifying the input (the bathroom ventilation fan) is not compatible with the desired output of increasing thermal efficiency through MVHR. Prescriptive standards give policymakers a degree of confidence in what practitioners should do that performance-based regulation might not offer, but if the current regulations, whose specific measures and materials, i.e. inputs rather than outcomes, do not deliver on achieving a reduction in energy demand and carbon emissions, then this is a risk that needs to be faced. Moving to measuring actual building performance also generates a new set of skills requirements, both in the building firms undertaking projects and in the advisors and building control regulators who approve works when complete.

Closely connected with technical standards is the need to create mechanisms for quality assurance. Quality assurance for projects has been clearly identified as an essential element to change UK construction performance [18]. Given the project-based nature of the industry, quality assurance needs to extend across practices and processes, not just products. Mandatory licensing could underpin better coordination of compliance checks with training and accreditation: a firm would support workers to undergo training because minimum qualifications would be needed to gain accreditation; accreditation would allow entry into the industry; and repeat compliance failures would carry a serious penalty, up to the withdrawal of accreditation. The exact details of trade licensing would need to be worked out, but such systems exist in many other countries, and the model is common in other professions.

5.3.4 The role of innovation in retrofit delivery

Innovation to enable and accelerate retrofit delivery will be in the products used, but, importantly, innovation is also needed in the practices and processes of project management and delivery.

For technology innovation, which focusses on a product but includes the enabling processes of design, installation, and use, we need to consider the attributes of technologies beyond their theoretical ability to change energy consumption or carbon emissions from energy consumption. Of course, the main attribute of any technology or technique is that it offers some relative advantage to the user and that it is effective, i.e. it actually delivers the promised advantage! In retrofit we have two categories of user – the installer and the building occupier – who may prioritise different forms of relative advantage and improving energy efficiency may not be top of that list, as indicated in the discussion of motivations and co-benefits above. Relative advantage is only one of the attributes of a technology that diffusion of innovation theory identifies as being important in the rate of technology adoption [27]. Another attribute which can accelerate take-up is the compatibility of the innovation with the existing system. In retrofit terms, from the builder's or installer's perspective, compatibility means that the technology must be compatible with other building components. It must also be compatible with the practicalities of installation; an insulation product that requires installers to wear full breathing apparatus will meet resistance to take-up across the industry. The issues of interfacing and detailing are the real issues of quality that retrofit must address. One of the challenges here is that this form of quality is not visible; it is difficult for the homeowner/bill payer to literally 'see' the quality of work undertaken. This often leads to an attractive aesthetic finish being used as a proxy for quality [16].

Further, technologies can only be adopted if they can actually reach the point of use, the RMI project, where they could be deployed. This leads us to look more closely at the supply chain systems in RMI. Wholesalers and builders' or plumbers' merchants are businesses with features that distinguish them from the businesses concerned with project management, building, and installation. Technology manufacturers certainly play a role in disseminating information about what is possible and supporting learning in the RMI sector, but merchants also play a vital mediating role in information dissemination [9,10]. Merchants' decisions about what to stock will also dictate the amount of what is installed in their local customer base. Merchants cannot stock all possible materials, so they 'choice edit' their stock for their customers, based on previous orders, an understanding of the local demand, and their own knowledge.

Innovation in building construction is often assumed to be concerned with new construction methods, such as off-site or modular construction, deploying more automation and seeking to develop an approach much closer to manufacturing than project-based working usually allows. Of course, there are opportunities to do more off-site production of components which fit together in effective, airtight ways, but work on existing buildings is inevitably bespoke; as soon as a building is occupied, then it is changed from the ideal, modelled energy performance that its designers imagined, one of the factors that creates the well-documented 'performance gap'

[28]. Altered through the very act of being lived in and used, each home will require a customised approach to energy retrofit, even if that customisation draws on a replicable menu of approaches and components. A new kitchen installation, a bathroom upgrade, a change to the windows or the heating system, as well as providing opportunities and trigger points for energy efficiency improvements – each of these projects will also alter how that building uses energy, and, in turn, each of those customisations will change the optimum set of retrofit activities which further need to be carried out.

5.4 Retrofit – an activity that changes over time

Thus far, we have tried to understand the nature of the RMI construction industry in terms of how it could shift to mainstreaming retrofit activity and outlined the range of interconnected systems that need to be aligned for that mainstreaming to be possible. For the final part of the chapter, we turn to considering how the 'over time' nature of retrofit affects what needs to be put in place for retrofit delivery. This is less a concern about the optimum way of arranging retrofit activities (as one off/whole house or room by room/over time [29]) and more a recognition that a large amount of retrofit activity in the housing stock will mean repeated interventions by a range of construction workers, both as the owner/occupier can afford and accommodate and as the potential to improve energy efficiency increases. Retrofit activities carried out now can only be a staging post in the ongoing work of maintaining a home so that it meets ever more demanding standards for energy use and associated carbon emissions.

All building projects – including retrofits – involve several steps including an initial concept, design, contract, installation, handover to the client, etc. In the UK, professional design and creation of buildings is guided by a seven-stage process, the RIBA Plan of Work [30]. While the RIBA Plan of Work starts with concept design, to be effective for retrofit projects, the Plan of Work needs additional stages. Retrofit, by definition, does not start with a blank sheet of paper (or area of ground); an initial baseline assessment and understanding of the building are needed before design thinking can start. Similarly, at the end of the construction process, an evaluation of performance under conditions of use is essential, so that future RMI works continue to improve building energy performance. Here, there is a real opportunity to capitalise on the structure of the RMI industry as it already exists, using the networks of small firms to share experiences and learning and disseminating knowledge through project delivery networks from job to job.

5.5 What needs to change in the retrofit supply chain to support high-quality, large-scale delivery?

Among the aspects of the RMI industry that need to change if large-scale, effective retrofit is to become a reality are: the ways in which work is organised at project and business/enterprise levels; the skills and professionalism that should be standard and unremarkable in the industry, and the procedures and institutions we use to capture and share knowledge and learning.

The linear representation of the plan of works does not work for retrofit delivery any more than it does for retrofit design, as outlined above. It is not possible to carry out a full baseline assessment for retrofit in advance of works starting. Some issues will only be revealed once plaster is off the walls or floorboards are lifted. The post-occupancy evaluation is only useful if there are mechanisms to tie the evaluation information into future plans. Feedback, feedforward, and learning are all essential elements to effective retrofit activity. Representing the plan as a 'retrofit wheel' can help focus attention on the need to connect the different stages and blur or share the responsibilities for successful project completion between different trades and roles [31]. Each stage in the retrofit cycle requires distinctive skillsets. Technical competence within narrow domains is not enough; there must be, as a minimum, an understanding of adjacent domains and, potentially, much greater development of multi-skilling so that individuals are not constrained to operate within the confines of one specific trade. In addition, the skills of communication, project management, and customer care are all skills that need to be deployed systematically throughout a project.

We must also face up to other challenges of how to organise retrofit, beyond the team processes and individual skills described above, in terms of the business models for enterprises undertaking retrofit projects. The fragmentation of the RMI sector is unlikely to change fundamentally, but perhaps there is the opportunity to work with the grain of the industry's existing mode, where local tradespeople often work together in different, overlapping teams and operate as virtual enterprises for different projects [15]. Different forms of enterprise, perhaps co-operatives or networks, are being tested, although not yet operating at the scale that the retrofit challenge and market transformation demand [32]. The industry will need to accommodate new partners, local authorities, social enterprises, or even community groups, who can act as brokers and, potentially, aggregators of demand. How retrofit activities are financed will also need to change. Currently, for owner-occupied or private rented housing, each dwelling is treated as an individual project and investment need. Building societies are offering green mortgages and other financial products to help individual homeowners. Social housing providers are able to develop programmes of activity across their entire housing stock and access programme-level funding to match. Grants to individual homeowners are not going to be sufficient for the challenge and, instead, large investors want to find ways to finance an area-based approach to retrofit, unlocking shared benefits beyond individual comfort and cost savings.

Once we accept that the programme that is easiest to model is not possible with the existing construction industry, i.e. that we cannot undertake all technically possible actions on an occupied dwelling at once, using a standard approach to labour and enterprise, across the whole housing stock, then 'whole house plans' emerge as a vital tool in ensuring that a long-term programme of housing energy upgrade is consistently moving in the right direction. In addition, all buildings require maintenance and repair, including those that have had energy efficiency improvements, as well as modifications in response to changing occupier needs. Retrofit is not a 'fit and forget' one-off activity; it is an ongoing cycle of change which needs to be tracked and understood, with information passed between owners as homes are sold. Creating and implementing whole-house plans, tracked by a 'building passport', listing RMI actions undertaken and what could happen next, a logbook which remains with the house, rather than its owner, generates another set of institutional and skills requirements [33].

5.6 Conclusion

In this chapter, we have been concerned with how to transform the existing market, which is already working effectively in supporting livelihoods and maintaining homes but which is not delivering on the low-carbon agenda, which ought now to be now inescapable for the construction industry. What we have set out is a 'market transformation' approach to RMI so that it delivers low carbon outcomes [8].

To achieve market transformation, we need to appreciate the socio-technical nature of the retrofit challenge, and in particular, we need to understand the small construction firms and micro-enterprises who are already busy making improvements to UK homes. The activities of these firms already provide opportunities to improve energy performance, and the projects they undertake are triggered by customer needs, which often have little to do with energy performance. Construction firms undertake work to deliver multiple benefits to their customers, and energy efficiency must be woven in alongside delivering those 'co-benefits'. However, the capabilities of those firms do not yet match the scale of the energy reduction challenge and there is much to be done to reduce risks and encourage continuous innovation and learning amongst firms, small firms in particular. Supporting standards, both technical standards and occupational standards, will be a vital part of developing robust quality assurance for energy performance of buildings. At the same time, there is no single uniform approach that will be effective on all buildings and a place-based approach that reflects more local norms and needs must also be developed.

In the RMI market, taking a systems approach means more than looking at the whole house or the whole energy system. A systems approach to tackling the challenge of transforming existing buildings means that both policy and practice need to move forward on skills, supply, standards, training, regulation, and customer experience *all at the same time*.

While this is all challenging, both to conceptualise and to implement, the opportunity that can be grasped if we are able to change the path of the existing RMI industry is significant. With more than one million people already working in the field and over £40 billion spent on housing RMI in 2023 [34], market transformation could achieve a significant amount of the carbon emissions reduction that comprehensive housing retrofit needs to make. But to rise to the challenge, we need to leave the apparent safety and certainty of technology-led models, abandon plans to create a new multi-billion-pound industry from scratch and instead work with the complex, messy reality of the RMI construction industry as it exists today.

References

[1] Killip G., Owen A., Topouzi M., Morgan M. 'Governance of low-carbon innovation in domestic energy retrofits in the UK'. *Proceedings of the 2017 International Refurbishment Symposium*. Dubai, UAE, March 2017. Cham: Springer; 2018, pp. 19–24

[2] Killip G., Owen A. 'The construction industry as agents of energy demand configuration in the existing housing stock'. *Energy Policy*. 2020;147:111816

[3] Janda K. B., Parag, Y. 'A middle-out approach for improving energy performance in buildings'. *Building Research and Information*. 2013;**41**(1): 39–50

[4] Simpson K., Janda K. B., Owen A. 'Preparing "middle actors" to deliver zero-carbon building transitions'. *Buildings and Cities*. 2020;**1**(1):610–24

[5] Simpson K., Murtagh N., Owen A. 'Domestic retrofit: Understanding capabilities of micro-enterprise building practitioners'. *Buildings and Cities*. 2021;**2**(1):449–66

[6] Owen A. M. 'Missing the point - The challenge of creating policies and programmes that tap into the motivations of the builders and installers'. *ECEEE 2015 Summer Study proceedings: First Fuel Now*. Hyres, France, June 2015. Paris: ECEEE; 2015

[7] Winch G. 'Zephyrs of creative destruction: Understanding the management of innovation in construction'. *Building Research & Information*. 1998;**26**(5): 268–79

[8] Killip G., Fawcett T., Jofeh C., Owen A. M., Topouzi M., Wade F. '*Building on our strengths: a market transformation approach to energy retrofit in UK homes'*. [online]. Federation of Master Builders/Centre for Research into Energy Demand Solutions; 2021. Available from https://www.creds.ac.uk/building-on-our-strengths-transformation-to-energy-retrofit/ [Accessed 16 June 2026]

[9] Killip G., Owen A., Topouzi M. 'Exploring the practices and roles of UK construction manufacturers and merchants in relation to housing energy retrofit'. *Journal of Cleaner Production*. 2020;251:219205

[10] Wade F., Shipworth M., Hitchings R. 'Influencing the central heating technologies installed in homes: The role of social capital in supply chain networks'. *Energy Policy*. 2016;95:52–60

[11] Energy Savings Trust. *Trigger Points: A Convenient Truth*. London: Energy Savings Trust; 2011

[12] Gram-Hanssen K. 'Retrofitting owner-occupied housing: Remember the people'. *Building Research & Information*. 2014;**42**(4):393–7

[13] Sorrell S., Dimitropoulos J., Somerville M. 'Empirical estimates of the direct rebound effect: A review'. *Energy Policy*. 2009;**37**(4):1356–71

[14] Murtagh N., Owen A., Simpson K. 'What motivates building repair-maintenance practitioners to include or avoid energy efficiency measures?

Evidence from three studies in the United Kingdom'. *Energy Research & Social Science*. 2021;**73**(1):101943

[15] Maby C., Owen A. *Installer Power: The Key to Unlocking Low Carbon Retrofit in Private Housing*. Leeds: University of Leeds; 2015

[16] Owen A., Mitchell G., Gouldson A. 'Unseen influence - The role of low carbon retrofit advisers and installers in the adoption and use of domestic energy technology'. *Energy Policy*. 2014;**73**:169–79

[17] WBCSD. *Transforming the Market: Energy Efficiency in Buildings*. [online]. Geneva: World Business Council for Sustainable Development; 2009. Available from https://www.wbcsd.org/Programs/Cities-and-Mobility/Resources/Transforming-the-Market-Energy-Efficiency-in-Buildings [Accessed 30 August 2021]

[18] Bonfield P. *Each Home Counts* [online]. London: BEIS; 2016. Available from https://assets.publishing.service.gov.uk/government/uploads/system/uploads/attachment_data/file/578749/Each_Home_Counts__December_2016_.pdf [Accessed 16 June 2026]

[19] Construction Leadership Council. *Greening our existing homes – National retrofit strategy consultative document* [online]. 2020. Available from https://www.constructionleadershipcouncil.co.uk/wp-content/uploads/2020/12/CLC-National-Retrofit-Strategy-final-for-consultation.pdf [Accessed 16 June 2026]

[20] Clarke L., Gleeson C., Winch C. 'What kind of expertise is needed for low energy construction?'. *Construction Management and Economics*. 2017; 35:78–89

[21] Zero Carbon Hub. *Closing the Gap Between Design and As-Built Performance, End of Term Report, July 2014*. London: Zero Carbon Hub; 2014

[22] Killip G. 'A reform agenda for UK construction education and practice'. *Buildings and Cities*. 2020;**1**(1):525–37

[23] Office for National Statistics. *Construction statistics annual tables Table 3.4* [online]. 2021. Available from https://www.ons.gov.uk/businessindustryandtrade/constructionindustry/datasets/constructionstatisticsannualtables [Accessed 16 June 2026]

[24] Gann D. M., Salter A. J. 'Innovation in project-based, service-enhanced firms: The construction of complex products and systems'. *Research Policy*. 2000;29:955–72

[25] Owen A. '*Factors that affect the diffusion and impact of domestic "green technology" and the role of "place"*'. PhD thesis, Leeds: University of Leeds; 2013

[26] Killip G. 'Products, practices and processes: Exploring the innovation potential for low-carbon housing refurbishment among small and medium-sized enterprises (SMEs) in the UK construction industry'. *Energy Policy*. 2013;62:522–30

[27] Rogers E. M. *Diffusion of Innovations*. New York: Free Press; 2003

[28] Zou P. X. W., Xu X., Sanjayan J., Wang J. 'Review of 10 years research on building energy performance gap: Life-cycle and stakeholder perspectives'. *Energy and Buildings*. 2018;178:165–81

[29] Fawcett T. 'Exploring the time dimension of low carbon retrofit: owner-occupied housing'. *Building Research & Information*. 2013;**42**(4):477–88

[30] RIBA. *RIBA Plan of Work* [online]. London: RIBA; 2013. Available from https://www.architecture.com/knowledge-and-resources/resources-landing-page/riba-plan-of-work [Accessed 16 June 2026]

[31] Topouzi M., Killip G., Owen A. M. 'Learning from "horror" stories: A plan of work to reduce the performance gap in deep retrofit'. *Passive Low Energy Architecture: Design to Thrive (PLEA 2017)* Edinburgh: PLEA; 2017, pp. 550–557

[32] Brown D. 'Business models for residential retrofit in the UK: A critical assessment of five key archetypes'. *Energy Efficiency*. 2018;11:1497–1517

[33] Fabbri M., De Groote M., Rapf O. *Building Renovation Passports – Customised Roadmaps Towards Deep Renovation and Better Homes*. Brussels: Building Performance Institute Europe; 2016

[34] ONS. *Construction statistics annual tables. Table 1.4.* [online]. 2023. Available from https://www.ons.gov.uk/businessindustryandtrade/constructionindustry/datasets/constructionstatisticsannualtables [Accessed 16 June 2026]

Chapter 6

Hard-to-treat dwellings: benefits and challenges of upgrading solid-wall buildings

Christopher Gorse[1], Kami Rakhshanbabanari[1] and Ali Saad[1]

6.1 Introduction

Hard-to-treat dwellings are those that cannot easily accommodate standard or cost-effective energy efficiency measures. According to government statistics, over half of the UK's building stock falls into this category, making these properties difficult to heat and particularly challenging to upgrade [1]. Conventional quick-fit thermal improvements often lack the flexibility needed for older buildings, where construction methods and materials are more complex and variable [2].

Recent retrofit programmes have shown that issues such as structural irregularities, limited material compatibility, and restricted access frequently disrupt installation, undermining the cost-effectiveness of retrofit work [3]. These subtleties limit the applicability of standardised solutions and highlight the need for bespoke, building-specific approaches.

Crucially, when thermal interventions are implemented without sufficient understanding of how solid walls behave, particularly their hygrothermal performance, unintended side effects, such as trapped moisture, mould, and material degradation can occur [4,5]. It is therefore vital to consider the thermal, moisture-absorbing, and structural properties of wall materials, their components, and how they are assembled. Additional factors such as the building's form, detailing, proximity, and connections to neighbouring structures further complicate retrofit design and specification, reducing the viability of off-the-shelf solutions. However, context-sensitive designs that are compatible with historic building fabric and tailored to specific materials can lead to more sustainable and resilient outcomes.

This chapter presents research insights into solid wall construction and key considerations for thermal retrofitting. It explores common solid wall types, their performance characteristics, and evidence-based retrofit strategies. Drawing on recent studies, this work evaluates the effectiveness of thermal upgrades in

[1]School of Architecture, Building and Civil Engineering, Loughborough University, Loughborough, UK

enhancing energy efficiency, while also addressing potential unintended consequences. It provides an overview of solid wall behaviour informed by contemporary research and practical retrofit initiatives.

As the UK advances toward decarbonising its building stock, retrofit strategies are evolving rapidly, with innovative solutions continually refining best practices [6]. However, a major challenge persists, ensuring that reliable, evidence-based information reaches householders, practitioners, and professionals alike [7]. Retrofit projects are costly and the services that support them are often fragmented [8]. Financing upgrades and making informed decisions remain complex tasks. Addressing these challenges requires sustained collaboration among researchers, policymakers, and industry experts to promote the adoption of best practices in both energy efficiency and heritage conservation.

6.2 The need to upgrade the building stock

While there are ongoing concerns about the quality and performance of newly built homes [9], the energy efficiency of recent developments has generally improved. Over 84% of new homes in England fall within Energy Performance Certificate (EPC) Bands A and B [10], although only a small fraction, just over 2%, achieve an A rating. Nevertheless, there is a consistent trend of newer dwellings outperforming older ones in terms of energy efficiency [11].

However, the contribution of new homes to national carbon reduction is minimal. For over two decades, the UK has not exceeded the construction of 200,000 new dwellings per year. As a result, new builds add less than 1% annually to the total housing stock, making their impact on the overall energy performance of UK housing negligible. With more than 70% of existing homes still in need of thermal upgrades [12], attention must shift toward improving the energy efficiency of the existing stock.

Meeting the UK's net zero targets requires urgent action to transform older, traditionally constructed homes, most of which are energy inefficient. Yet progress has been slow. Underperforming dwellings are rarely demolished, and the rate of new high-performing replacements is insufficient. Both social and private landlords have been encouraged to retrofit existing stock, but take-up has been limited. Fiscal incentives designed to stimulate energy efficiency improvements have had only a limited impact, particularly when it comes to solid wall properties. Since 2015, only 852,000 properties have benefitted from government-funded energy efficiency measures [13]. Funding efforts have generally focused on relatively straightforward interventions, such as loft insulation, cavity wall insulation, and boiler replacement measures that are easier and cheaper to install [14]. In contrast, solid wall insulation, which is more technically demanding and costly, has seen low levels of take-up.

Recent UK fiscal measures aimed at stimulating retrofit activity include:

- **The Green Homes Grant (2020–2021)**: A £2 billion scheme intended to support homeowners and landlords in installing insulation and low-carbon heating. The programme was withdrawn after just six months due to administrative issues and low take-up, reaching fewer than 10% of its target households [15,16].

- **ECO4 (Energy Company Obligation, 2022–2026)**: The latest phase of the ECO scheme, which mandates energy companies to fund energy-efficiency upgrades for low-income and vulnerable households. ECO4 includes support for more complex interventions, such as solid wall insulation, but still faces challenges in delivery and scale.
- **The Home Upgrade Grant (HUG, Phase 2: 2023–2025)**: Targeted at off-grid homes, this £630 million programme aims to support whole-house retrofits, including fabric-first upgrades and low-carbon heating [17].
- **Local Authority Delivery (LAD) Scheme and Social Housing Decarbonisation Fund (SHDF)**: These initiatives provide funding to local councils and housing associations to improve energy performance in social housing, with a particular focus on properties rated EPC D or below [18,19].

Despite these initiatives, fiscal measures have yet to significantly improve the performance of the hardest-to-treat homes, especially solid-wall buildings, which comprise a large portion of the UK's ageing housing stock.

6.3 Building form and characteristics

Approximately 8.5 million homes in the UK are believed to have solid wall construction [13]. However, this broad classification conceals significant variation in wall type, construction technique, and material composition. Effective thermal upgrades must be appropriate to each building's unique characteristics, especially regarding the substrates and construction details to which insulation is applied. Solid walls and their distinct behaviours must be carefully understood before retrofitting. Inappropriate interventions can lead to serious unintended consequences, such as surface and interstitial condensation, mould growth, and the premature degradation of wall or insulation materials. Interfaces between solid walls and other building elements, such as floors, roofs, windows, and doors, are also vulnerable, where they are at risk of becoming a significant cold bridge where condensation and mould can form if not properly integrated into retrofit designs.

Different materials and components within these buildings transmit heat and moisture at varying rates. Elements that act as thermal bridges, components that conduct heat more readily, can cause neighbouring materials to become colder and damper in winter. Therefore, understanding the makeup and assembly of traditional walls is essential to inform suitable retrofit strategies. Some materials may be insulated, isolated, or sealed, while others need to remain breathable or may require no treatment at all. Traditional buildings, particularly those with solid wall construction, commonly have substantial variation in form and fabric. These variations are more pronounced than in cavity wall construction, partly because solid walls have been used for much longer. Solid wall construction in the UK spans centuries, from Roman times through to the early twentieth century, and often reflects distinct regional materials and building practices [20,21]. As walls age, this diversity increases, making it difficult to apply universal retrofit solutions. Nevertheless, where solid wall behaviour is consistent, tried-and-tested interventions may be replicated.

The UK has one of the oldest housing stocks in the world and the oldest in Europe. Approximately 37.8% of homes were built before 1949, 39.7% between 1946 and 1980, 15.6% between 1980 and 2000, and only 6.9% after 2000 [22]. Some houses built as early as the twelfth century are still occupied today [23]. The evolution of housing types has largely mirrored economic and social trends, and understanding these peaks in construction can offer valuable insights into building form and construction techniques.

6.3.1 Demand and mass delivery of solid wall housing

The history of mass housebuilding in the UK has been shaped by fluctuating demand, driven by societal needs, and constrained by the socio-economic conditions of the time. For instance, following periods of conflict, particularly after the World Wars, there was a surge in the provision of cost-effective, off-site housing to accommodate low-income populations [24]. Similarly, during times of rapid industrialisation and economic growth, as people migrated to urban centres for work, the need for affordable housing led to widespread construction of solid-wall back-to-back houses and closely spaced terraced housing. Much of the solid wall housing stock built during the nineteenth and early twentieth centuries, particularly in industrial towns and cities, remains in use today [25,26].

Each surge in housebuilding brought with it an influx of new materials, technologies, and skills, which influenced construction practices and the forms of housing delivered. These transitions often introduced both improvements and challenges. While design preferences were shaped by the regulations of the time [27], the actual construction practices largely depended on the builders' familiarity with those regulations and the resources available. As a result, the quality and performance of buildings varied significantly, shaped by prevailing knowledge, skills, and materials.

The assumptions and practices of the era left a lasting impact, with gradual shifts in construction methods subtly affecting long-term performance. Occasionally, however, changes were dramatic and industry-wide. One key example is the transition from solid to cavity wall construction, which began in the early twentieth century and virtually ended the widespread use of solid wall systems. Another major shift was the introduction of prefabricated trussed rafters, replacing traditional hand-cut roof frames. While not directly related to wall construction, this shift illustrates how quickly building practices can change, even in a sector often viewed as slow to innovate.

Before cavity walls became widespread in the 1930s, solid wall construction varied considerably across the UK, reflecting local materials and craftsmanship. Some walls were built entirely of stone, while others combined a brick interior with a stone exterior to reduce costs. Examples of solid wall construction are shown in Figure 6.1. Where solid brick walls were used throughout, a range of bonding patterns such as

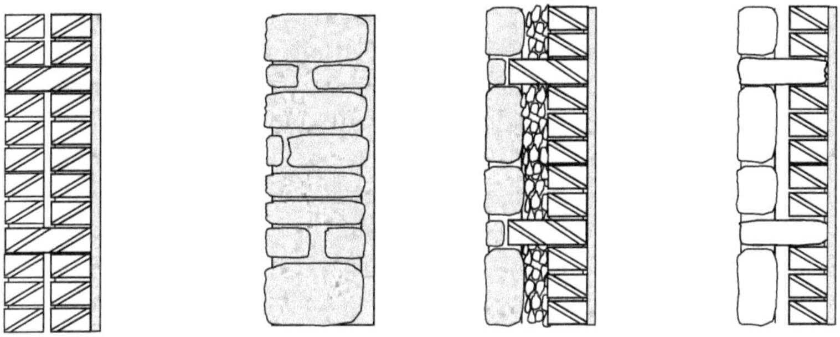

a. **Solid wall construction**	b. **Stone walls**	c. **Stone walls with rubble**	d. **Uninsulated stone /brick with small finger cavity**
Typical construction in 1800s. Solid wall or wall with a 15-25mm break to prevent moisture passage. Solid lime plaster applied to the face of the wall. The cold internal face makes the construction prone to surface condensation. Approximate U value 1.2 – 2.4 W/m^2 K	Solid stone walls are very dependent on local materials and practice, their construction and performance varies considerably based on the local materials used. Approximate U-values 1.2 – 2.0 W/m^{2}K (stone) 1.5 – 2.0 W/m^{2}K (sand stone) 1.4 – 2.3 W/m^{2}K (granite)	Cavity filled creating a solid wall. External stone is porous, but water that penetrates drips down the unconnected rubble cavity rather than crossing through. Thicker walls, with sealed finish tend to have lower U values Approximate U value 1.2 – 2.3 W/m^{2}K 0.9 – 2.4 (Baker, 2011)	Typical construction The thin cavities are often not sealed, leading to internal and external air channels through floor joists and voids, service penetrations etc. Approximate U value 1.4 to 1.5 W/m^{2}K

Figure 6.1 Examples of solid wall construction

English, Flemish, Sussex, and Monk bond and their garden-wall derivatives created subtle variations in both appearance and performance. Many solid walls feature stretcher bonds with occasional header courses to tie the masonry leaves together. Despite being classified as 'solid', these walls may contain construction voids, finger gaps between bricks, or rubble-filled cavities [28].

In cases where header bricks span the full width of the wall, they can form direct pathways for heat and moisture transfer. Cracks, voids, and interconnected gaps can facilitate the movement of moisture from the exterior to interior surfaces, especially under the influence of gravity, wind pressure, or capillary action. This is particularly concerning in light of climate change, which is expected to increase the frequency and intensity of wind-driven rain [29]. Solid walls are inherently permeable, so best practice recommends that they be allowed to 'breathe': rainwater should be resisted externally, but any moisture within the wall should be able to evaporate and escape.

The diversity of solid wall forms and materials means that some are more vulnerable to thermal and moisture transfer than others. As these buildings age, natural weathering, decay, and historic repairs further affect their performance [30]. While general retrofit strategies may be helpful in some cases, upgrading older solid wall buildings often requires a tailored, case-by-case approach. Both heating and ventilation systems must be designed in harmony with the building's materiality and thermal characteristics [31–34]. A sound understanding of these influencing factors is essential for retrofit designers seeking to improve energy performance without compromising the building's integrity.

Despite numerous government attempts to encourage system-built housing as a means of meeting demand [24], masonry construction has remained the dominant building method in the UK. Solid wall masonry was the prevalent form until the 1940s, after which cavity masonry walls became standard [35,36]. Understanding the historical evolution and enduring legacy of solid wall housing is, therefore, central to any meaningful discussion on energy retrofitting and housing resilience in the UK.

6.3.2 Solid wall and thermal performance

Solid wall construction varies significantly in its thermal performance due to the wide range of materials, construction techniques, and local building traditions used over time. Factors such as wall thickness, type of masonry, bonding pattern, presence of voids or inclusions, and the condition of the materials all contribute to this variability. As a result, solid walls cannot be treated as a uniform building element when assessing energy performance or planning retrofit measures; however, some assumptions on the thermal properties should be made and informed through research, survey, and measurement.

A list of the thermal performance of walls is provided in Table 6.1. Rather than being treated as a definitive guide, the list should be treated with some caution, as it is based on limited data. It is always prudent to undertake a full survey of properties prior to retrofit in order to determine the wall's construction and anticipated performance. Further work is essential in this area to build up a greater appreciation of the variance that does exist between walls and the different thermal properties that manifest. The list also introduces a sample of walls that have been upgraded with thermal insulation, demonstrating a level of energy efficiency that can be achieved with simple upgrades.

Further work undertaken by Glew *et al.* [41] conducted a sizeable study for BEIS on the use of internal wall insulation and the benefits achieved. A comparison is offered in the following tables of the performance calculated, Table 6.2, and measured, Table 6.3. The breathable thin insulants offer less risk, with some significantly improving the thermal performance. In the study, the material and installation cost of phenolic foam were found to be the most expensive, reasonably comparable with aerogel, followed by cork render, latex foam, expanded polystyrene (EPS), and polyisocyanurate (PIR), which again were in a similar price range. Thermal reflective paint was the least expensive but showed no measurable improvement. When evaluating performance, it is essential to consider how readily and safely the thermal insulation can be applied.

Table 6.1 Thermal properties of traditional wall construction [37–39]

Wall construction	Measured *U*-value (from [37])	Assumed Sap value (from [51])	Measured *U*-value (from [39])
Masonry with Granite	1.4 W/m^2K (for 520 mm construction)	2.3 W/m^2K (for 500 mm construction)	1.1 ±0.2 W/m^2 K (for 600 mm – Baker) 0.8–0.9 W/m^2 K (Field measurements Baker)
Masonry with Sandstone	1.5 W/m^2K (for 475 mm construction)	2.0 W/m^2K (for 500 mm construction)	as 1.1 ±0.2 W/m^2 K (for 600 mm - Baker) 0.8–1.6 W/m^2 K (Field measurements Baker) 1.7 W/m^2 K (for 600 mm – EST)
Masonry with brick	1.4 W/m^2K (for 220 mm construction)	2.1 W/m^2K (for 220 mm construction)	
Cob	0.7 W/m^2K (for 600 mm construction)	0.8 W/m^2K (for 550 mm construction)	0.4–0.8 W/m^2 K (Field measurements Baker)
Timber frame	2.0 W/m^2K (for 135 mm construction)	2.5 W/m^2K (for 150 mm construction)	
Concrete blocks, both leaves with 65 mm uninsulated cavity			1.1 W/m^2 K (Field measurements Baker)
Concrete blocks, both leaves with 65 mm insulated cavity			0.6 W/m^2 K (Field measurements Baker)
Bricks, both leaves with uninsulated cavity		1.44. W/m^2 K (Anderson from Baker) 50 mm cavity	1.3. W/m^2 K (Field measurements Baker) 65 mm cavity, 300 mm construction
Bricks, both leaves with 65 mm insulated cavity		0.3 W/m^2 K (EST) brick/block cavity wall	0.3 W/m^2 K (field measurements, 300 mm construction, Baker)

6.3.3 Classification of walls

It may be useful to classify buildings in order to align different walls with different thermal upgrades; however, classification systems can be too general, although it is useful to glean an insight into the classification systems.

Most of the houses constructed before 1930 tend to be solid walls, making up 28–29% of the UK building stock [13,42]. Also, many of the buildings constructed after this period are assembled in a manner that is not considered to be of

Table 6.2 Attributes of internal wall insulation and U approx. U-value when notionally applied to a solid brick wall of 2.09 W/m²K [41].

Insulation product – for internal wall insulation	Thickness mm	λ value (W/mK)	Approximate U-value	breathable
Phenolic, PUR, PIR, EPS, etc. foam	12.5–100	0.018–0.040	0.19–1.04	No
Aerogel blankets	10–60	0.014	0.21–0.72	Yes
Cork insulating render	10–75	0.037–0.058	0.40–1.79	Yes
Latex foam rolls	10–20	0.019	0.66–1.01	No
Mineral wool blankets or slabs	25–100	0.035–0.038	0.3–0.87	Yes
Vacuum Insulated Panels (VIPs)	20–40	0.0036–0.008	0.20–0.36	No
Wood fibre board (inc. cement)	20–100	0.037–0.048	0.36–0.92	Yes
Calcium silicate board	30–50	0.059	0.8–1.06	Yes
Thermo-reflective aerogel paint	1	0.014	2.02	Yes

Table 6.3 Measured performance field results based on the application of internal wall insulation to solid-wall buildings [41]

Insulation product – for internal wall insulation (applied to the internal surface)	Thickness of insulation	Original solid wall U-value	Measured U-value	U-value reduction
PUR	70	2.11	0.30	83%
PIR	27	2.11	0.78	63%
Aerogel boards	14	2.11	0.76	64%
EPS	22	2.01	0.98	49%
Corks render	20	2.01	1.06	32%
Latex rolls	10	2.10	1.49	38%
Thermal paint	1	2.10	2.25	n/a

conventional cavity wall construction [43]. While buildings can be grouped by their age and their wall construction, such as solid and cavity construction, their actual wall assembly is not so easily characterised and often does not follow a conventional type. Thompson's methodology on classifying cavity walls is of interest, noting some of the variations found and useful when considering the challenges of retrofitting. While the descriptions and methodology of classification by Thompson go beyond those listed below, the following classifications, shown in Table 6.4, provide some indication of the varying nature of walls; many of the unconventional walls will also be considered to be of a solid wall nature.

6.3.4 *Different construction forms and assemblies within a single wall and misdiagnosis*

While the descriptions offered in Table 6.4 go beyond that of the solid wall classifications, it is important to understand that the wall construction can vary within a dwelling or within a single elevation. Where traditional buildings have been altered

Table 6.4 Broad classification of wall types. Adapted from [43].

Wall type	Description
Solid walls	**Solid masonry walls** – typically brick, stone or in situ concrete with no significant cavity. **Solid walls with cladding** – often the cladding comprises timber or steel frame studs for the hanging of cladding, weatherboard, mineral sheets or tiles – in some instances, the gap or cavity within the studwork is insulated, but where uninsulated cavities exist, these could potentially be insulated. **Solid wall with lath and plaster** – solid mass wall with lath and a 40 mm cavity, which could be insulated.
Cavity walls	**Narrow cavities** (masonry walls with less than a 50 mm wide cavity) **Cavity walls that have been fully filled** (cavities of various widths that cannot be filled any further with insulation) **Partially filled masonry cavity walls** (cavity with fill and residual cavity typically 30–50 mm).
Unconventional walls	**Narrow masonry cavity** (various, nondescript cavity construction comes in many forms) **Prefabricated concrete** construction with cavity **Metal frame** construction with cavity **Random stone walls** – where there is a cavity, the width is uneven. **Mixed wall type** – where the wall is not of a single wall type, including dwellings where the outer face is completely or partially faced with cladding or hung tiles.
Timber frame	**Insulated or partially insulated** studwork without a masonry cavity.

or benefitted from an extension, combinations of cavity and solid walls can be common within one building. It is not wrong that the different wall forms are used within one building or elevation, but it is important that the different wall types and their contribution to whole house behaviour are evaluated. Where walls are of different construction, they are likely to have different hygrothermal properties and behaviours. Where different wall types contribute to the enclosure of a single room, those walls that are poorer thermal insulators are likely to be colder in the winter and more susceptible to condensation. If insulation is applied to the building, additional care is needed to ensure that the properties of the wall and changes in the wall's behaviour post-insulation are properly considered. Different elevations and walls within the building may need different methods of insulation. A failure to properly consider the construction and insulation substrate can lead to latent defects.

In King's [44] review of defects found in insulated solid walls, many of the risks occur as a consequence of not giving due consideration to the different interfacing materials and building elements. Inadequate identification of construction form and early construction techniques used in the UK, such as snapped headers (a half brick that appears but does not extend across two leaves of construction) and rat-trap bond (bricks bedded on their narrow edge, which results in a cavity between two skins), were considered to result in misdiagnosis of the wall's construction form. Ultimately, the misdiagnosis of the wall led to wrongly specified and designed thermal upgrades [44]. While the form of construction will favour certain thermal upgrades more than others, some mixed forms of construction will prove challenging and require further consideration.

Where insulation has been installed without due consideration for the characteristics of the wall, the choice of insulating products and positioning of the insulation may result in latent defects leading to unintended consequences, such as damp, mould, and degradation of materials [32,40,41]. Where inappropriate insulation products are installed, the thermal benefits gained may be short-lived. Incorrect use of insulation products may result in water ingress, condensation forming, or other construction defects. Saturated insulation and materials in close proximity may need to be replaced or refurbished. Indeed, King's [44] review of solid walls that had benefitted from thermal upgrades found 126 different unintended consequences that had not been effectively considered during the upgrade. While many of the consequences are relatively superficial, some are detrimental to the fabric, structure, and/or impact on the health of the occupants. The general observations from this study are useful when considering the choice and design of refurbishment and thermal upgrade to solid-walled properties (Table 6.5).

Table 6.5 Common defects found following retrofit and potential latent defects [32,41,44]

Defect	**Consequence – latent defect**
Incorrectly designed, combined, or installed	When materials or systems are poorly researched or designed, they may not perform as expected or required. For example, insulation can become saturated with moisture, lead to mould growth, reducing its effectiveness and energy-saving ability. Poor thermal performance can exacerbate economic deprivation and fuel poverty.
Air tightness increasing radon risk	Increasing air tightness without proper ventilation can trap radon gas inside a building, especially in high-radon areas. Buildings must include measures to prevent radon entering and provide adequate ventilation to safely disperse it.
Increased formaldehyde, VOCs, CO and CO_2 levels (reduced indoor air quality)	Introduction of materials, changes in airtightness and possible derogation of finishes can reduce air quality. Improvements in airtightness may require increased purpose-provided ventilation to prevent the build-up of VOCs. VOCs are often initially high following construction works. This is generally associated with new products installed and the VOCs can be high – the vulnerable occupants may be particularly at risk. CO and CO_2 levels in the building can rise in an airtight structure if it is not sufficiently ventilated.
Cracks or failure in external wall finish allowing moisture ingress. Gaps, channels between interconnecting surfaces	Moisture that enters the fabric can freeze during the winter months, expanding and spalling and cracking clays, concretes and mortars. Where metal brick reinforcement is present, corrosion can take place, the oxidation causing expansion and cracking around the reinforcement. Capillary action and winds can also lead to water tracking upwards, rather than solely being driven by gravity.

(Continues)

Table 6.5 (Continued)

Defect	**Consequence – latent defect**
	Moisture that enters the fabric can dissolve soluble salts; the crystals can form and expand as the water evaporates near or at the surface. Expanding salts can contribute to surface cracks and dusting of masonry and discolour the surface of walls.
Interstitial moisture (condensation within walls)	When insulation prevents walls from "breathing," moisture can become trapped inside the wall. As temperatures drop, water condenses within the structure. This increases dampness, encourages mould growth, and accelerates material decay through freeze-thaw cycles. Proper vapour control layers and breathable materials are needed to manage this.
Internal moisture	Changes to the thermal properties of the external wall can lead to cold spots on the surface of the wall; the increase in air tightness can result in an increase in relative humidity. Cold building elements, parts of the structure and other cold fabric offer areas where condensation from saturated air can form.
Thermal (cold) bridging	Certain areas—such as joints, corners, service entries, and junctions—are difficult to insulate properly and have a higher proportion of materials that readily allow thermal exchange. These areas become colder than the surrounding structure, increasing the likelihood of condensation forming, which can lead to damp and mould problems.
Pollution risk to external walls	Damp or cold walls are more vulnerable to airborne pollutants. Pollutants (especially acidic or alkaline substances) can react with wall materials, causing staining, surface damage, or long-term deterioration.
Premature failure of building systems	When moisture enters the building or insulation alters temperature conditions, materials may deteriorate faster. External wall elements may stay colder and wetter (due to internal insulation), making them less able to dry out. This increases the risk of damage from freeze-thaw cycles and weathering, shortening the lifespan of building components.

6.3.5 Choosing which solid wall buildings to retrofit: the fuel poverty vs. high emitters dilemma

Achieving a resilient, net zero housing stock will require significant upgrades to all solid-wall properties, including significant energy efficiency and comfort improvements – likely requiring a combination of changes to the building fabric, heating systems, ventilation, and energy supply. Retrofitting solid wall dwellings is rarely inexpensive and must be approached with care to ensure long-term thermal, environmental, and health benefits.

Homes where solid walls form the majority of the building envelope and where energy bills are high stand to gain the most from solid wall insulation. Comprehensive upgrades to such properties can reduce energy consumption and fuel bills by as much as 75%, with walls playing a significant role in these savings [45]. Detached, semi-detached, and end-terrace houses, which typically have greater exposed wall surface area, benefit most from reductions in heat loss through external walls. Even in homes with lower energy bills, reducing heat loss through solid walls can improve comfort and reduce risks of condensation and mould.

However, retrofitting must be done with caution. Incorrect installation of insulation can increase the risk of internal moisture issues. While thermal upgrades generally offer health benefits, some interventions, particularly those that reduce ventilation, can adversely affect indoor air quality [46]. Material selection and system design must, therefore, prioritise both performance and occupant health.

Households experiencing fuel poverty may not have the financial means to adequately heat their homes and may not see significant financial returns from energy-saving measures. Residents in this group often have little choice in their housing options, and those renting are dependent on landlords, private or social, for maintenance and upgrades [47]. Fuel-poor households often live in thermally inefficient homes with inadequate comfort levels, which contributes to poorer health outcomes [32].

In practice, retrofit funding is often prioritised for buildings with the highest emissions, which tend to be larger homes occupied by wealthier households. Yet many of the hardest-to-treat homes, such as older, solid wall dwellings, are occupied by low-income residents, who are not high energy users but still live in thermally inadequate conditions. This creates a dilemma: should retrofit efforts target the highest emitters to cut carbon or the most vulnerable to address fuel poverty? A balanced strategy is needed, one that considers both environmental impact and the socio-economic needs of occupants.

6.4 Progress on retrofitting hard-to-treat homes: too fast or too slow?

The UK's Clean Growth Strategy set a goal to upgrade all homes to EPC Band C by 2035, 'where practical, cost-effective, and affordable', and to achieve this standard in fuel-poor and most rented homes by 2030 [47,48]. Yet progress remains slow. Only around 30% of homes currently meet EPC bands B or C, and just 2% fall within the top-performing A–B bands [10,12]. Despite policy commitments, the terms 'practical', 'cost-effective' and 'affordable' have yet to be clearly defined [49].

Solid wall properties, many of which are classified as difficult to treat, are often excluded from large-scale retrofit programmes due to their perceived impracticality. However, these homes house approximately 30% of the UK population, including many fuel-poor households [47,50]. Retrofit activity in this segment of the housing stock has been inconsistent and underwhelming.

Data from schemes such as the Energy Company Obligation (ECO) and the Green Deal (GD) suggest that the rate of solid wall insulation installations has significantly declined from around 65,000 installations in 2014 to approximately 18,000 in recent years [13]. In total, fewer than 800,000 of the UK's estimated 8.5 million solid wall homes have received internal or external wall insulation [13]. Furthermore, these figures may overstate the actual retrofit rate, as even partial upgrades to a single wall are recorded as solid wall insulation [51].

Despite slow take-up, National Energy Efficiency Data (NEED) shows that solid wall insulation, when applied as a single measure, can yield greater gas savings than cavity wall or loft insulation, offering annual reductions in gas consumption exceeding 18% [11]. This suggests substantial potential for impact, provided upgrades are carried out effectively.

6.5 Summary

Long-term research on the performance of solid wall insulation remains limited. While some properties have experienced issues such as damp and condensation due to poorly designed or installed upgrades, others have demonstrated marked improvements in both occupant health and financial outcomes. More robust research, both in the field and in the laboratory, is needed to better understand the nature of solid wall construction and to identify the most reliable and effective upgrade strategies.

The introduction of PAS 2035: Retrofitting dwellings for improved energy efficiency aims to establish a coordinated framework for assessment, design, and delivery of retrofit works, tailored to both the building and its occupants. However, the success of this framework depends on a robust and consistent evidence base, particularly regarding solid wall performance, which is still developing.

Decarbonising the UK's building stock is central to addressing climate change. Effective thermal upgrades to solid-wall buildings can significantly reduce heating demand and improve overall energy efficiency. As the UK transitions to renewable energy sources, managing and conserving energy at the household level becomes increasingly important. Solid wall buildings represent a substantial portion of the housing stock and must be upgraded – but with care, precision, and a strong evidence-led approach.

References

[1] Department for Energy Security and Net Zero and Department for Business, Energy and Industrial Strategy (DESNZ/BEIS). *Building energy efficiency technical research: Collection of government retrofit research* [online]. 2025. Available from https://www.gov.uk/government/collections/buildings-energy-efficiency-technical-research [Accessed 08 July 2026]

[2] Historic England. *Energy efficiency and historic buildings: Advice for practitioners* [online]. 2024. Available from https://historicengland.org.uk/advice/technical-advice/retrofit-and-energy-efficiency-in-historic-buildings/ [Accessed 08 July 2026]

[3] Department for Communities and Local Government (DCLG). *English Housing Survey Energy Efficiency of English Housing – 2012 Annual Report on England's Housing Stock*. London: DCLG; 2014

[4] BSI Standards Publication. *PAS 2035: Retrofitting Dwellings for Improved Energy Efficiency: Specification and Guidance*. London: BSI; 2023

[5] Fletcher M., Thomas F., Miles Shenton D., *et al. Deterioration of Retrofit Insulation Performance (DRIP): Phase 1*. London: Department for Energy Security and Net Zero; 2024

[6] UK Green Building Council. *Decarbonising the UK's Building Stock: Policy Recommendations and Industry Trends – Impact Report 2023–2024*. London: UK Green Building Council; 2024

[7] Wise F., Gillich A., Palmer P. 'Retrofitting homes at scale for net zero: Lessons from the Social Housing Decarbonisation Fund Demonstrator'. *Energy Research & Social Science*. 2025;119:103866

[8] Samson H., Hall J. *The Role of Non-Profit Organisations in the Provision of Retrofit Services in England*. Bristol: Centre for Sustainable Energy; 2025

[9] Wilson W. *New-Build Housing: Construction Defects – Issues and Solutions (England)*. House of Commons Library, No 07665, 20 August 2020

[10] Ministry of Housing, Communities and Local Government (MHCLG). *Energy Performance of Buildings Certificates Statistical Release: Q1 2018 – England and Wales*. London: MHCLG; 2018

[11] Department for Business, Energy and Industrial Strategy (BEIS). *National Energy Efficiently Data-Framework (NEED) Summary of Analysis, Great Britain*. London: BEIS; 2020

[12] Green Alliance. *The Route to Clean Growth: Using Local Industrial Strategies to Drive Change*. London: Green Alliance; 2019

[13] Department for Business, Energy and Industrial Strategy (BEIS). *Household energy efficiency national statistics, detailed report 2018*. London: BEIS; 2019

[14] Department for Business, Energy and Industrial Strategy (BEIS). *Household energy efficiency detailed release: Great Britain data to December 2019*. London: BEIS; 2020

[15] National Audit Office (NAO). *Green Homes Grant Voucher Scheme [online]*. London: National Audit Office; 2021. Available from https://www.nao.org.uk/reports/green-homes-grant-voucher-scheme/ [Accessed 08 July 2026]

[16] Department for Business, Energy & Industrial Strategy (BEIS). *Green homes grant statistics* [online]. 2021. Available from https://www.gov.uk/government/statistics/green-homes-grant-statistics [Accessed 18 July 2025]

[17] Department for Energy Security and Net Zero (DESNZ). *Home upgrade grant phase 2: Guidance for local authorities* [online]. 2023. Available from

https://www.gov.uk/government/publications/home-upgrade-grant-phase-2-hug2 [Accessed 18 July 2025]

[18] Department for Energy Security and Net Zero (DESNZ). *Net zero strategy: Delivery update – Local authority schemes* [online]. 2024. Available from https://www.gov.uk/government/publications/net-zero-strategy-delivery-update-2024 [Accessed 18 July 2025]

[19] House of Commons Library. *Social housing decarbonisation fund and home upgrade grant: Funding retrofit measures. Briefing Paper No. 9768* [online]. 2023. Available from https://researchbriefings.files.parliament.uk/documents/CBP-9768/CBP-9768.pdf [Accessed 18 July 2025]

[20] Labate V. '*Roman Walls'*. World History Encyclopedia [online]. 32 August 2016. Available from https://www.worldhistory.org/article/942/roman-walls/ [Accessed 08 July 2026]

[21] Souza E. *How were the walls of roman building constructed* ArchDaily [online]. 19 March 2020. Available from https://www.archdaily.com/935423/how-were-the-walls-of-roman-buildings-constructed [Accessed 08 July 2026]

[22] Nicol S., Roys M., Ormandy D., Ezratty V. *The cost of poor housing in the European Union*. Watford: BRE; 2016

[23] Martinscroft T. *12 buildings in use today that were around when King Richard III was on the throne* [online]. n.d. Available from https://www.abroadintheyard.com/12-buildings-today-around-when-richard-iii-on-throne/ [Accessed 08 July 2026]

[24] Hashemi A. *Review of the UK Housing History in Relation to System Build*. London: The Bartlett School of Architecture, University College London; 2013

[25] Harrison J. 'The origin, development and decline of back-to-back houses in Leeds, 1787–1937'. *Industrial Archaeology Review*. 2017;**39**(2):101–16

[26] Historic England. *A brief introduction to Terraced Housing* [online]. 15 November 2019. Available from https://heritagecalling.com/2019/11/15/a-brief-introduction-to-terraced-housing/ [Accessed 08 July 2026]

[27] Coles T., Dinan C., Warren N. 'Climate change mitigation and the age of tourism accommodation buildings: A UK perspective'. *Journal of Sustainable Tourism*. 2015;**23**(6):900–21

[28] Smith M., Gorse C. *Building Surveyor's Pocket Book*. Oxon: Routledge. 2021

[29] Sanderson M.G., Eastman M., Lowe J. A. *Projected Wind-Driven Rain in the United Kingdom*. London: Department for Energy Security and Net Zero; 2024

[30] Gorse C., Highfield D. *Refurbishment and Upgrading of Buildings, 2nd ed.* London: Spon Press Taylor and Francis; 2009

[31] Glew D.W., Smith M., Miles-Shenton D., Gorse C. 'Assessing the quality of retrofits in solid wall dwellings'. *International Journal of Building Pathology and Adaptation*. 2017;**35**(5):418–33

[32] Gorse C., Glew D., Johnston D., *et al*. *Core Cities Green Deal Monitoring Project Leeds*. London: BEIS; 2017

[33] Zhao J., Du Y. 'A study on energy-saving technologies optimization towards nearly zero energy educational buildings in four major climatic regions of China'. *Energies*. 2019;12:4734

[34] Cho K., Cho D., Kim T. 'Effect of bypass control and room control modes on fan energy savings in a heat recovery ventilation system'. *Energies*. 2020; 13:1815

[35] Department for Communities and Local Government (DCLG). *English Housing Survey HOMES 2010*. London: HMSO; 2012

[36] Bailey J.D. '*A performance evaluation of mainstream timber framed and traditional masonry housing in the UK*'. PhD Thesis, University of Nottingham; January 2015

[37] Rye C. *U values and retrofitting traditionally built walls* [online]. 2014. Available from https://www.wtbf.co.uk/newsite/cymraeg/_blog/app/web/upload/3%20Caroline%20Rye%20U-values%20%26%20Retrofitting%20Traditionally%20Built%20Walls.pdf [Accessed 08 July 2026]

[38] Department of Energy and Climate Change (DECC). *The Governments Standard Assessment Rating of Dwellings SAP 2009 edition incorporating RdSAP 2009, Table S6:Wall U values England and Wales RdSAP, Appendix S*. London: HMSO; 2011

[39] Baker P. *U-values and traditional buildings in situ measurements and their comparisons to calculated values. Technical Paper 10*. Edinburgh: Historic Scotland; 2011

[40] Gorse C., Smith M., Brooke-Peat M., *et al*. 'Thermal performance' in: Smith M., Gorse C. (eds.) *Building Surveyor's Pocket Book*. Oxon: Routledge; 2021, pp. 169–219

[41] Glew D., Fylan F., Farmer D., *et al*. *Thin Internal Wall Insulation, Measuring Energy Performance Improvements in Dwellings using Thin Internal Wall Insulation. Summary Report*. London: BEIS; 2021

[42] National Energy Foundation *Solid Wall Insulation: Internal Vs External Benefits and Disadvantages* [online]. 2024 Available from https://nef.org.uk/internal-vs-solid-wall-insulation-which-is-right-for-my-property/ [Accessed 08 July 2026]

[43] Thompson R. *Review of the Number of Cavity Walls in Great Britain. Methodology*. Department of Energy and Climate Change; [online]. 2012. Available from https://assets.publishing.service.gov.uk/government/uploads/system/uploads/attachment_data/file/48433/5620-review-of-the-number-of-cavity-walls-in-great-brit.pdf [Accessed 08 July 2026]

[44] King C. *Solid Wall Heat Losses and the Potential for Energy Saving. Consequences for Consideration to Maximise SWI Benefits: A Route Map for Change*. Department of Energy and Climate Change, BRE; [online]. 2016. Available from https://www.bre.co.uk/filelibrary/pdf/projects/swi/UnintendedConsequencesRoutemap_v4.0_160316_final.pdf [Accessed 08 July 2026]

[45] Farmer D., Gorse C., Swan W., *et al.* 'Measuring thermal performance in steady-state conditions at each stage of a full fabric retrofit to a solid wall dwelling'. *Energy and Buildings*. 2017;155:158–67
[46] Maidment C.D., Jones C.R., Webb T.L., Hathway E.A., Gilbertson J.M. 'The impact of household energy efficiency measures on health: A meta-analysis'. *Energy Policy*. 2014;65:583–93
[47] Housing and Regeneration Research. *Evidence Review: The Lived Experience of Fuel Poverty in Scotland*. Edinburgh: Scottish Government; 2020
[48] HM Government. *The Clean Growth Strategy: Leading the Way to a Low Carbon Future* [online]. October 2017. Available from https://assets.publishing.service.gov.uk/government/uploads/system/uploads/attachment_data/file/700496/clean-growth-strategy-correction-april-2018.pdf [Accessed 08 July 2026]
[49] Wilson W., O'Donnell M., Bellis A., Barton C. *The Cost of Unhealth Housing to the National Health Service*. CDP-2019-0046, London: House of Commons Library; 2019
[50] Hills J. *Getting the Measure of Fuel Poverty: final report of the Fuel Poverty Review*. Centre for Analysis of Social Exclusion, Report 71, 2012
[51] Department of Energy and Climate Change. *Methodology Note accompanying statistical releases on estimates of home insulation level in Great Britain* [online]. 2013. Department of Energy & Climate Change. Available from https://assets.publishing.service.gov.uk/government/uploads/system/uploads/attachment_data/file/245591/methodology_note_estimate_home_insulation_levels_gb.pdf [Accessed July 08 2026]

Chapter 7

Standardized approaches for systemic retrofit in buildings

Chiara Dipasquale[1] *and Roberto Fedrizzi*[1]

7.1 Introduction

The share of energy consumption of residential buildings and the high percentage of European buildings built before 1990 require the promotion of large-scale energy efficiency retrofits of the existing building stock. However, this process is limited by high investment costs; low acceptance of the owners or tenants; unclear economic, energy, and comfort advantages; design effort; and the difficulty of intervention works in occupied buildings, as well as the overall definition of the retrofit solution.

Some of these barriers can be overcome by selecting the most appropriate solution for the building typology, avoiding component oversizing, and having clear expected energy and economic savings. Commonly, retrofit solutions in multi-occupancy buildings are defined by standard layouts and one generation unit that simplifies the installation and management of the whole system but does not exploit renewable energy nor optimize the system's function. Moreover, a standard configuration often implies that components may be oversized and, consequently, have high installation and running costs, inefficient system operation, and user discomfort.

As such, a better understanding of suitable retrofit solutions for specific building typologies, with expected energy savings and operation costs, would help identify solutions and, therefore, improve the renovation rate. Here, retrofit packages are analyzed through dynamic modeling for four representative building typologies of multifamily houses situated in four locations throughout Europe. The retrofit solutions investigate interventions on the envelope following the regulatory standards of envelope performance at the time of construction, with the main focus being on three heating, cooling, and ventilation (HVAC) system layouts coupled with renewable energy technologies; solar thermal collectors and photovoltaic (PV) systems. A database of model-based results identifies the energy performance of the studied cases, calculated through dynamic simulations, where the buildings

[1]EURAC Research, Institute for Renewable Energy, Bolzano, Italy

with the proposed renovation packages are modeled and simulated in four reference locations to explore the impact of differing climates.

7.2 Building study background

In this section, we explore the modeling of the buildings prior to energy-efficient retrofit. The modeling analyzes the different levels of intervention, assessing the energy savings, the economic impacts, and the building performance across four EU climates, recognizing the wide range of environmental conditions buildings must perform in.

7.2.1 Approach to the simulations

The approach adopted in this study uses dynamic models to analyse the influence that renovation interventions to an existing building have on energy, environmental, and economic performance. The developed methodology consists of the definition of reference buildings located throughout Europe, to which retrofit packages are applied. The retrofit packages include envelope insulation, window replacement, heating and cooling generation efficiency improvements, and adoption of renewable energy sources, as shown in Figure 7.1.

The reference buildings are multifamily houses, as they represent more than half of the heated spaces in some European countries [1], with 80% being built before 1990. For this reason, there is a large potential for decarbonisation of the residential sector if retrofit solutions are applied to these building typologies in the EU. Taking into account different building shapes and the related heating demands, four building typologies are used with different surface-to-volume ratios (S/V). The four reference building typologies are: low-rise building, high-rise building, small multifamily house, and large multifamily house. Details on the building typologies are outlined in Section 7.2.3.

The demand variability and external conditions are also considered by running dynamic simulations in four different climates through Europe: Oceanic for a cold-mild northern climate, Continental for a European cold climate, Southern Dry and Mediterranean for warm climates, both dry and humid.

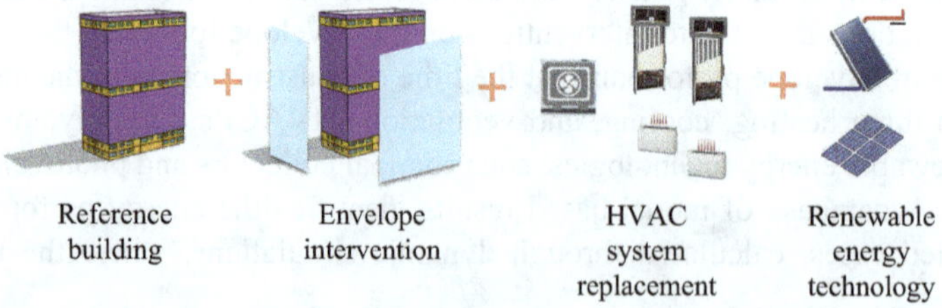

Figure 7.1 Methodology for the study of retrofit protocols

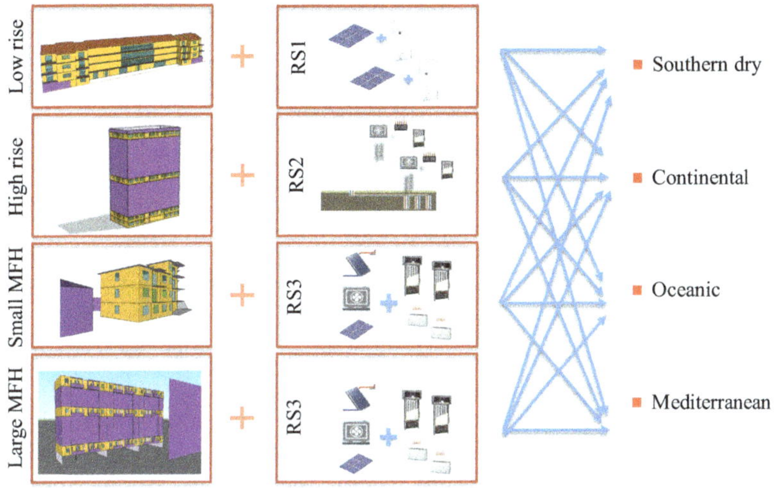

Figure 7.2 Combination of reference buildings and retrofit packages in different climatic conditions

The starting point of the approach is the definition of the reference buildings. They are assumed to have been built in the period 1980–1990. Therefore, the average thermal characteristics for the building envelope are referred to this period. To these reference buildings, interventions to the envelope are applied, reducing the building's energy demand by following the national requirements of thermal transmittance in renovation works.

In order to improve the energy efficiency of the existing buildings and contribute to the reduction of CO_2 emissions, the retrofit also considers the replacement of the existing HVAC system with a more efficient one. The analysis of the possible retrofit interventions applies renewable energy technologies to further reduce the use of imported energy. Depending on the building typology and available external surface, each retrofit solution, at dwelling or building level, may use photovoltaic or solar thermal systems, or a combination of both. The reference cases with applied different retrofit packages are modeled in TRNSYS, a dynamic environment [2], in four climatic conditions to assess the building's energy behavior under different working conditions (see Figure 7.2).

For each retrofit solution (RS) package, performance indicators on energy, environment, and economy have been studied and collected in an online database [3]. The main key performance indicators are explained in Section 7.2.1.2.

7.2.1.1 Climate variants

The energy performance of reference buildings with retrofit packages was assessed for four different climate conditions, representative of the whole of Europe, including the UK. The representative climates are used to define the building's

Table 7.1 Reference climatic conditions and heating degree days

Reference climate	Country	City	Heating degree days	Average temperature [°C]
Continental	Germany	Stuttgart	2902	9
Oceanic	United Kingdom	London	2452	12
Southern Dry	Spain	Madrid	1993	14
Mediterranean	Italy	Rome	1400	16

thermal characteristics before and after renovation, as well as the weather conditions that influence renewable energy production. A country, representative of the climatic zone, is used for the definition of the building transmittance characteristics, based on that country's regulations, while a city is used for the external weather file. Table 7.1 summarizes the four climates with the associated reference countries and cities. The fourth column of the table reports the Heating Degree Days (HDD) calculated on an 18 °C basis for 3 years (2017–2019) [4]. As the four climates differ by approximately 500 HDD each, they are, therefore, representative of a wide range of countries. The table also provides the average yearly temperature, which characterizes the climate and the ground temperature.

7.2.1.2 Key performance indicators

The performance of the reference buildings, with applied retrofit solutions, is evaluated by using energy, environmental, and economic key performance indicators. Energy performance indicators quantify energy required by the building to maintain the comfort conditions at 20 °C during wintertime and 25 °C during summertime. When considering the same boundary conditions between different climates, the comfort temperature has been assumed during a 24-hour period. Definitions of the energy KPIs are as follows.

Energy Demand (ED) - Energy demand denotes the energy required to maintain the comfort temperature during the year for space heating, space cooling, and domestic hot water (DHW) needs. During wintertime, this energy is positive, meaning that energy is added to the thermal zone, maintaining the set point of 20 °C, while during summertime, energy is extracted from the zone, maintaining the set point of 25 °C. DHW is assumed to be supplied at 45 °C. This quantity is indicated in terms of energy over a unit of heated area, represented in $kWh/m^2 \cdot y$.

Final Energy (FE) - Final energy represents the energy used to meet the building energy demand. This is dependent on the heating source. In the case of a gas boiler, final energy is the quantity of energy generated by natural gas through combustion. For electric resistances, heat pumps, pumps, fans, and split units, final energy refers to the quantity of electricity used.

Seasonal Performance Factor (SPF) - The SPF represents the effectiveness of the whole system and is defined as the ratio between the net energy demand to the user and all the electricity used in producing that energy. In contrast to the Seasonal Coefficient of Performance, this indicator refers to the useful energy effectively used by the building, net of system thermal losses, and includes all electricity consumption and auxiliaries and distribution devices, such as fan coils.

$$SPF_{el} = \frac{Energy\ demand}{Electric\ consumption}$$

Due to this definition, SPF can be linked to one energy source only. In the following, SPF refers to electricity. This indicator can be calculated for all the building uses or for each use, such as space heating, space cooling, and DHW.

Solar contribution - The contribution of the solar thermal system to space heating or DHW is accounted for in the percentage it represents as the share of space heating (SH) or DHW covered by solar energy.

$$SC = \frac{Solar\ thermal\ energy\ for\ SH\ or\ DHW}{Total\ thermal\ energy\ for\ SH\ or\ DHW}$$

PV use - The contribution of a PV system to the reduction of a system's electric consumption is assessed in terms of PV self-consumption. PV has a number of sub-indicators that indicate the consumption and energy flows. Energy may be consumed within the building or exported to the grid. The following definitions break this consumption down into the relevant categories.

The *PV self-use fraction* is the percentage of electric energy produced by PV and directly used by the HVAC system.

$$PV_{use} = W_{\mathrm{PV,used}} / W_{consumption}$$

where $W_{\mathrm{PV,used}}$ is the self-consumed PV energy and $W_{consumption}$ is the total electricity consumed. This quantity is calculated hourly and takes into account the simultaneity between PV production and electricity consumption.

PV self-consumption is energy produced by the PV system and used by non-HVAC devices that are consuming electricity, such as lighting or small power.

$$PV_{self} = W_{\mathrm{PV,used}} / W_{produced}$$

where $W_{\mathrm{PV,used}}$ is the self-consumed PV energy and $W_{produced}$ is the total electricity produced by PV. This quantity is calculated hourly and takes into account the simultaneity between PV production and electricity consumption. *PV production* represents the electricity produced by the PV system and then used by the system or sent to the grid.

Finally, *electricity to the grid* is the surplus energy produced by the PV system, which is fed into the grid.

Environment indicators refer to Primary Energy consumption for each energy source and to CO_2 emissions.

Primary Energy - Primary Energy consumption is an indicator that allows a comparison between different energy sources and considers energy consumed by electricity or gas. Primary Energy multiplies Final Energy for a Primary Energy Factor that quantifies the nonrenewable Primary Energy used to provide the final energy, including the energy used for construction of the electric grid and power plants, or the embodied energy. This indicator accounts for the Primary Energy from fossil, nuclear, and primary forest resources, such as original forests that are destroyed and replaced by farmland, defined in terms of Primary Energy to final energy—kWh_{PE}/kWh_{FE}.

$$PE = FE * \mathrm{PEF}$$

Since the provenance of electrical energy at the plug varies widely from country to country, due to their power generation and import mixes, the adopted values are reported in Table 7.2 and refer to the "Primary Energy Factors and Members States Energy Regulations" document [5].

CO_2 Emissions - The emission of CO_2 is another indicator that gives an estimation of the impact that the proposed solutions will have on the environment. When calculating this value, we must calculate the emission factor for each unit of electricity or gas production, following. the emissions factors outlined in the "Covenant of Mayors — Technical Annex to the SEAP template instruction document: The emission factors" [6]. Table 7.3 outlines the European average values to be used:

Table 7.2 Values of Primary Energy factors

Energy source	**PEF ([kWh_{PE}/kWh_{FE}])**
Electricity	2.3
Main gas	1.1

Table 7.3 Values of CO2 emissions factor

Energy source	**Standard emission factor [t CO_2/MWh_e]**
Electricity	0.460
Main gas	0.202

7.2.2 Reference buildings and HVAC systems

The reference buildings have been defined on the basis of geometric characteristics, with the aim of addressing the most common typologies in the EU and, at the same time, having a variety of cases. Table 7.4 summarizes the main reference buildings' geometric characteristics. The four typologies differ in the number of floors, the horizontal and vertical aspect, and the total heated area. Looking at the form factor, denoted by the surface-to-volume (S/V) ratio, the chosen multifamily house typologies range from 0.15, for a very compact building, to 0.45, for a smaller building.

As previously discussed, the building envelope's thermal characteristics depend on the specific climate. For the existing pre-retrofit case, average walls, roof, ground, and windows transmittance for a building built between 1980 and 1990 were adopted. Wall constructions can differ, depending on the building typology, but the total transmittance has been assumed to be the same for each climate. These values refer to the study conducted during the FP7 project iNSPiRe [7] and are reported in Table 7.5.

Thermal transmittances for each of the retrofit cases are designed to achieve the minimum requirements defined in each reference country when renovating. These are reported in Table 7.6 and implemented when retrofit packages are applied to the reference building. We do not specifically look at the technical solutions here, as the focus is on HVAC retrofit solutions, but we assume an

Table 7.4 Reference building characteristics

Building typology	**Number of floors**	**Number of dwellings per floor**	**Area tot**	**S/V**
	[n]	[n]	[m^2]	[-]
Low rise (LRMF)	4	15	4055	0.34
High rise (HRMF)	17	6	7140	0.15
Small MFH (SMFH)	4	2	712	0.45
Large MFH (LMFH)	8	2	7120	0.27

Table 7.5 Transmittance values for reference buildings built between 1980 and 1990

	U_{walls} [W/m^2·K]	U_{roof} [W/m^2·K]	U_{ground} [W/m^2·K]	U_{window} [W/m^2·K]
Continental	0.65	0.42	0.69	2.92
Oceanic	0.76	0.59	1.07	4.36
Mediterranean	1.00	1.24	1.51	4.03
Southern Dry	1.56	1.33	1.07	3.49

Table 7.6 Transmittance values defined by current regulations for buildings retrofitted

	U_{walls} [W/ $m^2 \cdot K$]	U_{roof} [W/ $m^2 \cdot K$]	U_{ground}[W/ $m^2 \cdot K$]	U_{window}[W/ $m^2 \cdot K$]
Continental [8]	0.25	0.19	0.23	1.10
Oceanic [9]	0.18	0.13	0.13	1.40
Mediterranean [10]	0.29	0.26	0.34	2.00
Southern Dry [11]	0.41	0.35	0.65	1.80

Table 7.7 Solar technology installations

Scenario	PV	STC
LRMF+ RS1	55.4 kWp (12% vertical, 88% at 30°)	–
HRMF + RS2	48 kWp (50% vertical, 50% at 30°)	–
SMFH + RS3	10.8 kWp – 10°	37 collectors, vertical, south facing
LMFH + RS3	54 kWp – 10°	37 collectors, vertical, south facing

upgrade to the fabric based on these minimum requirements for the purposes of the model.

Depending on the building typology and available building external surfaces, such as the roof or external walls oriented to the south, east, or west, each retrofit solution is coupled with a photovoltaic (PV) system only or an additional solar thermal collector (STC) system. PV covers part of the electricity used by heat pumps or auxiliary devices, while STC provides a share of Domestic Hot Water (DHW) production. The size of the adopted PV and STC systems is shown in Table 7.7.

As mentioned previously, each reference building has been simulated in the four countries to assess the performance of the intervention. In each energy model, internal gains are identified as the contribution from occupancy, lighting, and appliance use. Different profiles of people's activity are used to account for the nonsimultaneous behavior of the tenants. The profiles of occupancy are stochastic [8] and consist of a time series of values denoting whether people are active, sleeping, or away from the building. Appliances and lighting use are defined according to the presence of occupants. Peak values for internal gains, set-point temperatures in winter and summer, infiltration, and ventilation rates assumed for the simulations are summarized in Table 7.8.

Each retrofit solution is assumed to replace an existing system. For each retrofit solution, the reference system is described and used to assess the savings in terms of energy, CO_2 emissions, and operational costs.

Table 7.8 Boundary conditions for the building model

Occupancy	Value
Number of people per dwelling	3
Sensible heat	65 W/person
Latent heat	55 W/person
Appliances and lighting	Value
Peak sensible heat	10 W/m^2
Standby sensible heat (min value)	2 W/m^2
Infiltration and ventilation	Value
Infiltration flow rate	0.1 vol/h
Mechanical ventilation flow rate	0.3 vol/h
Set temperatures	Value
Schedule space heating	0–24 h
Heating set temperature	20 °C
Cooling set temperature	25 °C

7.2.2.1 Reference building 1: low-rise building

A low-rise building is characterised by a total number of floors less than four and it extends horizontally. The reference building for this typology has four floors, including the ground floor. It is north-south exposed. The building is composed of five blocks with three dwellings each: two 85 m^2 dwellings facing north and south and one 50 m^2 dwelling facing south. In total, the building has 53 dwellings, 8 on the ground floor and 15 on the other floors (Figure 7.3). Due to its shape, the building has an S/V ratio of 0.34, meaning that it is quite compact.

Solutions for reducing heating and cooling demand consist mainly of external wall insulation and window replacement. The target U value to be achieved through the renovation works is reported in Table 7.6.

7.2.2.2 Reference building 2: high-rise building

A high-rise building is characterised by a larger vertical dimension than horizontal one; therefore, it has a large number of floors. The reference building for this typology has 17 floors with 6 apartments on each. The total number of apartments is 102. The facades of the building are oriented at 45° with respect to the cardinal directions. Figure 7.4 shows the exposure of the single apartments on a typical floor. All the apartments have a regular shape and no balconies are present. The building has a compact shape and, consequently, has a low S/V ratio (0.15). The floor area of apartments 1, 2, 5, and 6 is 80 m^2, while apartments 3 and 4 are 56 m^2.

7.2.2.3 Reference building 3: small multifamily house

The small multifamily house is composed of a small number of floors and two to three dwellings per floor. The reference building for this typology has four floors and eight apartments, each having a floor area of 89 m^2. The apartments are east- and west-oriented. The south façade is assumed to be external, while the northern

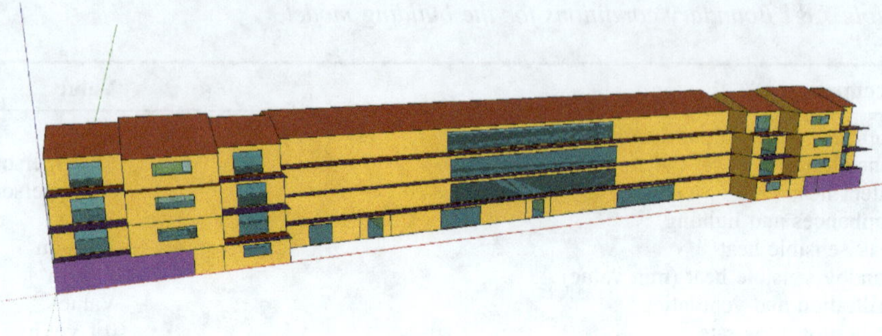

Figure 7.3 3D model—view of the South façade

Figure 7.4 High-rise building—orientation and geometry of apartments in a typical floor (left) 3D model view (right)

one links with another building. The building has an S/V ratio of 0.45, indicating a non-compact multifamily house (Figure 7.5).

7.2.2.4 Reference building 4: large multifamily house

The large multifamily house is characterised as having more than four floors and, generally, includes more than one staircase. The reference building for this typology has five staircases (see Figure 7.6), each serving 16 apartments on 8 floors. The total number of apartments is 80, with the same geometry and a floor area of 89 m^2. The apartments' exposure is mainly toward the east and west. The apartments belonging to the north and south staircases also have facades exposed toward these directions, without windows. The first floor is not adjacent to the ground; it is raised, and the floor slab is exposed to external air. The shape of the apartments is quite compact, apart from the recesses of the balconies, and the S/V ratio is 0.27.

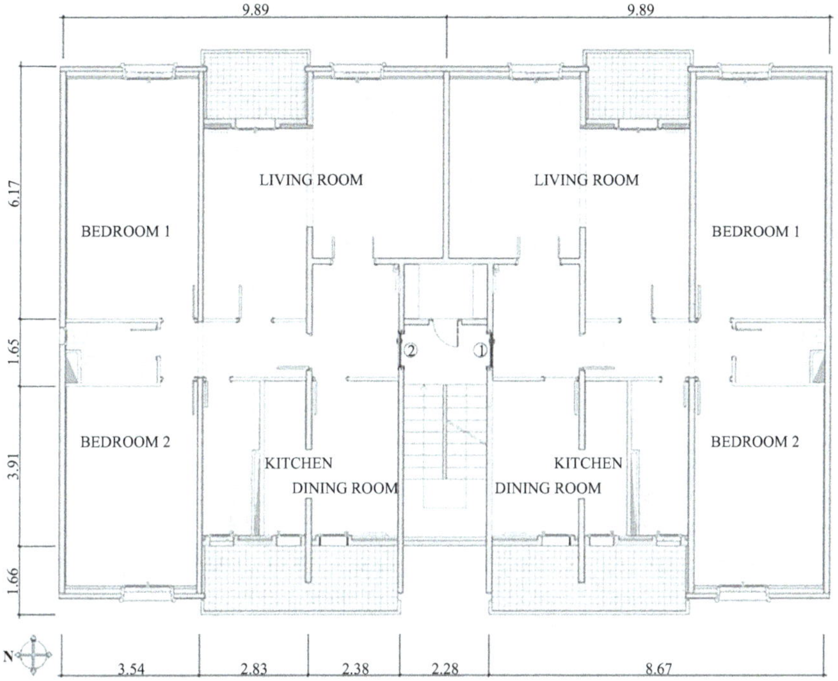

Figure 7.5 Small multifamily house—orientation and geometry of a typical apartment

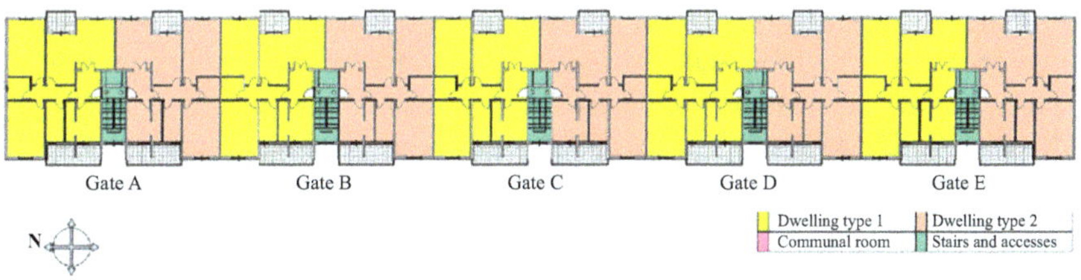

Figure 7.6 Large multifamily house—orientation and geometry of the apartments and shape of the building plant

7.2.3 Retrofit solutions

The following section outlines the proposed retrofit solutions for the reference buildings. These start by identifying the base case, the starting point for a specific aspect of the building, and then outline a proposed retrofit solution to improve the performance of the building. As stated previously, the envelope retrofit is assumed to be new windows and wall insulation to the standards outlined in Table 7.6. The main focus of this study is the HVAC systems that are part of the retrofit package.

7.2.3.1 Retrofit solution 1: decentralised HVAC system

The reference base case, or the pre-improved case, for Retrofit Solution 1 (RS1) is a decentralised system composed of a gas boiler installed in each dwelling to deliver space heating and DHW. The assumed efficiency for this system is 0.85. Space cooling is covered by a non-performant split unit system with an efficiency, or coefficient of performance, of 2.5. In the base case, a mechanical ventilation system is not present (Figure 7.7).

The HVAC system proposed within RS1, outlined in Figure 7.9, is a three-in-one heat pump plus mechanical ventilation, able to cover space heating and cooling loads, DHW production, and ventilation. Commercial units with this range of applications are usually composed of a mechanical ventilation unit with heat recovery, which takes fresh air from outside and blows it into the buildings directly as ambient air if there is no need for air conditioning. When the ambient air needs to be heated up, fresh air is mixed with recirculation air and is warmed through the condenser of the pump, which brings supply air to the optimal temperature to achieve the comfort temperature. The evaporator of the heat pump takes energy from the exhaust air before it is expelled outdoors. When space cooling is needed, the heat pump cycle is inverted. In this case, instead of expelling warm air to the ambient, condenser heat is used to preheat domestic hot water.

Typical units have 90–150 litre storage tanks for DHW preparation. For the calculations presented here, we have selected a unit with a 90-litre DHW storage tank; the air flow rate during ventilation-only mode is 100 m^3/h, in heating/cooling mode, the air flow rate is 400 m^3/h, and in free cooling mode, the air flow rate is 200 m^3/h. The machine has a heating capacity of approximately 3 kW and a cooling capacity of 2 kW. The coefficient of performance is assessed to be 3.83, resulting in a power capacity of less than 1 kW.

RS1 also includes the use of renewable energy by adopting a PV field of 1 kWp for each dwelling. Depending on the available surface, five panels are installed on the roof with an inclination of 30°, or eight are positioned on the east and west façades with a 90° slope. RS1 includes an inverter that transforms PV energy for use by the heat pump. In the case of PV overproduction, a heating

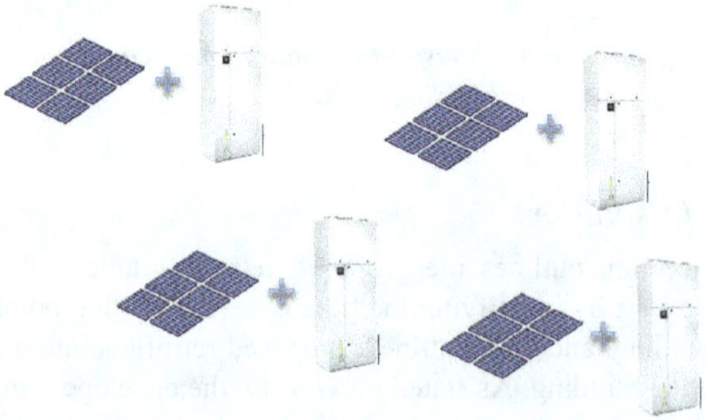

Figure 7.7 Schematic of the RS1 HVAC system

element is switched on in the DHW tanks to store excess energy in the form of hot water.

7.2.3.2 Retrofit solution 2: decentralised generation and centralised water loop

In the reference pre-retrofit scenario for Retrofit Solution 2 (RS2), the heating system for space conditioning and DHW production is a centralised gas boiler. The assumed efficiency for this kind of system, including thermal losses, is 0.80. Energy distribution in the apartments is provided by radiators working at high temperatures (70 °C). Space cooling is covered with low-performance split units with an efficiency of 2.5 and there is no mechanical ventilation.

The RS2 identifies the installation of a ground-source heat pump (GSHP) of 6 kW for each dwelling for space heating and space cooling and it is coupled with a 120-litre thermal store for DHW use. The GSHPs are connected to a brine loop at source side, composed of four risers with a circulation pump in each, which, in turn, are connected to a bore field (see Figure 7.8).

Each apartment has six radiators of different sizes, depending on the apartment's floor area. Radiators have a maximum design supply temperature of 45 °C for outdoor temperatures lower than −5 °C, which changes according to a climatic curve, providing weather compensation. The minimum supply temperature is 30 °C

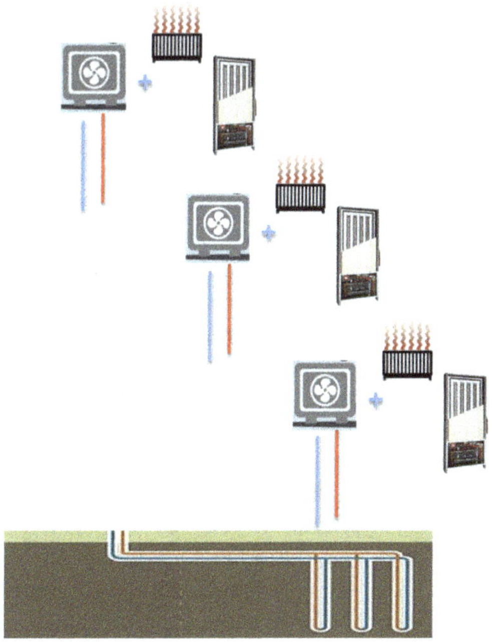

Figure 7.8 Layout of the system at building level. Schematic of the decentralised heat pumps and bore field.

for outdoor temperatures greater than 15 °C. This climate control, plus the inverter-driven heat pump, allows the system to adapt the emission power to the climate and building load.

Two kinds of renewable energy sources are used with this retrofit solution, a geothermal heat source and a solar source using photovoltaic modules.

The geothermal source is particularly interesting due to the temperature that the ground can offer, which is more advantageous when compared to the external air during both winter and summer months. The ground temperature annual profile is less affected by the location than the air temperature profile.

The PV field is composed of 210 panels installed on the roof with a slope of 30° and 120 panels installed on the southwest façade. The PV panels are mono-crystalline, with a peak power of 200 W. The two apartments on the southwest façade have four panels each, covering part of the electric load, while the other dwellings, plus the two on the ground floor, can benefit from three panels each, installed on the roof.

7.2.3.3 Retrofit solution 3: centralised HVAC system

The reference pre-retrofit scenario for Retrofit Solution 3 (RS3) is a centralised system composed of a gas boiler that delivers space heating. The assumed efficiency for this kind of system is 0.8. DHW production relies on individual electric boilers, with one in each apartment. Space cooling demand is met by low-performance split units with an efficiency of 2.5 and there is no mechanical ventilation.

RS3 is applied to both the large and small multifamily houses (LMFH and SMFH). When applied to the LMFH, each of the five staircases has an independent HVAC system based on a centralised air-to-water heat pump (Figure 7.9). Each heat pump can cover heating, cooling, and DHW demands for the apartments of the respective block.

The south-facing block has solar thermal panels positioned vertically on the south façade, which can make a significant contribution to DHW production. This contribution accounts for the whole building. There are 8 apartments per staircase in the SMFH and 16 apartments per staircase in the LMFH. In this second case, the share of renewable energy per m^2 of heated area is lower.

The heat pump generation system on the roof includes a DHW storage tank of 1000 litres and a buffer tank of 500 litres for heating and cooling distribution. All of the dwellings have a prototype HVAC system that includes a 140-litre water storage tank for DHW, a mechanical ventilation unit with heat recovery, and a hydraulic module connected to the fan coil circuit for space heating and cooling. In case of excess heat from the solar thermal field, the storage tank can be overheated.

During summer, the heat pump can recover waste heat for DHW production. For this reason, the control system also includes an operation mode in which the evaporator side absorbs heat from the water for space cooling and the condenser side transfers heat to the DHW storage. The size of the HP depends on the building typology (SMFH or LMFH) and on the climate.

Each staircase block of the LMFH and the SMFH has a PV field installed on the roof with the same size as the available surface. The PV field has 36 panels with

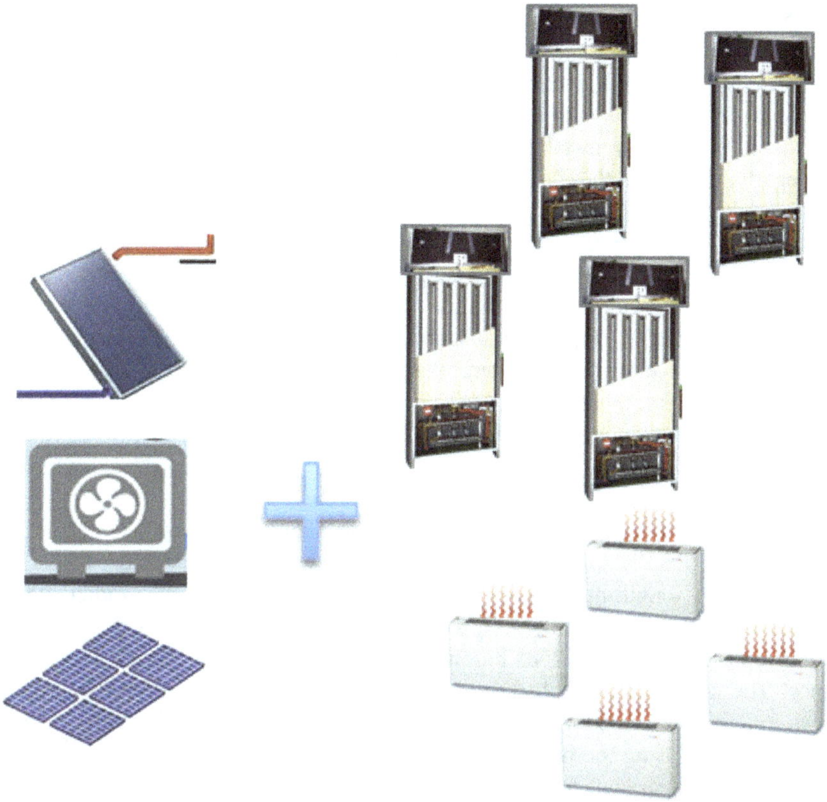

Figure 7.9 Schematic of the HVAC system in RS3

a total power capacity of 10.4 kW. Electricity produced by the photovoltaic field is used to run the heat pump and auxiliaries of the centralised system.

The solar thermal collector field is integrated into the south façade and has 37 panels for both building typologies, with a total installed area per building of 85 m^2. The solar system stores harvested energy in a thermal energy tank of 1500 litres.

With each of the proposed retrofit solutions identified for each of the types, we will apply these solutions to the reference buildings to evaluate their performance across the identified key performance indicators.

7.3 Results

The presented retrofit solutions have been applied to the reference buildings. Each combination of reference building, plus retrofit solution, has been evaluated in each climate, meaning a total number of 16 cases (Figure 7.2). The justification for combining a solution to a reference case is outlined below.

RS1 and RS2, the decentralised solutions, are not influenced by the shape of the building but by the load of the single apartment. Considering different apartment shapes in different locations guarantees a variety of loads. RS2 is usually adopted in cases with a high number of users and, therefore, the behaviour of the centralised water loop is taken into consideration in a building with a higher number of floors. For these reasons, RS1 and RS2 are applied to one reference building only, RS1 to the low-rise building and RS2 to the high-rise building. Due to the different dwelling sizes and studied climates, the results could give an overview of the retrofit packages' energy, environmental, and economic performance in different working conditions.

The solution of a centralised system, RS3, is characterised by thermal losses higher than a decentralised system. Plant configuration and control strategies influence this effect, therefore, the energy performance of such a system depends on the building typology. Buildings that are wide or high and, therefore, require longer pipes will have higher thermal losses through the pipework. For this reason, it is suggested to adopt the centralised system solution in those building typologies with a limited number of floors or that serve one staircase at a time. Due to these considerations, RS3 is applied to the small and large multifamily houses.

The energy, environmental, and economic performance of the studied RSs is reported in terms of the influence that the intervention on the envelope, HVAC system replacement, and use of renewables have on the whole building's performance. The influence that interventions on the envelope have on the total energy consumption is calculated, as is the energy consumed to cover the building demand after retrofit, by using the heating and cooling system of the existing case.

All the comparisons and assessments of energy savings, unless otherwise specified, are calculated with respect to the reference case.

7.3.1 Southern dry and continental climate

In this section, we explore the results for the Southern Dry and Continental climates. This links to the BuildHEAT project, which had case study buildings within these specific climates that were monitored. The complete modeling results can be found in the final report [12]. Figure 7.13 and 7.14 refer to the MED and the CON climates, respectively, and show the Primary Energy consumption in the reference case (first column — PRE), Primary Energy with implemented intervention on the envelope and HVAC system (second column — ENV/HVAC), and the total Primary Energy consumption when the PV and STC, renewable energy, are implemented (third column — REN EN). Each column is divided by energy use: space heating, space cooling, DHW demand, and ventilation. On the right axis, it is possible to read the PV self-consumption and STC contribution.

Despite the variety of building typologies, renovation solutions, and climates, we can note that the upgrading of the envelope's thermal characteristics to the current regulatory requirements for each country, outlined in Table 7.6, and the use of more efficient HVAC systems can reduce Primary Energy consumption by a range of 60–65%, when compared to the pre-renovation case. The only exception is the HRMF building, which is characterised by a large number of apartments, with only one external surface and located on an intermediate floor. As a consequence of this, an intervention on the envelope is effective in reducing heating demand, but not the cooling demand. The envelope retrofit leads to an increase in cooling demand once the intervention is implemented, especially in warmer climates. However, covering this load with a heat pump leads to a reduction of PE consumption in the order of 25% in the Mediterranean climate, where cooling demand is higher, and 50% in a northern climate. To increase these savings, renewable energy might further contribute, depending on the installation surface availability and control logics adopted for managing energy production (Figure 7.10).

In the Mediterranean climate, the annual PV self-consumption is 33% in the LRMF+RS1 scenario, 15% in both HRMF+RS2 and SMFH+RS3 scenarios, and

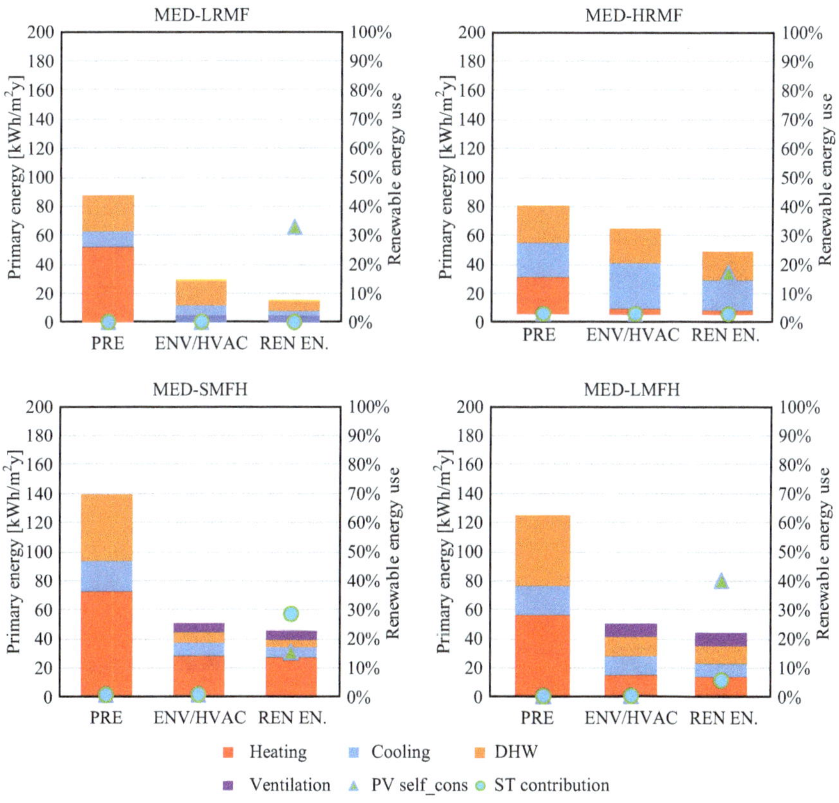

Figure 7.10 Primary energy decrease for each intervention and by energy use, PV self-consumption, and STC contribution in the Mediterranean climate

38% in the LMFH+RS3 scenario. In the Continental climate, the PV self-consumption is 27% in the LRMF+RS1 scenario, 12% in the HRMF+RS2, 11% in the SMFH+RS3 scenario, and 28% in the LMFH+RS3 scenario. The increase of PV self-consumption in warmer climates is due to the more frequent match between consumption and PV production, which in turn is due to the greater demand for cooling.

Solar thermal collectors contribute 28% and 11% of the building's total DHW production in the Mediterranean and the Continental climates in the SMFH +RS3 scenario. This is reduced to 5% and 2% in the LMFH+RS3 scenario. The very low ST contribution in the SMFH+RS3 scenario is due to the small south-oriented surface where STC is installed (Figure 7.11).

The overall results present a predicted reduction in Primary Energy consumption for each of the retrofits in the identified climates, indicating large potential savings. The following discussion revisits the key performance indicators previously outlined and considers the results in the light of these wider outcomes.

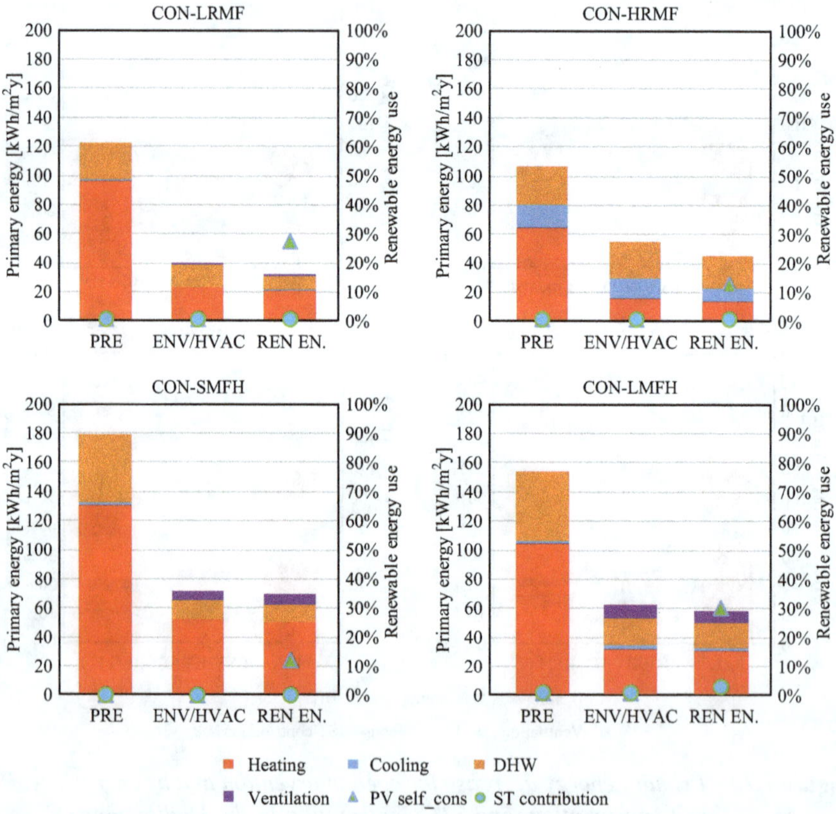

Figure 7.11 Primary energy decrease for each intervention and by energy use, PV self-consumption, and STC contribution in the Continental climate

7.4 Discussion

In understanding the retrofits, we reflected on a wider set of key performance indicators. Rather than considering only the Primary Energy, it is important to understand energy, environmental, and economic impacts. This gives a broader picture of the impact of the proposed retrofit solutions.

7.4.1 Energy performance

The proposed retrofit packages address the most common renovation solutions in different multifamily house typologies and climates throughout Europe. Due to the variety of building typologies and retrofit solutions, a comparison between the different cases is not straightforward. However, energy, environmental, and economic considerations can be reflected upon.

The proposed retrofit packages deliver energy savings by reducing the building energy demand, mainly by improving the energy plant efficiency and by adopting renewable energy. Interventions on the envelope are defined based on the minimum regulatory thermal transmittance requirements, as defined by individual countries when refurbishing a building. This means that for each country, the target external surfaces' performance is the same regardless of the building typology.

Figure 7.12 shows the final electric energy consumption for all the case studies. The dots indicate energy demand for space heating, space cooling, and DHW. DHW demand per square meter has been assumed to be the same for all the climates and buildings. While final energy depends on the system efficiency, energy demand is related to the building shape and thermal characteristics.

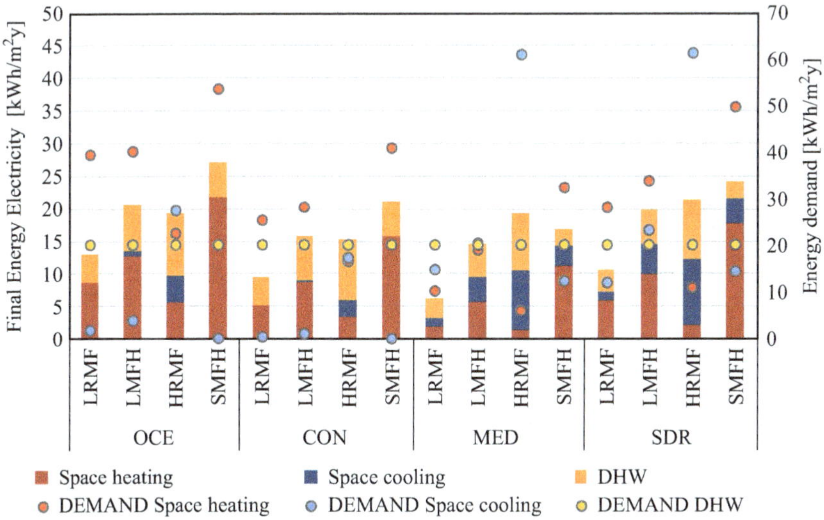

Figure 7.12 Final energy consumption and energy demand per building typology and climate

Figure 7.12 shows that, with each country having the same external surfaces' thermal transmittance, the building shape factor and the use of the building influence heating demand by 60–80%, depending on the climate. The variance between the high-rise building and the small multifamily house is 30 kWh/m^2·y in the Oceanic climate, 24 kWh/m^2·y in the Continental, 27 kWh/m^2·y in the Mediterranean, and 39 kWh/m^2·y in the Southern Dry. The difference between the building typologies for space cooling is higher than for space heating due to the decrease in walls' transmittance and replacement of windows that reduce thermal losses, avoiding cooling the ambient temperature during nighttime. As a consequence, space cooling between the buildings with the lowest and highest S/V ratio varies by 27 kWh/m^2·y in the Oceanic climate, 17 kWh/m^2·y in the Continental, and almost 50 kWh/m^2·y in Mediterranean and Southern Dry climates.

With the proposed interventions, the system efficiency is increased due to the adoption of generation devices that are able to exploit renewable energy, namely air-source and ground-source heat pumps, leading to a higher coefficient of performance. As a consequence, the SPF of the analyzed cases ranges between 2.7 and 7, depending on the climate and RS. The highest values are found in RS1 due to the use of a heat pump, together with heat recovery, air recirculation, and the PV system. The decentralised system with ground-source heat pumps shows SPFs for the whole system around 3–3.5 for all the different climates due to the less variable temperatures in the ground boreholes. Centralised systems (RS3) have an SPF for all the uses in a range of 2.7–2.8 in the coldest climates and 3.1–3.4 in the warmest ones. This lower performance with respect to the other analyzed cases is due to the higher variability of the heat pump coefficient of performance. This is due to external air temperatures and thermal losses through the pipework. Despite this, the advantages of centralised systems lie in the installation costs and the lower

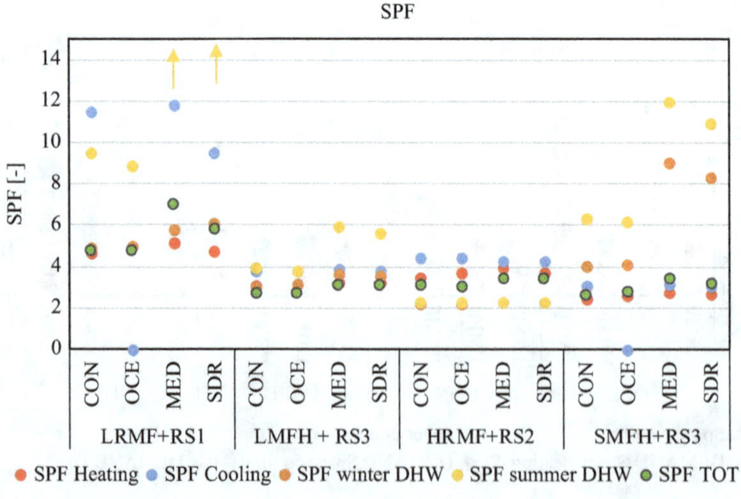

Figure 7.13 SPF values for each energy use, building typology, and climate

disruption of the works, which can be essential when considering the adoption of retrofits.

Looking at the SPF for each energy use, the lowest system performance is generally for DHW production, as the heat pump will generally perform less well when required to generate higher temperatures. However, if DHW is produced with waste heat from cooling production, or by electrical resistance exploiting excess PV production, then the SPF for DHW during the summer season becomes higher. The zero value for SPF refers to a case where there is no cooling production, while the yellow arrows indicate high values of SPF for DHW production in summer for the reasons presented above. The green dot indicates the total SPF referred to the whole system.

The impact of renewable energy on the total energy consumption depends on the installed PV capacity, the simultaneity between PV energy availability and the load, and the PV field orientation. Figure 7.14 shows the total PV production for each case divided into self-consumed energy (red column), energy fed into the grid that is not exploited due to overproduction (yellow column), and energy required from the grid (gray column). The quantities are weighted over the surface area of the PV, as these results are strictly related to the installed capacity per square meter of living area. For the LRMF, the PV field is 11% of the heated area; in LMFH, the share between installed PV and lived area is 4%; in HRMF, it is 5%; and in SMFH, it is 8%.

The highest values of PV share are observed in the LRMF, due to the improved control strategy that exploits PV overproduction for DHW uses. However, in the other cases, the PV share, that is, PV-generated energy used over the total

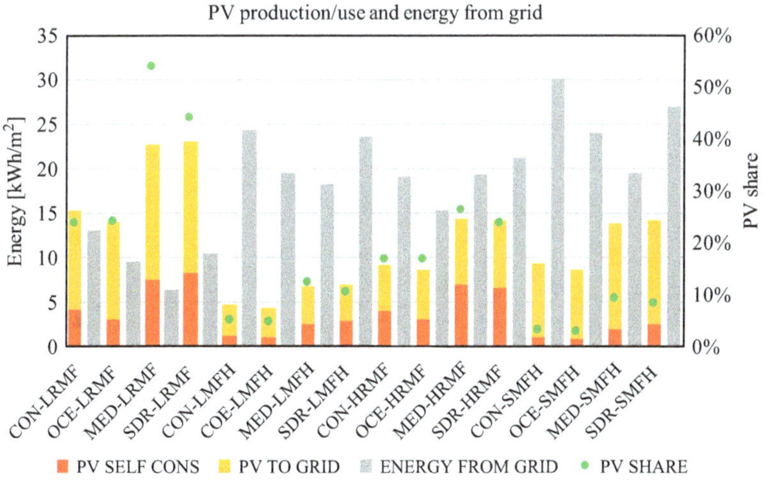

Figure 7.14 Summary of energy produced by a renewable source, self-consumed energy, and energy taken from the grid for each analyzed case

consumed electricity, ranges between 3% for the northern climates with a centralised system and 27% in the southern climates with decentralised systems. It should be noted that the implementation of optimized control strategies to increase the use of renewable energy may improve this share.

The red/yellow columns, indicating self-consumed energy with respect to the total PV production, are around 25–35% in LRMF case, between 30% and 40% in LMFH, between 35% and 50% in HRMF, and between 10% and 20% in SMFH.

7.4.2 Environmental performance

Figure 7.15 shows the Primary Energy consumption of the cases before the implementation of the retrofit packages and the achieved consumption after each of the interventions. The three darker green columns represent the consumed Primary Energy after the installation of insulation and replacement of the windows (Envelope), the replacement of the HVAC system (Env. + HVAC sys.), and the use of renewable energy (Env. + HVAC sys. + Renew. En.). The darkest green column, therefore, indicates the final Primary Energy of each case after renovation. The proposed solutions are able to reduce energy consumption below 50 kWh/m^{2}y in almost all cases. The exception is the SMFH in cold climates, as this typology has a higher heating demand.

Considering Primary Energy consumption and savings before intervention, the proposed interventions on the envelope reduce Primary Energy consumption by up to 60%. Only in two cases, the high-rise buildings in the two warmer climates, is

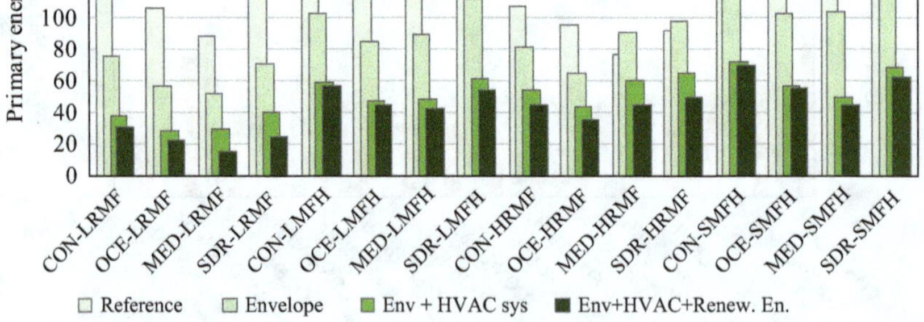

Figure 7.15 Primary Energy consumption reduction and final Primary Energy consumption post-intervention for each case, thanks to the adoption of the analyzed retrofit solutions

the increase in space cooling higher than the reduction of space heating. However, the replacement of the existing gas boiler-based heating system and non-efficient split units with centralised, or decentralised, heat pump systems can cover the increase in building energy demand and bring additional Primary Energy savings. In particular, efficient HVAC systems applied to renovated buildings can reduce Primary Energy consumption of an existing building in the range of 20–74%. Finally, the use of renewable energy, specifically solar thermal collectors for centralised systems and PV systems, adds a Primary Energy consumption reduction of around 10–20%, depending on the climate and building typology.

CO_2 emissions are calculated based on the conversion factors defined in Table 7.3. In Figure 7.16, CO_2 emissions of the cases post-intervention, shown in the dark green column, range between 3 and 14 $kgCO_2/m^2 \cdot y$ depending on the building typology, RS implemented, and climate. As already mentioned, for high-rise buildings in warmer climates, interventions on the envelope increase energy consumption for space cooling, with consequently higher CO_2 emissions. However, in the other cases, envelope insulation and window replacement can reduce CO_2 emissions by up to 50%. The enhancement of the HVAC systems gives an additional CO_2 emissions reduction in the range of 20–40%, being slightly higher for the decentralised systems. The use of renewable energy can also make an important contribution, achieving a CO_2 emissions reduction of up to 20% in the southern climates, where solar radiation is higher and cooling demand occurs simultaneously with solar energy availability.

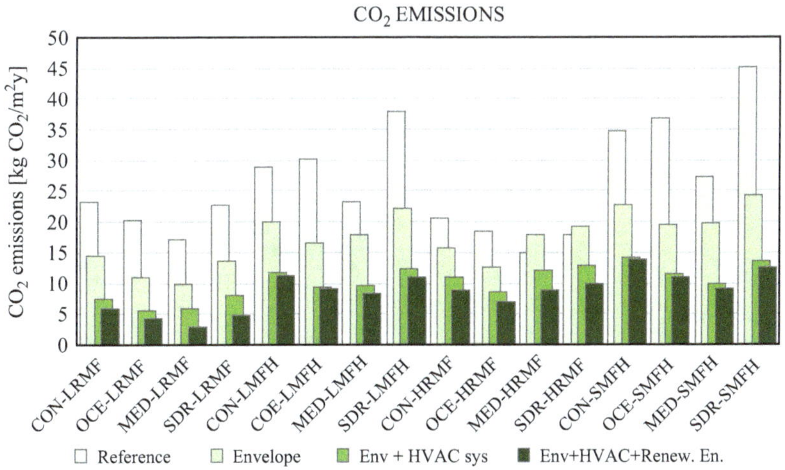

Figure 7.16 *CO_2 emissions of each studied solution before the renovation (whole column), contribution to the reduction by each intervention on the envelope, HVAC system, and renewable energy, and final CO_2 emission after renovation (dark green part)*

7.4.3 Economic performance

The annual operating costs of each reference case and retrofit solution are calculated based on the EU27 average cost of energy for both natural gas and electricity. EUROSTAT data for household customers in 2023 and 2024, including taxes and levies, have been taken as a reference, with natural gas priced at 116 €/MWh and electricity at 288 €/MWh.

Every country has implemented a variety of policies to either incentivize electricity exported to the grid or to favor physical and virtual renewable electricity self-consumption. Additionally, there have been frequent changes to these policies. For this reason, this section reports on the economic savings derived from renewable electricity physical self-consumption, while other economic benefits are disregarded due to the complexity for each country. This is a conservative approach to calculating savings, as larger benefits would be expected in most cases due to the various incentives in specific countries.

The total operating costs (C_{tot}) in the pre- and post-RS are calculated based on the costs of the vectors derived from the natural gas and electricity grids. The results are shown in Figure 7.17 and are normalized to reflect the building floor area. The orange columns represent operating costs of the reference system without any renovation intervention; the yellow columns refer to the operating costs after renovation. The annual cost reduction ranges from 37% to 81%, largely driven by the building aspect ratio, with climate playing a minor role. High-rise buildings achieve lower savings, as the amount of surface available for PV field installation is smaller; likewise, small multifamily homes featuring a high aspect ratio perform less well due to higher heat losses. Low-rise multifamily homes can benefit the most from the retrofit measures considered.

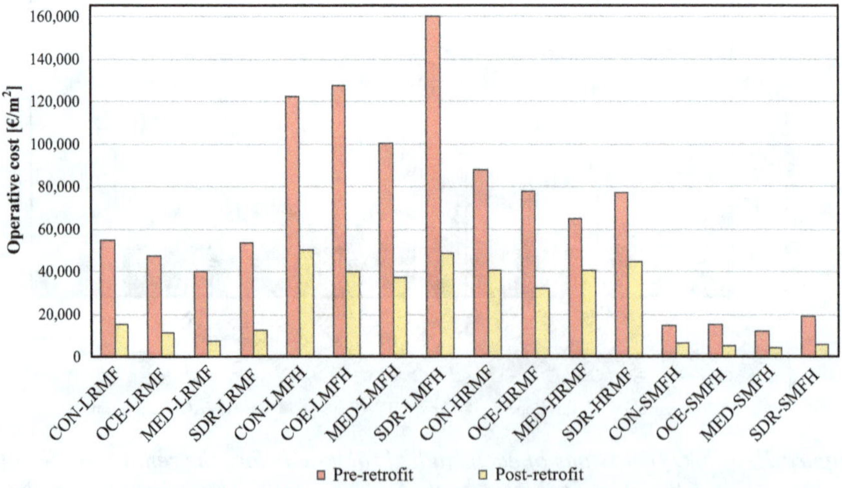

Figure 7.17 Operating costs in the pre-RS and in the post-RS with respect to the building archetypes and climates concerned

7.5 Conclusion

In this study, a set of renovation measures for multifamily houses was analysed, considering a variety of building typologies, retrofit solutions, and climates. Simulation-based results identify that the adopted interventions can lead to a significant Primary Energy demand reduction in all climates, while improving the comfort condition for the occupants.

For a consistent reduction of CO_2 emissions and decarbonisation of the residential sector, the most common solution of insulating the building envelope and replacing the windows can produce only partial results. The study demonstrates that coupling improvements to the envelope to reduce a building's energy demand, undertaken together with the replacement of existing HVAC systems, specifically ground- and air-source heat pumps, can significantly cut Primary Energy consumption and CO_2 emissions of multifamily houses. The combination of these solutions with technologies that exploit solar energy for electrical or thermal energy production achieves Primary Energy savings of approximately 40–80%. Similarly, CO_2 emissions reductions range from 40%, in those cases with the increase in cooling demand due to the higher insulation, to 85%, in cases with a decentralised system coupled with a PV array and control strategies that improve the use of solar energy.

Depending on the building geometry, weather conditions, and type of intervention, different renovation solutions can be implemented. The energy demands of buildings with same thermal characteristics are influenced by the building shape, resulting in lower heating demand in compact buildings but higher cooling demand in warm climates. The increase of cooling demand after envelope insulation can be significant. However, the use of an efficient HVAC system with heat pumps that exploit waste heat from cooling production for DHW uses, or coupled with a photovoltaic system, can overcome the increase in energy demand and show significant energy consumption savings.

The impact of solar energy technologies on Primary Energy reduction depends on the space availability for installation and can be limited when compared to the contribution of envelope renovation and HVAC replacement. However, it does contribute to the PE consumption of the non-heating and cooling uses of 5% and up to 20–30% for DHW production. The integration of photovoltaics with heat pumps becomes central to the renovation strategy, especially in those cases where cooling demand is not negligible.

Better exploitation of renewable energy use can be achieved by implementing control strategies that maximize the match between energy production and energy consumption and reduce the level of surplus energy exported to the grid. This is also highlighted by an analysis of the operating costs, indicating that a support scheme based on net-metering can lead to the complete compensation of the cost of electricity when total electricity consumption is close to the overall PV production. While the economic analysis conducted does not focus on the renovation investment cost and payback time, a significant reduction of the annual operating cost in all the scenarios shows the potential profitability of the interventions.

Acknowledgments

The research leading to these results has received funding from the European Union's Horizon 2020 research and innovation program under the project BuildHeat Standardized approaches and products for the systemic retrofit of residential BUILDings, focusing on HEATing and cooling consumption attenuations, Grant Agreement No. 680658.

References

[1] Heiselberg, P.K., 'Survey and simulation of energy use in the European building stock', *CLIMA 16 Proceedings of the 12th REHVA World Congress*; Aalborg, Denmark, May 2016. Aalborg: Aalborg University; 2016

[2] Klein SA, Beckman A, Mitchell W, Duffie A. *TRNSYS17 – A transient systems simulation program*. Madison: Solar Energy Laboratory, University of Wisconsin; 2011

[3] Dipasquale, D., Fedrizzi, R., Bee, E. *BuildHeat - Retrofit solutions performance dataset* [online]. Available from https://doi.org/10.5281/zenodo.3734970 [Accessed 11 June 2024]

[4] Degree Days.net. *Custom Degree Day Data*. [online] Available from: http://www.degreedays.net [Accessed 16 October 2024]

[5] Hitchin R, Thomsen KE, Wittchen KB. *Primary energy factors members states energy regulations*. [online] 2018. Available from https://epbd-ca.eu/wp-content/uploads/2018/04/05-CCT1-Factsheet-PEF.pdf [Accessed 11 June 2024]

[6] Covenant of Mayors *Technical annex to the SEAP template instruction document: The emission factors*. [online]. 2016. Available from https://www.eumayors.eu/IMG/pdf/technical_annex_en.pdf [Accessed 16 June 2024]

[7] Ministry of Housing Communities and Local Government. *The Future Home Standard 2019. Consultation on Changes to Part L (Conservation of Fuel and Power) and Part F (Ventilation) of the Building Regulations for New Buildings* [online]. London: MHCLG; 2019. Available from https://assets.publishing.service.gov.uk/government/uploads/system/uploads/attachment_data/file/852605/Future_Homes_Standard_2019_Consultation.pdf [Accessed 17 May 2024]

[8] ANIT (Associazione Nazionale per l'Isolamento Termico e Acustico). *Mini Guida ANIT: Appendice A Allegato I DM 26/6/2015. Descrizione dell' edificio di riferimento e parametri di verifica* [online]. Milan: ANIT; 2018. Available from https://anit.it/wp-content/uploads/2018/03/miniGuidaANIT_gennaio2018.pdf [Accessed May 16 2024]

[9] Gobierno de España. *Codigo técnico de la edificación. Guía de aplicación DB HE* [online]. 2019. Available from https://www.codigotecnico.org/pdf/GuiasyOtros/Guia_aplicacion_DBHE2019.pdf [Accessed 29 May 2024]

[10] Dipasquale C, Bee E, Fedrizzi R. *D4.5 Retrofit protocols. BuildHeat project. EC H2020 Grant Agreement No. 680658* [online]. 2020. Available from https://zenodo.org/record/3734970#.X0S7x-gzY2w [Accessed 21 October 2026]
[11] Clivet. *Elfopack heat pump* [online]. Available from https://clivet.com.ua/wp-content/uploads/2020/04/CPAR-XIN.pdf [Accessed 21 February 2021]
[12] Banja M, Jégard M, Monforti-Ferrario F, *et al. Renewables in the EU: The support framework towards a single energy market* [online]. 2017. Available from https://op.europa.eu/en/publication-detail/-/publication/83d9ab2f-647d-11e8-ab9c-01aa75ed71a1/language-en [Accessed 21 October 2025]

[10] [illegible] C, [illegible] L, Esch [illegible] [illegible] grows [illegible] COVID-19 [illegible] 2020. Available from: [illegible] [Accessed 28 October 2020]

[11] [illegible] Available from: [illegible] [Accessed 2? February 2021]

[12] [illegible] N., Mo[illegible] for the [illegible] 2017. Available from: [illegible] [Accessed [illegible] 2020]

Chapter 8

An overview of the Salford Energy House

Grant Henshaw[1], Will Swan[1], Richard Fitton[1], David Farmer[1] and Bill George[1]

8.1 Introduction

In 2011, the University of Salford launched the Salford Energy House. This is a complete pre-1919 Victorian terraced property, a UK archetype that covers approximately 20% of the UK housing stock in terms of building typology [1]. As a retrofit problem, it is also defined as a hard-to-treat (HTT) property [2], meaning it represents one of the most complex challenges in delivering a successful retrofit. The property sits within an environmental chamber, meaning research and testing regimes can be carried out under controlled conditions. While climatic chambers are widely used for elemental [3] and vehicle research [4], their application has not previously been applied to buildings. Over the last decade, the facility has undertaken a series of tests that range from whole-house retrofits [5–7] to simple energy efficiency solutions such as curtains [8,9] Here, we outline some of the work that has been developed over the years to identify the lessons that have been learned in terms of retrofit. The breadth of the work identifies the complexity of retrofit, as it has developed from fabric energy efficiency as the focus of the solution to embrace wider technical solutions such as energy flexibility, where the house is more fully engaged in the wider energy system [10]. Since its inception, the mission of the research group has been focused on both energy efficiency and net zero, working in close collaboration with industry and policymakers. Given the recent announcement of the Warm Homes Plan [11] in the UK, which has shifted perspective to less fabric-oriented technical solutions, understanding the technical implications for proposed approaches is essential and requires addressing at pace.

The work of the Salford Energy House sits within a wider body of work in the UK and internationally to understand 'what works' in terms of retrofit. The aim of Salford Energy House is to focus on whether an individual intervention, or whole system change, can deliver energy or emissions savings when used as intended. This generally does not consider issues around poor installation or use, but it is

[1]Energy House Labs, University of Salford, UK

clear that delivering a wide array of technical projects will invariably lead to, sometimes hard, lessons for innovators who are looking to deliver their products, systems, and services to the market.

Here, we outline the research context for the development and operation of the facility, the technical details of the facility itself, the operating principles, and the key research that has been undertaken during the last 10 years. The Salford Energy House operates in a wider context of technical retrofit research which incorporates modelling, stand-alone product and materials tests, larger-scale research facilities, and field trials. It does not replace any of these approaches but can provide valuable insights that are not possible within the current methods in terms of detail and time frames in which work is delivered. As with any method, it has its limitations which must be known and understood if the data is to be properly analysed and interpreted.

8.1.1 Retrofit: a technical perspective

Retrofit, by its very nature, is a socio-technical issue [12,13]. When a retrofit is delivered in the field, it is linked with a whole series of issues that can drive or constrain its delivery. Researchers taking a more theoretical view of the wider impacts rightly point to issues such as skills, legislation, standards, user practices, supply chains, and finance as essential elements that must be considered if retrofit is to be successfully delivered and perform for consumers and occupants. Researchers may look at retrofitting through sociological theoretical lenses such as Actor Network Theory [14] or Assemblage Theory [15]. While this is essential to understand how we deliver retrofit within wider communities, or as part of national programmes, the work within the Salford Energy House considers retrofit from a technical perspective, focusing on performance. This points to a positivist stance that takes data as the foundation of understanding the performance of the retrofit, as well as a consideration of any unintended consequences. However, we also recognise that for a retrofit to be a success, it must be aware of the wider socio-technical context in which it sits. In many respects, it provides an argument for the socio-technical traditions to more widely embrace the technical as well as the social aspects if progress is to be made.

8.1.2 Defining retrofit

To understand the gap that the Salford Energy House was designed to fill, it is useful to consider the pre-existing approaches to how the technical understanding of retrofit was established. In doing this, it is useful to have a working definition of retrofit. Retrofit has a number of different definitions, and there is complexity in establishing one that is universally agreed upon [16]. If we consider some of the early definitions, where retrofit was first identified as a policy issue to address energy efficiency, fuel poverty, and carbon emissions, modernisation, retrofit and refurbishment were often used interchangeably [17–22]. Target-led definitions, such as those from Renovate Europe, the Superhomes network, or the EnerPHit standard, identify process- and metrics-led standards with a focus on a defined 'deep' low-carbon retrofit with targets such as a 60% reduction in emissions [16]. Safri and Beagon [23] identify that retrofits must sit outside basic repair and maintenance and use a 2014 definition,

> *'Sustainable retrofit can be defined as the upgrading of the building fabric, systems and/or controls to improve the energy performance of the property.' (Brown, Swan and Chahal 2014) [24]*

This points to an issue around intentionality; why is the retrofit work being done? If the intention of the retrofit must go beyond the simple upgrade of the property to maintain its fabric or improve its aesthetics, then we may consider the underlying reasons for retrofit that sit outside of these issues. This is not to say that this entirely resolves the problem of retrofits that may be done for aesthetic or maintenance reasons that then deliver the co-benefits of energy efficiency or comfort, such as the installation of double glazing, which is a common home improvement upgrade that can deliver energy savings of £235 per annum [25], but may be adopted to replace existing poor-quality windows.

This argument, to some extent, is driven by the perspective of 'whole house retrofit', a view that the house should be considered a system [26], citing failures of the elemental approach from the supplier obligation delivery of small-scale interventions to reduce carbon emissions, particularly the Carbon Emissions Reduction Target, since replaced by various iterations of the Energy Company Obligation (ECO). ECO was criticised in the recent National Audit Office report [27], which highlighted a stark picture of failure in the delivery of solid-wall insulation. However, financial barriers often create a need to stage retrofit activities. This issue has been identified in the UK standard PAS2035, which takes a whole-house position but recognises that this might not reflect the financial reality for many owner-occupiers or private sector landlords. The staged approach is managed through a whole-house plan, which retains a whole-house perspective but recognises these interventions may be piecemeal to reflect the ability of the household to finance the retrofit [28].

Given the difficulties in defining retrofit around intention, targets, scale, and scope, it may be useful to consider stepping back and taking a more technical perspective to reflect the nature of the research under discussion. Salford Energy House has considered the following main issues; energy efficiency, carbon dioxide emissions, and costs as related to fuel poverty. If a specific intervention contributes to the delivery of these issues, from a technical perspective, intentionality is less of an issue. Where heating and fabric issues are under discussion, it also considers issues of comfort and ventilation. With the increase of the smart meter infrastructure and related smart energy assets, such as batteries, renewables, smart controls, electric heating systems, and electric vehicles, allowing households to not only consume but also generate, store, and trade both energy and control to the network through demand-side response, this should also be considered as potentially supporting low carbon, energy savings, and cost savings. Taking a technical perspective, specifically focused on the discussion of the research within the Salford Energy House, we take a more inclusive definition of technical retrofit.

> *'Technical retrofit is defined as any product or service intervention that can reduce carbon dioxide emissions, energy consumption, or reduce cost to an occupant or asset owner.'*

While this definition may be arguable, the difficulty in defining a specific definition for retrofit more widely, perhaps suggests that it may shift depending on the research question being asked. As the Salford Energy House has no occupants, it is difficult to ascribe an intentionality to a decision-making process. By taking the outcomes of cost, energy saving and emissions, we can perhaps more successfully bind the work into a technical question of 'what works'. This is not to say we do not recognise the socio-technical context; rather, we recognise the boundaries of our work.

8.2 Technical retrofit research

Taking a purely technical perspective, we can identify a body of research that focuses on the technical performance of retrofit interventions. This can be viewed through three specific perspectives. The first is one of scale, or unit of analysis, where the research may focus on the understanding of an individual product. Multiple products can be combined to make systems, which may be evaluated. Scaling up further, we have whole houses, ranging from individual properties to large-scale field trials. The second is the scope of the research questions being asked. This might be by exploring a single metric, such as establishing heat-transfer performance of a building element, to more all-encompassing studies with multiple metrics, such as heat-transfer coefficients, energy consumption, and air tightness, as well as more qualitative technical approaches, such as thermography or physical surveys, which remain within the technical family of studies. Finally, is the question of setting. This means the physical context that the study is being undertaken in, which may range from simple test rigs or bench tests to understand individual products. These may be located within buildings, such as environmental chambers or hot boxes, or externally located, such as Modcell test boxes. At the whole house scale, we may have the quasi-realistic settings such as the Salford Energy House, to homes in an external context, such as unoccupied houses or twin houses in external environments, and finally to fully occupied properties in field trials, which need to take into account people interacting with their homes.

The test context must also take into account the different methods that might be applied. When assessing fabric performance, methods such as the establishment of heat transfer or U-values can be applied in all contexts, with variations around a core method. Some methods such as the establishment of the heat-transfer coefficient (HTC), are only applicable at the whole-house level. These individual methods create a suite of analytical tools that can be applied to answer the question of technical performance. In addition, additional tools such as modelling can also be applied to understand the performance of buildings. However, this approach can have limitations when applied without appropriate measurement. Models calibrated with measured data to address assumptions within the model are much closer to establishing reliable simulations [29].

Fundamentally, Salford Energy House sits between two traditions of evaluating technical solutions. The first is elemental testing and the second is externally located full properties and field trials that are subject to the weather.

8.2.1 Elemental testing and rigs

Testing of building components to established thermal and air-permeability standards is supported by an international network of specialist laboratories capable of evaluating performance from material to façade scale. These assessments span multiple methodologies, including calibrated and guarded hot box testing (e.g. ISO 8990, EN 1934, EN ISO 12567-1/2) [30–33] for whole-element thermal transmittance, guarded hot plate and heat flow meter techniques (ISO 8301, ISO 8302, EN 12664, EN 12939) [34–37] for insulation conductivity, glazing calorimetric and hot box methods (EN 673, EN 674, EN 675, ISO 10077-1/2) [38–41] and laboratory air permeability testing of components and façades (EN 12114, EN 1026, EN 12153) [42–44].

The IEA EBC Annex 58 Subtask 2 decision tree provides a structured framework for navigating this test landscape, acting as a practical tool to identify appropriate laboratory or in situ methods for quantifying different aspects of a building's thermal properties, from material conductivity through to whole-façade heat transfer and airtightness performance [45].

In the UK, material-level testing is conducted at facilities such as the University of Salford's Thermal Measurement Laboratory using guarded hot plate and heat flow meter apparatus aligned with ISO and EN conductivity standards [46]. Lucideon operates specialist hygrothermal testing laboratories that evaluate moisture transport, vapour permeability, sorption behaviour, and coupled heat-moisture performance of construction materials and systems under controlled environmental conditions [47].

Across Europe, facilities such as the Fraunhofer Institute for Building Physics and CSTB operate accredited guarded hot box and climatic façade chambers, capable of determining U-values of opaque and glazed assemblies [48,49]. If Rosenheim and FIW Munich provide certified fenestration and insulation thermal testing [50,51], with RISE offering additional large-scale envelope and conductivity laboratories [52].

Façade air permeability, watertightness, and curtain wall performance testing are further delivered by specialist envelope laboratories such as Peutz, Wintech Testing & Certification, and Applus+ Laboratories, operating full-scale façade rigs aligned with EN 12153 and related standards [53,54]. Internationally, guarded hot box testing is delivered at facilities such as the National Research Council Canada and Oak Ridge National Laboratory [55,56], while commercial laboratories, including Intertek and UL Solutions, provide accredited hot box, fenestration, and air permeability testing [57,58]. CSIRO and BRANZ focus primarily on insulation conductivity and resistance measurement using heat flow meters and guarded plate techniques [59,60].

8.2.2 External houses and field trials

The use of external houses ranges from the use of twin-paired houses through small collections of case study houses to large-scale field trials. Twin pairs are used to compare a close, like-with-like house. They are co-located and allow comparison under the same weather conditions. Often, one property will be unchanged while another might be subject to an intervention to allow a direct comparison. Examples of these types of installation are located at Fraunhofer [61] and Loughborough [62]. Another approach is to have a 'free running' experimental property. This may be an existing house type or represent a new type of construction. These may be similar in construction to allow twin-pair-type experiments, but often they will be of different types to allow the exploration of different archetypes. Examples of this may include the experimental eco-homes constructed at Nottingham University [63] or the historic ConstructThor facility at Leuven [64]. These approaches provide useful case study data, particularly on novel systems. However, they lack the repeatability of the Salford Energy House, which can extend the time experiments take to deliver and the comparability of results of different staged interventions.

Field trials are complex. There are identified risks in establishing samples, ensuring data collection is appropriate and, as these are occupied houses, managing residents' expectations and continued engagement [65]. A good early example of the complexities of field trials is the Pennyland Project in the UK [66], which details, in an exceptionally transparent report, the complexities of sampling and controls, data collection, and equipment. This work, undertaken in the 1970s, was exceptionally labour-intensive, although it led to some good initial findings on the impact of insulation in a period when studies of housing and energy were in their infancy. Retrofit for the Future was another major UK field trial: 119 dwellings were identified in the sample, with 86 having some data available, emissions data for 45, 36 with indoor temperatures and 23 where the critical pre- and post-retrofit data were available [67]. This clearly shows the complexity of monitoring for even a national flagship project. Given the failure rate, the balance between risks and costs of field trials need to be carefully considered. This is not to discount field trials, merely to highlight their complexity and resource-intensive nature. Field trials, when well-designed and managed, have contributed significantly to our understanding of issues such as insulation, zonal heating controls, and heating system performance. The growth of IoT and smart metering creates new, lower-cost opportunities to help us understand the built environment, with lower cost and disruption than previous approaches. The Smart Energy Research Lab (SERL) at UCL uses smart meter data to understand the performance of more than 13,000 homes in the UK [68], which can give detailed data on a population.

Both elemental testing and field trials are an essential part of understanding the energy performance of the domestic built environment. They both have limitations, as do all research methods. When conceiving the development of the Salford Energy House, the consideration was largely focused on the methodological gap that might span these two approaches and make a contribution to understanding retrofit solutions in a different way.

8.3 About the Salford Energy House

The Salford Energy House is a pre-1920s solid-wall end-terrace dwelling, constructed using reclaimed materials and traditional building techniques representative of approximately 20% of the UK's historic housing stock. Built in 2011 at the University of Salford, the property is located within a large-scale environmental chamber that enables both steady-state testing and repeatable dynamic temperature profiles, with external air temperatures ranging from −12 °C to +30 °C (±0.5 °C). The chamber is additionally equipped to simulate rain (up to 200 mm per hour), wind (up to 10 m/s), snow, and solar gains through fixed simulation rigs, allowing external boundary conditions to be imposed in a controlled and repeatable manner (Figure 8.1).

8.3.1 House construction

The Salford Energy House is a full-scale replica of a pre-1919 end-terrace dwelling, constructed using reclaimed materials and traditional building techniques. The

Figure 8.1 The Salford Energy House

external walls comprise 222.5 mm solid brick, finished internally with 13 mm dense gypsum plaster. The property adjoins Pemberton House (formerly the Conditioning Void, or Guard House) via a shared party wall, enabling adjacent heating conditions to be controlled so as to replicate those of an occupied neighbouring dwelling [69].

8.3.2 Heating, Ventilation and Air Conditioning (HVAC) and weather control

The ability to maintain well-defined external boundary conditions is fundamental to the experimental value of the Salford Energy House. Unlike field environments, where weather variability introduces noise that can obscure causal relationships, the Energy House operates within a large-scale environmental chamber that allows temperature and weather-related variables to be specified, controlled, and reproduced with a high degree of precision. This capability underpins both steady-state testing, where stability and repeatability are essential, and dynamic experiments designed to examine system response, transient effects, and methodological performance under changing conditions.

Thermal control of the external environment is achieved through a combination of HVAC systems, each operating over a defined temperature range. A blast chiller system is used to deliver low-temperature conditions, enabling external air temperatures in the range of approximately −12 °C to −5 °C to be maintained. For milder and higher-temperature conditions, an air-handling system provides controlled operation over a broader range, extending from around 0 °C to 30 °C. Together, these systems allow the facility to represent a wide spectrum of UK-relevant winter and shoulder-season conditions, as well as warmer scenarios where overheating or system response under elevated temperatures is of interest. Both systems are integrated into the central building management system, allowing temperature conditions to be held constant, or varied, according to predefined schedules.

The integration of HVAC plant with the building management system allows for precise control over temporal profiles, as well as absolute setpoints. External temperature can therefore be ramped, stepped, or cycled in a controlled manner, enabling the investigation of transient behaviour such as warm-up and cool-down dynamics, control response, and the interaction between building fabric and services. This level of control is particularly important for the validation of dynamic measurement techniques and simulation models, where known and repeatable boundary-condition changes are required to distinguish between method behaviour and building response.

In addition to air-temperature control, the environmental chamber is equipped to simulate a range of weather-related phenomena that influence building performance. Solar gains can be replicated using an array of 84 middle-wavelength infrared heaters (λ_{peak} = 3034 nm), distributed across the three external façades of the Energy House and the roof plane. These heaters are individually controllable via an ENTTEC ODE MK2 Ethernet-DMX control system, allowing spatially and temporally variable radiant inputs to be applied. In practice, this enables the simulation of dynamic solar conditions such as sunrise and sunset, as well as asymmetric loading of façades, effectively allowing the orientation of the dwelling with respect to solar gains to be altered without physically moving the building.

Precipitation effects can also be introduced within the chamber. A water-delivery system, supplied from a dedicated storage tank of 2500 litre capacity, allows controlled application of rainfall to the building envelope at intensities of up to 200 mm per hour. This capability supports studies concerned with moisture exposure, surface wetting, and the interaction between rainfall and thermal performance, as well as combined scenarios where rain is applied in conjunction with low external temperatures. Wind effects can be approximated through the use of large fans, which can be positioned within the chamber to generate air movement of up to 10 m/s across the building façades. While such arrangements do not replicate the full complexity of atmospheric boundary-layer flow, they provide a means of increasing convective heat transfer and examining the sensitivity of fabric performance and airtightness-related heat loss to external air movement under controlled conditions.

Snow conditions can be reproduced through coordinated operation of the temperature-control and water-delivery systems, allowing precipitation to be applied at sub-zero temperatures. This enables exploratory investigation of snow accumulation and melt processes and their influence on heat loss, surface temperatures, and moisture behaviour. Although simplified relative to natural snowfall, this approach provides a level of experimental control that is otherwise unattainable in the field.

Collectively, the HVAC and weather-control systems allow the Salford Energy House to operate across a wide envelope of environmental conditions, supporting both simple, single-variable experiments and more integrated scenarios that combine multiple weather drivers. The ability to impose repeatable and well-characterised boundary conditions is central to the facility's role in supporting measurement-led investigation of building performance and retrofit interventions.

8.3.3 Sensor arrays and additional methods

A defining characteristic of the Salford Energy House is the extent to which its operation and measurement regimes can be controlled and observed in a systematic and repeatable manner. In contrast to field environments, where occupancy, user interactions, and weather introduce substantial uncertainty, the Energy House is instrumented and automated such that both boundary conditions and internal disturbances can be deliberately applied, monitored, and reproduced. This capability is essential to the facility's role as an experimental platform: it enables controlled manipulation of variables such as ventilation, internal gains and system schedules, supporting the generation of datasets that are suitable for high-resolution diagnostics and method validation (Table 8.1).

Automation within the facility is coordinated through a central building management platform, where the Cylon system provides supervisory control over both the environmental chamber and the house systems. This includes the HVAC plant that governs external chamber conditions, as well as building-level automations

Table 8.1 Energy House monitoring equipment

Measurement	Details
Air temperature	15 locations in each room – three heights in each corner and in centre Conditioning void – each zone Environmental chamber – mid-storey on each elevation Underfloor void and loft
Ground floor slab temperature	Surface Below ground at three depths
Black globe temperature	Centre of each room
CO_2 concentration	Ground Floor and First Floor of Energy House Conditioning void Environmental chamber
Relative humidity	Centre of each room in Energy House Each zone in conditioning void Environmental chamber – mid-storey on each elevation
Heat flux density	Each thermal element (including party wall) Ten locations on ground floor slab
Electricity consumption	Each circuit Electric heating supply (e.g. IR) to each room Selected sockets
Natural gas	Volume Flow rate
Central heating (gas and ASHP)	Boiler heat metering (power, energy, flow rate, volume, flow and return temperatures) Radiator heat metering (power, energy, flow rate, volume, flow and return temperatures) Corrosion rate and pressure monitoring (Resus RisyCor) OpenTherm monitoring

used to emulate occupant-driven events and typical domestic operation. In practice, these automations include the controlled opening and closing of windows and doors, operation of lighting systems, activation of water outlets, and the scheduling or simulation of occupancy patterns. The intent is not to recreate the full behavioural complexity of real households, but to allow specific, well-defined internal disturbances to be introduced in a way that is consistent between tests. In doing so, the facility supports both steady-state testing, where disturbance minimisation is desirable, and dynamic regimes, where controlled perturbations are essential for evaluating system response and methodological performance.

The measurement strategy within the Energy House reflects the need to capture both whole-house performance outcomes and the underlying spatial and temporal mechanisms that drive those outcomes. The facility has a substantial permanent sensor network, supplemented by deployable instrumentation selected according to the research question being addressed. At the environmental chamber scale, temperature, and humidity sensors are distributed around the house's external façades to characterise boundary conditions and confirm the uniformity and stability of the imposed external environment. This provides the essential reference against which heat transfer and energy balance calculations are made. Within the dwelling itself, high-density internal monitoring is used to characterise the thermal environment in a manner that exceeds typical field practice, reflecting the need to quantify spatial heterogeneity and to support both comfort analysis and validation of modelling assumptions.

Within Energy House 1, the room-level instrumentation has been designed to provide a detailed representation of air and radiant conditions. Each room is equipped with an array of air-temperature sensors, positioned to capture stratification and spatial gradients. The arrangement includes measurements at multiple heights and at multiple horizontal locations, such that both perimeter effects and the geometric centre of the space are represented. In parallel, black globe temperature sensors are deployed to capture mean radiant effects and to support thermal comfort interpretation, with sensors positioned at mid-height in corners and at multiple heights through the centre of the space. This combination is important because in solid-wall dwellings, discomfort is often driven not simply by air temperature but by radiant asymmetry and cold-surface effects, which can persist even where mean room temperature appears acceptable. High-resolution measurement, therefore, supports a more complete interpretation of comfort outcomes, particularly when comparing pre- and post-retrofit conditions (Figure 8.2).

In addition to the thermal environmental monitoring, the Energy House includes metering infrastructure required to quantify energy flows and system performance. Heat meters are installed on primary heating systems and at distribution points to enable the calculation of delivered heat, system efficiencies, and

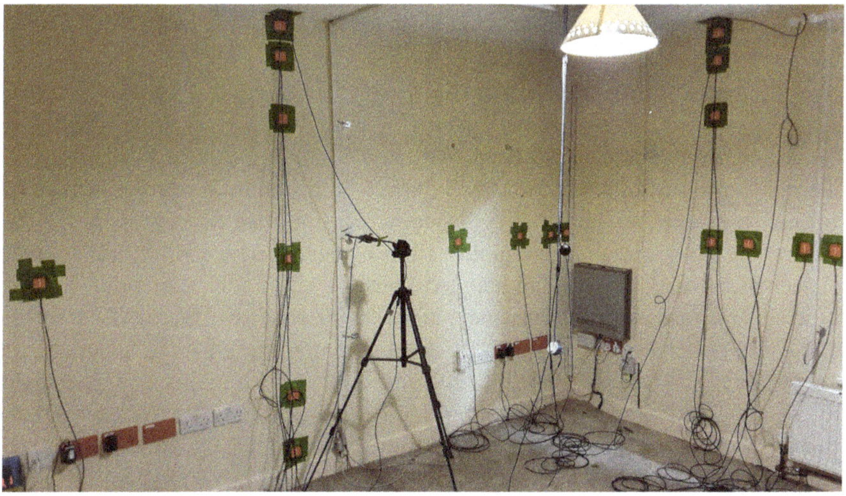

Figure 8.2 Salford Energy House measurement equipment

radiator-level outputs and to support the analysis of both space-heating and domestic hot water demand, where relevant. Electrical power and energy are monitored at whole-house level and at individual circuits, enabling the separation of end uses and the identification of system-level changes arising from interventions or control strategies. This provides a basis for relating thermal performance changes to energy outcomes in a way that is compatible with both experimental analysis and model calibration.

The permanent instrumentation is supplemented by additional sensors deployed as needed to address specific research objectives. These include heat-flux plates for in situ U-value assessment and the direct quantification of heat transfer through building elements, as well as surface temperature sensors to characterise thermal bridging and to support hygrothermal risk assessments. Instrumentation associated with thermal comfort can be expanded beyond globe thermometry where required, and airflow measurement devices, such as anemometers, can be used to identify air movement patterns that influence convective heat transfer and perceived draughts. The ability to add such instrumentation within a stable, repeatable environment is central to the facility's diagnostic capability: it enables point measurements to be interpreted within a broader spatial and system context and allows causal mechanisms, such as convective bypass, emitter-driven plumes, or localised stratification, to be identified with greater confidence than is typically possible in occupied field studies.

Overall, the sensor and automation infrastructure should be understood as an enabling layer rather than an end in itself. Its purpose is to ensure that experimental conditions are controllable, that measurements are sufficiently resolved to capture relevant mechanisms, and that results can be reproduced across test campaigns. In combination, these capabilities support the Energy House's broader mission of providing measurement-led evidence on building performance and retrofit outcomes, while also offering a platform for the development and validation of emerging diagnostic and modelling methods.

8.4 Lessons from the Salford Energy House

The Salford Energy House has been used as a controlled testbed to interrogate a wide range of retrofit interventions and diagnostic methods. While the facility has produced a substantial body of evidence across multiple programmes, the key contribution is not simply the quantification of energy savings but the ability to isolate mechanisms, identify unintended consequences, and distinguish between predicted and measured performance under repeatable conditions. This section collates the principal lessons arising from the programme of work undertaken to date, grouping the findings into three themes: fabric interventions, energy systems and services, and testing methodologies. In doing so, it seeks to draw out the transferable learning that is relevant to retrofit delivery beyond the laboratory setting.

A recurring feature across the work is the extent to which retrofit outcomes are shaped by factors that are difficult to control or even observe in the field. These include workmanship, detailing at junctions, convective bypass, interactions between fabric and heating systems, and the limitations of common modelling assumptions. The Energy House does not replace field trials, which are essential for understanding installation variability, occupant practices, and long-term persistence. However, it can provide a level of diagnostic certainty that is rarely possible in practice, enabling both the validation of emerging technologies and the refinement of methods used to measure performance. Taken together, the findings highlight that 'what works' in retrofit must be considered not only as an energy question, but also in relation to durability, comfort, health risk, and the robustness of evidence used to justify intervention.

8.4.1 Fabric

The Energy House programme has consistently shown that fabric measures remain central to reducing heat loss in traditional solid-wall dwellings, but that performance is not simply a function of the nominal insulation specification. Instead, it is shaped by installation quality, detailing decisions at junctions, and the extent to which an intervention changes both conductive and convective heat-transfer pathways. A key lesson is that the measured heat-loss benefit of a given measure can diverge meaningfully from prediction and that this divergence is often attributable to physical mechanisms, such as thermal bypass or incomplete air-sealing, rather than uncertainty in the measurement method. This reinforces the role of controlled testing in distinguishing between 'design intent' and delivered performance.

Across staged retrofit studies, external wall insulation (EWI) emerges as the dominant driver of whole-house heat-loss reduction for this archetype. Within the DEEP 5.01 programme [5], a sequence of piecemeal fabric upgrades reduced the whole-house heat-transfer coefficient (HTC) by approximately half, with EWI accounting for the majority of the improvement. A similar staged retrofit study [70] demonstrated a full-package HTC reduction of around two-thirds, again with EWI responsible for roughly three-quarters of the change. The convergence of these results across independent studies is important, as it provides empirical support for the prevailing view that wall heat loss is the principal fabric challenge in solid-wall homes. However, the studies also demonstrate that even when measures are correctly targeted, their delivered performance can be materially reduced by practical issues, with measured underperformance observed relative to prediction for both the whole-house package and the EWI system in isolation.

The mechanisms underpinning this underperformance are instructive. The controlled steady-state environment enables coheating-derived HTC estimates and in situ U-value measurements to be undertaken under conditions where confounding influences, such as solar gains, wind variability, and occupancy, are minimised. In this setting, U-value measurements have highlighted gaps between expected and measured thermal resistance, for both internal and external wall insulation systems, indicating that thermal performance limitations can be embedded within the system as installed, rather than simply arising from modelling error. Blower-door testing undertaken in the chamber further demonstrates that convective bypass, driven by poor edge sealing of mechanically fixed insulation boards, can create a pathway for heat loss that is not captured by simple one-dimensional calculations. This reinforces a key lesson for retrofit: nominal thickness and material properties do not guarantee performance if junctions and interfaces are not treated as primary design elements.

The same staged retrofit evidence also highlights that reductions in heat loss do not necessarily translate into proportional improvements in airtightness. Across the DEEP 5.01 sequence, the measurable change in airtightness was shaped more by specific floor-related upgrades than by the accumulation of other measures. Where airtightness improved, it did so through interventions that either deliberately addressed leakage paths or inadvertently created sealing effects. In contrast, subsequent best-practice 'whole-house' detailing did not generate measurable additional airtightness benefits beyond those already achieved through piecemeal upgrades. This finding is important, as whole-house retrofit is often framed as a route to improving airtightness by addressing continuity of the air barrier. The Energy House evidence suggests that, at least for this archetype, airtightness improvements are not automatically delivered by fabric upgrades unless leakage paths are explicitly identified and treated and that some parts of the building fabric, particularly at the ground floor, may dominate the achievable change.

The comparison between piecemeal practice and a whole-house approach further illustrates that the value of best-practice detailing may lie less in incremental energy savings and more in risk mitigation. After piecemeal upgrades were installed, additional detailing measures were applied consistent with whole-house principles, including insulating below the damp proof course (DPC), aligning openings with the insulation line, extending the eaves and improving bay-window detailing. These measures produced a negligible additional reduction in whole-house heat loss, with the cumulative effect amounting to around a further one per cent reduction in HTC. In purely energy terms, this is a modest return on investment. However, the same interventions delivered a substantial reduction in condensation and mould risk at thermal bridges. Baseline testing indicated that a large proportion of junctions were at risk; piecemeal retrofit reduced this risk but also introduced new risks at specific junctions, suggesting that interventions can shift hygrothermal vulnerability rather than simply reduce it. Under whole-house detailing, risk was largely eliminated, with residual issues concentrated at junctions where installation is intrinsically challenging. The implication is that a whole-house retrofit may be justified primarily through durability and health outcomes, rather than through incremental reductions in steady-state heat loss.

This creates a broader lesson regarding how retrofit is appraised and funded. If best-practice junction detailing delivers risk reduction more than energy reduction, then appraisal frameworks based solely on predicted kilowatt-hour savings will undervalue its contribution. The Energy House evidence supports the need for funding and policy mechanisms that recognise avoidance of damp and mould risk as a legitimate outcome of retrofit, particularly in hard-to-treat solid-wall dwellings where thermal-bridge conditions can be severe. It also reinforces the role of diagnostics in determining when enhanced detailing is warranted. In practice, the additional cost of whole-house detailing can be substantial, and therefore, the ability to identify where it is most needed becomes central to delivering cost-effective retrofits that protect both occupants and the building fabric.

Beyond insulation systems, the Energy House programme has also shown that 'soft' or low-disruption measures can deliver measurable fabric-related benefits, but that these benefits are often under-represented in conventional guidance and models. For suspended timber floors, the replacement of worn floor coverings with higher-quality carpets and underlays produced small but consistent reductions in whole-house HTC, alongside measurable reductions in underfloor heat flux [71]. While the magnitude of the HTC change is modest compared with major fabric measures, the mechanism is revealing: professional installation created perimeter sealing effects that reduced air movement through floor-edge leakage paths. Airtightness testing supported this interpretation, with the installation of both old and new coverings producing measurable reductions in air permeability, and new coverings yielding a larger effect. In a controlled context, these changes translated into a measurable reduction in daily energy consumption. The broader lesson is that seemingly minor measures can combine thermal resistance changes with convective control and that their performance depends on installation practice as much as on material properties.

The suspended timber floor studies also provide a cautionary lesson regarding the heterogeneity of fabric heat loss and the limits of low-resolution measurement [72]. High-density heat-flux monitoring across the Energy House floor has demonstrated that point U-values can vary by a factor of two across a single uninsulated floor, with elevated heat loss concentrated at perimeters and near ventilation openings. Area-weighting of high-resolution measurements produced a whole-floor U-value substantially higher than values predicted by common methods used in International Standards Organisation (ISO), Standard Assessment Procedure (SAP), and Chartered Institute of Building Services Engineers (CIBSE) guidance. This difference is consequential, not only for the estimation of energy savings, but also for the calibration of building models and the prioritisation of interventions. The study further showed that one- or two-point measurement strategies risk producing misleading results, with a large proportion of two-point combinations failing to represent the whole-floor performance within uncertainty bounds. The implication is that both modelling and measurement methods can systematically under-estimate floor heat loss in this archetype and that robust characterisation requires either careful sensor placement strategies or higher-resolution measurement approaches than are commonly deployed.

The window-covering studies illustrate similar themes: measurable performance improvement exists, but the magnitude depends on product design, installation fit, and interaction with heating dynamics. Controlled steady-state testing has demonstrated that blinds, shutters, and curtains can significantly reduce heat loss through glazing, with performance improvements ranging from modest reductions for free-hanging roller blinds to substantial reductions for fitted systems incorporating low-emissivity layers or improved perimeter sealing [9]. These findings reinforce that the principal mechanisms are the restriction of convective air movement within the window-covering cavity and the reduction of radiative heat transfer through low-emissivity surfaces. The work also demonstrates that baseline glazing performance can deviate from expected specification values due to real-system effects, highlighting the importance of empirical measurement when validating industry tools and product claims.

However, the Energy House evidence also shows that steady-state fabric metrics do not fully capture how window measures perform in dwellings with real heating systems. Tests comparing steady-state conditions with operation under a gas central heating system revealed that the apparent benefit of window coverings depends strongly on emitter placement [8]. Where radiators are located beneath windows, a common configuration in the UK, the convective plume can increase heat loss through glazing, and window coverings can, therefore, deliver larger local benefits than models based on homogeneous steady-state assumptions might indicate. Conversely, in rooms without direct emitter–window interaction, savings were smaller and more variable. The implication is that window-covering performance cannot be understood purely as an incremental increase in thermal resistance; it is also shaped by the dynamics of the heating system and air movement in the room. This points to a broader lesson that fabric interventions should be evaluated in relation to system operation where plausible interactions exist.

The work on innovative blind systems extends this argument by coupling controlled measurement with archetype modelling to infer annual heating impacts. By measuring product-induced U-value reductions under controlled conditions and embedding these changes within a dynamic simulation of a representative dwelling, the study quantified potential whole-year energy benefits that would be difficult to establish through short-term field monitoring alone [73]. The results suggest that advanced blinds can deliver meaningful reductions in heating demand relative to conventional roller blinds, positioning them as a low-disruption intervention where glazing replacement may be constrained. While such modelling necessarily relies on assumptions, the lesson here is methodological: controlled product characterisation can be used to parameterise simulations in a way that is grounded in measured performance, allowing more credible assessment of emerging technologies than reliance on manufacturer data alone.

Finally, the analysis of thermal paint, identified as an insulant, provides an important counterpoint, demonstrating the value of rigorous laboratory validation in challenging low-cost retrofit claims [74]. Detailed microscopy and UKAS-accredited steady-state testing showed that thin coating systems containing ceramic or aerogel-based additives provide only marginal increases in thermal resistance and do not exhibit the nano-porous, vacuum-retaining behaviour often implied by marketing narratives. Emissivity measurements indicated that most coatings are similar to conventional finishes, offering no meaningful radiant-barrier effect. When translated into whole-house modelling, the predicted energy savings were extremely small and associated with implausibly long payback periods. The lesson is that superficially plausible retrofit products may not deliver meaningful performance improvements when assessed through established physical measurement and that the Energy House programme benefits from integrating laboratory materials testing with whole-house performance evaluation to avoid misattributing performance outcomes.

Taken together, the fabric studies demonstrate a set of consistent themes. Major fabric measures such as EWI can deliver large reductions in heat loss, but only if installation quality prevents convective bypass and junction detailing is treated as an integral component of performance. Whole-house retrofit principles may deliver limited incremental energy savings beyond staged upgrades yet provide substantial benefits in reducing condensation risk and improving durability. Minor measures, such as floor coverings and window treatments, can deliver measurable benefits that depend strongly on fit, sealing, and interactions with heating dynamics. Across all measures, the Energy House evidence reinforces that credible performance assessment requires both a whole-house outcome metric and diagnostic tools capable of revealing the mechanisms behind that outcome.

8.4.2 Energy systems and services

The Energy House programme demonstrates that fabric retrofit materially alters the operational behaviour of heating systems, changing not only energy demand but also controllability, cycling behaviour, and the distribution of comfort conditions within the home. This is an important lesson for retrofit delivery, as much of the policy and practice discourse treats fabric and services as separable packages. In reality, changes in heat loss shift system loads, alter the effectiveness of existing controls, and reveal opportunities, sometimes necessities, for rebalancing, downsizing, and re-commissioning. If these system-level adjustments are not considered, there is a risk that measured energy savings will be lower than technically achievable or that comfort outcomes will be sub-optimal even in an upgraded building fabric.

Within the DEEP 5.02 programme [75], controlled testing of the Energy House heating system across pre-retrofit, piecemeal retrofit, and whole-house retrofit stages showed that reductions in HTC translate into shorter burner-on periods, reduced daily gas consumption, and faster attainment of setpoint temperatures. Temperature mapping indicated that fabric improvements moderated internal temperature swings and improved spatial uniformity, particularly in zones previously affected by cold-surface losses. This finding is consistent with a view that retrofit should be evaluated not only in terms of energy use but also in relation to thermal comfort metrics, such as radiant asymmetry, surface temperature gradients, and mean radiant temperature. In pre-1919 solid-wall homes, cold-surface effects are a prominent driver of discomfort, and the Energy House evidence suggests that fabric upgrade reduces these effects in ways that may not be fully captured by energy metrics alone.

The same work also highlights that retrofit can reveal shortcomings in heating controls and system configuration. Under reduced heating loads, poorly configured control strategies can lead to inefficiencies, including unnecessary cycling and sub-optimal distribution of heat. The Energy House evidence, therefore, supports the argument that services retrofitting should be considered as an integral element of whole-house planning, even where the primary intervention is fabric-led. The implication is not that all systems must be replaced during a fabric upgrade, but that the heating system should be assessed for suitability under the post-retrofit load, with commissioning and optimisation treated as essential steps in realising the full value of fabric improvements.

The controlled evaluation of a passive deaerator provides a complementary lesson regarding add-on heating-system technologies [76]. In the field, products of this type are frequently promoted on the basis of theoretical performance arguments and anecdotal reports, but direct before-and-after evidence is challenging to obtain due to variability in weather, occupancy, and system operation. By running two matched 24-hour tests under controlled chamber conditions, the Energy House enabled a like-for-like comparison with high confidence. The results suggest that any energy impact of the passive deaerator is marginal and falls within the uncertainty bounds of common metering, with no clear evidence of the substantial gains sometimes claimed. Moreover, the test revealed that system-side issues, such as micro-leaks and valve performance, can influence outcomes and may be exposed through interventions that change system hydraulics or air content. This reinforces a broader lesson: retrofit add-ons should be subjected to controlled evaluation before they are incorporated into policy, incentives, or widespread practice, and commissioning quality remains a key determinant of whether claimed mechanisms can plausibly operate in real systems.

The systems studies collectively point to a consistent conclusion: the performance of energy services is not static but shifts as the building fabric changes. Retrofit, therefore, needs to be understood as an integrated system change in which fabric, heating plant, emitters, and controls interact. This suggests that retrofit guidance and funding models should place greater emphasis on post-retrofit commissioning, control strategy optimisation, and the alignment of service capacity with reduced loads. It also implies that comfort should be treated as a co-equal outcome alongside energy and emissions, particularly for hard-to-treat dwellings where cold-surface discomfort can be a dominant driver of occupant behaviour and heating practices.

8.4.3 Testing methodologies

A distinctive contribution of the Salford Energy House programme is the development and validation of testing methodologies that aim to quantify building performance more accurately, more rapidly, or with greater diagnostic resolution than is typically possible in the field. The methodological work is important, not only for understanding the Energy House itself, but also because many retrofit decisions rely on methods, such as in situ U-value measurement, coheating, airtightness testing, and model calibration, that carry implicit assumptions and practical limitations. By providing a stable and repeatable environment, the Energy House enables these methods to be tested, refined, and compared against reference conditions, supporting more robust interpretation when they are applied in real dwellings.

The validation of the QUB method illustrates the potential for rapid whole-building heat-loss measurement [77]. QUB seeks to infer a building heat-loss coefficient from a short dynamic test involving controlled heating followed by free cooling, rather than the multi-week steady-state coheating approach. The theoretical foundation is that, after sufficient time, the building's thermal response becomes dominated by a single effective time constant, allowing the system to be approximated with a reduced-order model. Controlled experiments in the Energy House, undertaken under highly stable chamber temperatures, demonstrated that QUB-derived heat-loss coefficients can align closely with steady-state reference measurements across differing roof insulation conditions. The key parameter governing accuracy, a dimensionless ratio reflecting the applied heat input relative to initial conditions, provides a practical control variable that can be targeted to maintain acceptable error bounds.

The broader methodological lesson of QUB is that speed and scalability in performance measurement may be achievable, but only if test conditions and input parameters are carefully constrained. Subsequent validations in other test facilities and in external environments indicate that errors remain larger than those achievable through coheating but are potentially acceptable for screening or large-scale deployment where traditional tests are impractical. This positions QUB as a complementary method rather than a replacement: it may enable rapid identification of buildings with anomalous performance or support pre/post-retrofit comparisons, while coheating remains a reference approach when highest precision is required. The Energy House contribution is therefore both empirical, demonstrating feasibility under controlled conditions and methodological, by clarifying the conditions under which the method is likely to be reliable.

The controlled evaluation of Heat3D provides a parallel lesson regarding spatial diagnostics [78]. Heat3D uses thermography to estimate heat flux and infer U-values over large areas, offering the potential to move beyond point-based heat-flux plate measurements. In the Energy House, Heat3D was tested against ISO 9869-1 heat-flux plates across multiple regimes, including steady-state and dynamic conditions. The controlled chamber environment enabled repeated surveys and a large, paired dataset, supporting robust comparison. Agreement for heat flux was strong, with most Heat3D estimates falling within combined uncertainty of reference sensors, suggesting that quantitative IR methods may be capable of providing rapid, spatially resolved heat-loss information under suitable conditions.

However, the same work also shows that translating heat-flux agreement into reliable instantaneous U-values is more challenging, particularly under rapidly varying external conditions, where sensor thermal lag and transient gradients influence results. This reinforces a broader lesson in measurement: methods that infer U-values from surface temperatures must contend with dynamic effects that may be negligible in steady-state but become significant in realistic conditions. The implication is not that such methods are unsuitable, but that their deployment requires careful understanding of boundary conditions, calibration practices, and uncertainty. In this respect, the Energy House functions as an intermediary step between laboratory development and field deployment, allowing emerging diagnostics to be stress-tested against reference instrumentation before they are used for retrofit decision-making or quality assurance.

The 'closing the gap' modelling study highlights the methodological importance of aligning simulation assumptions with measured reality [29]. By comparing an uncalibrated dynamic model populated with standard SAP/CIBSE assumptions against high-precision coheating and in situ U-value measurements, the study demonstrated a substantial prediction error in whole-house heat loss. Crucially, the source of the error was not a construction defect, or unusual building behaviour, but the use of generic assumptions that overstated heat loss through solid brick walls and mis-specified airtightness. Stepwise substitution of measured airtightness and U-values reduced the modelling error to a small residual, bringing model predictions into close agreement with measured HTC. The lesson is that model sophistication does not inherently reduce error if the input assumptions are poorly aligned with the specific building and that measurement-led calibration is essential if models are to be used credibly for retrofit appraisal, design, or policy evaluation.

This has wider implications for the performance gap debate. In many contexts, model–measurement discrepancies are attributed to occupant behaviour, construction quality, or operational factors. The Energy House evidence suggests that, at least for this archetype, a material portion of the gap can arise from conservative or generic fabric assumptions embedded within modelling defaults. This shifts the emphasis toward improving the empirical basis of those defaults or providing accessible means to replace them with building-specific measurements. It also supports the argument that the measurement and modelling communities need closer integration, as calibration workflows require both high-quality measurement data and modelling practices that can assimilate that data systematically.

Finally, the work on temperature sensor placement in in situ U-value measurement provides a practical lesson that is directly relevant to field protocols [79]. By repeatedly calculating U-values under controlled steady-state and dynamic conditions, while varying the horizontal placement of internal and external air-temperature sensors, the study showed that internal sensor position had little effect on calculated U-values within the tested ranges, suggesting robustness to internal stratification effects for this dwelling. In contrast, external sensor placement produced small but consistent differences, particularly under dynamic external conditions, implying that local external temperature measurement near the façade may reduce bias relative to more distant 'environmental' measurements. Although the observed differences were generally within uncertainty, the pattern highlights the importance of careful external sensor specification and provides an evidence base for refining measurement guidance.

Across the methodological studies, a consistent lesson emerges; uncertainty in building performance measurement is not only a function of instrumentation accuracy but also of boundary-condition control, sensor placement, dynamic effects, and the appropriateness of method assumptions. The Energy House provides a platform where these factors can be isolated and interrogated, supporting the refinement of both rapid diagnostic tools and established reference methods. In doing so, it contributes to the broader goal of enabling more reliable comparison between predicted and measured performance, which underpins credible retrofit appraisal, policy design, and quality assurance.

8.4.4 Summary of retrofit research

The lessons from the Salford Energy House can be summarised as follows. First, fabric retrofit can deliver substantial reductions in heat loss in solid-wall terraces, but delivered performance is contingent on installation quality and the management of thermal bypass and junction effects. Second, whole-house detailing may deliver limited incremental energy savings beyond staged upgrades yet provide significant benefits in reducing condensation and mould risk, implying that retrofit appraisal should incorporate health and durability outcomes. Third, energy systems and services respond dynamically to reduced loads, and the benefits of fabric retrofit are mediated by controls, commissioning, and system suitability. Finally, measurement and modelling methods underpinning retrofit decision-making require careful validation and refinement; controlled testbeds can play a critical role in improving both the speed and robustness of performance assessment.

In combination, these findings support a view of retrofit as an integrated technical intervention in which fabric, services, and evidence methods must be treated as mutually dependent. The Energy House programme demonstrates that the ability to isolate mechanisms under controlled conditions is valuable not only for research but also for improving the credibility and effectiveness of retrofit delivery at scale.

8.4.5 Limitations

Research conducted using facilities such as the Energy House does have some significant benefits. However, it can never be a replacement for elemental testing or field trials, which answer different types of questions. A major limitation is observed by Lomas *et al.* [80], in that the weather is 'synthetic', so potentially an oversimplification of real weather. This would require comparable work to be undertaken on two dwellings, both under controlled conditions and in the field, to understand the potential differences this makes. The second issue is that it is a single case study. This means it is not representative of a whole population. Work can be undertaken to attempt to scale, but for a fuller understanding, large-scale field trials would be required to complete the picture. However, the Salford Energy House is well-placed to de-risk field trials by addressing whether an innovation should be taken to field trial, as an impact that is not measurable in the facility is unlikely to make a measurable difference in the field, given the increased uncertainty in measurement in field trials. Additionally, measurement methods can be pre-tested before large-scale deployment. This type of approach is relatively new, so further work on the application of the method is welcome.

8.5 Future research

The future research agenda for the Salford Energy House is shaped by two converging drivers. The first is the evolving national policy landscape, most notably the UK Government's Warm Homes Plan, which places increasing emphasis on systems-based retrofit and household-scale energy integration. The second is the expansion of Energy House Labs, from the original Salford Energy House (EH1) to Energy House 2.0 and the forthcoming Retrofit Centre, which together provide an experimental platform capable of addressing a far broader range of research questions than was previously possible. This section separates these two strands before drawing them together in a synthesis that positions the Energy House programme as a key empirical bridge between policy ambition and technical evidence.

8.5.1 Future policy context: the Warm Homes Plan and systems-led retrofit

The recently released Warm Homes Plan [11] signals a significant shift in the framing of domestic retrofit within the UK. While fabric efficiency remains an underpinning element, the plan places increased emphasis on systems-level interventions as central mechanisms for reducing household energy costs, cutting carbon emissions, and improving resilience. The programme commits £15 billion of public investment with the stated ambition of upgrading up to five million homes by 2030 and lifting around one million households out of fuel poverty. In contrast to earlier policy initiatives that were strongly fabric-first, the Warm Homes Plan foregrounds the deployment and integration of technologies, such as solar photovoltaics, battery storage, low-carbon heating, electric vehicles, and smart control systems.

A defining feature of the plan is its focus on energy generation, storage, and management within the home. Solar PV and batteries are positioned not only as tools for decarbonisation but also as mechanisms to reduce exposure to volatile energy prices and to enable households to participate more actively in a flexible, low-carbon energy system. Smart systems and digital controls are presented as essential enablers, coordinating generation, storage, and demand under increasingly complex tariff structures. While insulation and basic efficiency measures remain part of the policy offer, particularly for low-income households, the narrative emphasis is clearly on how homes operate as integrated energy systems rather than as passive energy consumers.

This shift raises important technical and evidential questions. Many of the projected benefits of systems-based retrofit depend on interactions between building fabric, heating systems, on-site generation, storage assets, and control strategies. Yet there remains limited empirical evidence on how these technologies perform together in real dwellings, particularly across different housing archetypes and retrofit stages. The Warm Homes Plan therefore, implicitly creates a demand for research that can move beyond isolated technology assessment and, instead, evaluate combined system behaviour, dynamic operation, and unintended consequences. Addressing these questions requires facilities that can impose controlled conditions, capture high-resolution data, and disentangle complex interactions, capabilities that are largely absent from field-only research approaches.

8.5.2 Research infrastructure: Energy House 2.0 and the Retrofit Centre

In parallel with these policy developments, the Salford Energy House programme has undergone a substantial expansion in experimental capability. Following the success of Energy House 1, which established the value of whole-house testing within a controlled climatic chamber, Energy House 2.0 was constructed in 2023 at a cost of approximately £16 million. The facility comprises two large-scale climatic chambers, each capable of accommodating two full-scale test houses, allowing a total of four dwellings to be constructed, instrumented, tested, and, ultimately, decommissioned or replaced. This represents a step change in experimental flexibility, throughput, and scope.

The first chamber within Energy House 2.0 has been configured with two dwellings designed to interrogate performance in the context of the forthcoming Future Homes Standard. This work focuses on new-build-relevant fabric and systems configurations, providing empirical evidence to test regulatory assumptions and to examine how compliant dwellings perform under controlled but realistic operating conditions. By enabling side-by-side testing within the same chamber, Energy House 2.0 allows direct comparison between alternative specifications, system choices, or control strategies, reducing uncertainty associated with weather variability and external confounding factors.

Construction is now due to commence on the Retrofit Centre within the second chamber of Energy House 2.0. This facility will house representations of a 1930s dwelling and a 1980s dwelling, capturing two highly prevalent and contrasting segments of the existing UK housing stock. The Retrofit Centre is explicitly designed to support staged and iterative retrofit research, enabling interventions to be applied, removed, and re-sequenced in a way that is not possible in occupied homes. This will allow systematic investigation of retrofit pathways, including partial upgrades, system substitution, and the interaction between legacy fabric and modern energy technologies.

Collectively, Energy House 2.0 and the Retrofit Centre extend the original Energy House concept from a single archetype into a modular research platform capable of addressing both new-build and retrofit questions across multiple construction eras. Importantly, the ability to demolish and rebuild test houses ensures that the facility can evolve alongside policy priorities and industry practice, maintaining long-term relevance.

8.5.3 Linking policy and infrastructure: from EH1 to integrated futures

Taken together, the Warm Homes Plan and the expansion of the Energy House facilities define a coherent future research agenda. Energy House 1 has already demonstrated the value of controlled whole-house testing for understanding fabric performance, system interactions, and measurement uncertainty in hard-to-treat dwellings. Energy House 2.0 builds on this foundation by enabling parallel testing, greater typological coverage, and explicit alignment with both new-build standards and retrofit needs. The Retrofit Centre, in particular, provides a direct experimental response to the Warm Homes Plan's focus on upgrading existing homes through combinations of systems and smart technologies.

The alignment between policy ambition and research capability is significant. The Warm Homes Plan envisages homes that generate, store, and manage energy in increasingly sophisticated ways, yet the technical evidence base for these integrated systems remains limited. By linking the detailed fabric and systems insights generated in EH1 with the expanded experimental capacity of Energy House 2.0, the Salford Energy House programme is well positioned to provide the measurement-led evidence required to underpin this transition. In doing so, it can help ensure that future retrofit pathways, whether driven by policy, industry, or consumers, are grounded not only in aspiration but in a robust understanding of what works, under what conditions, and for which parts of the housing stock.

References

[1] MHCLG. *English Housing Survey 2023–2024: Headline report – Chapter 1: Profile of households and Dwellings, Annex tables* [Online]. Ministry of Housing, Communities and Local Government, London, UK, November, 2024. Available from https://assets.publishing.service.gov.uk/media/6983411113622473b51ca9d6/2023-24_EHS_Headline_Report_Chapter_1_Profile_of_households_and_dwellings_Annex_Tables_.ods [Accessed 17 July 2026]

[2] Loveday D. L., Vadodaria K., Haines V., *et al.*, *Refurbishing the UK's "hard to treat" dwelling stock: Understanding challenges and constraints* [Online]. Available from https://repository.lboro.ac.uk/articles/conference_contribution/Refurbishing_the_UK_s_hard_to_treat_dwelling_stock_understanding_challenges_and_constraints/9338432 [Accessed 17 July 2026]

[3] Ferrari S., Zanotto V. 'The thermal performance of walls under actual service conditions: Evaluating the results of climatic chamber tests'. *Construction and Building Materials*. 2013;43:309–16

[4] Kucybała P., Gawlik A., Pobędza J., Walczak P. 'The usage of a thermo-climatic chamber for technoclimatic tests of special vehicles and mobile machines'. *Journal of KONES*. 2018;25 [Online]. Available from https://yadda.icm.edu.pl/baztech/element/bwmeta1.element.baztech-aadf2034-4fe3-480c-8cfb-7dd1c22dd69a [Accessed 17 July 2026]

[5] Farmer D., Henshaw G., Tsang C., Roberts B., Fitton R., Swan W. *DEEP report 5.01: Salford energy house – Fabric performance testing* [Online]. Department for Energy Security & Net Zero, London. October 2024. Available from https://assets.publishing.service.gov.uk/media/6717be32e 319b91ef09e3857/5.01_DEEP_Energy_House_Fabric.pdf [Accessed 17 July 2026]

[6] Farmer D., Henshaw G., Burrows R., Roberts B., Fitton R., Swan W. *DEEP report 5.02: Salford energy house – Heating systems testing* [Online]. Department for Energy Security & Net Zero, London; October 2024. Available from https://assets.publishing.service.gov.uk/media/6717bf3938149ce9d 09e3840/5.02_DEEP_Energy_House_Services_and_Comfort.pdf [Accessed 17 July 2026]
[7] Alzetto F., Farmer D., Fitton R., Hughes T., Swan W., 'Comparison of whole house heat loss test methods under controlled conditions in six distinct retrofit scenarios'. *Energy and Buildings*. 2018;168:35–41
[8] Fitton R., Swan W., Hughes T., Benjaber M. 'The thermal performance of window coverings in a whole house test facility with single-glazed sash windows'. *Energy Efficiency*. 2017;**10**(6):1419–31
[9] Henshaw G., Farmer D., George B. *Salford energy house thermal performance of blinds and shutters*. [Online]. 2023. Available from https://bbsa.org.uk/wp-content/uploads/2023/03/March-2023-BBSA-Initial-Report-v2.pdf [Accessed 17 July 2026]
[10] Paraskevas I., Alan D., Sitmaldis A., *et al.*, 'The critical role of IoT for enabling the UK's built environment transition to net zero'. *Energies*. 2025; 18(21):2025
[11] Department for Energy Security and Net Zero. '*Warm homes plan*', [Online]. 2026. Available from https://assets.publishing.service.gov.uk/media/696f8a3ec0f4afaa9536a0c4/warm-homes-plan-standard-print.pdf [Accessed 17 July 2026]
[12] Tweed C. 'Socio-technical issues in dwelling retrofit'. *Building Research & Information*. 2013;**41**(5):551–62
[13] Swan W. 'Retrofit innovation in the UK social housing sector' in Swan W. and Brown P. (eds.), *Retrofitting the Built Environment*. Chichester: Wiley & Sons Ltd; 2013. pp. 36–52
[14] Padilla Gil H. M., Moore T., Willand N., Horne R. 'Beyond the walls: How materiality shapes social processes of housing insulation retrofit'. *Energy Research & Social Science*. 2025;127:104233
[15] Ince R., Marvin S. 'Constructing domestic retrofit as a new urban infrastructure: Experimentation, equitability and contested priorities'. *Local Environment*. 2019;**24**(9):825–42
[16] Fawcett T. 'Exploring the time dimension of low carbon retrofit: Owner-occupied housing'. *Building Research & Information*. 2014;**42** (4): 477–88
[17] Swan W., Ruddock L., Smith L., Fitton R. 'Adoption of sustainable retrofit in UK social housing'. *Structural Survey*. 2013;**31**(3):181–93
[18] Hong S. H., Gilbertson J., Oreszczyn T., Green G., Ridley I. 'A field study of thermal comfort in low-income dwellings in England before and after energy efficient refurbishment'. *Building and Environment*. 2009;**44** (6):1228–36
[19] Jenkins D. P. 'The value of retrofitting carbon-saving measures into fuel poor social housing'. *Energy Policy*. 2010;**38**(2):832–9

[20] Reeves A., Taylor S., Fleming P. 'Deep carbon emission reductions in existing UK social housing: Are they achievable, and how can they be funded'. Presented at *ECEEE Summer Studies*. Act! Innovate! Deliver!; Ile Saint-Denis, France, 2009

[21] Bell M., Lowe R. 'Energy efficient modernisation of housing: A UK case study'. *Energy and Buildings*. 2000;**32**(3):267–80

[22] Kelly M. J. 'Retrofitting the existing UK building stock', *Building Research & Information*. 2009;**37**(2):196–200

[23] Saffari M., Beagon P. 'Home energy retrofit: Reviewing its depth, scale of delivery, and sustainability'. *Energy and Buildings*. 2022;269:112253

[24] Brown P., Swan W., Chahal S. 'Retrofitting social housing: Reflections by tenants on adopting and living with retrofit technology'. *Energy Efficiency*. 2014;**7**(4):641–53

[25] Energy Saving Trust. *Windows and doors advice* [Online]. Available from https://energysavingtrust.org.uk/advice/windows-and-doors/ [Accessed 17 July 2026]

[26] Jones P., Li X., Perisoglou E., Patterson J. 'Five energy retrofit houses in South Wales'. *Energy and Buildings*. 2017;154:335–42

[27] National Audit Office. *Energy efficiency installations under the Energy Company Obligation* [Online]. National Audit Office, London; 2025. Available from https://www.nao.org.uk/reports/energy-efficiency-installations-under-the-energy-company-obligation/ [Accessed 17 July 2026]

[28] Killip G., Fawcett T., Jofeh C., Owen A., Topouzi M., and Wade F. *Building on Our Strengths: A Market Transformation Approach to Energy Retrofit in UK Homes*. Centre for Research into Energy Demand Solutions; 2021

[29] Marshall A., Fitton R., Swan W., *et al*. 'Domestic building fabric performance: Closing the gap between the in situ measured and modelled performance'. *Energy and Buildings*. 2017;150:307–17

[30] International Organization for Standardization. *ISO 8990:1994 Thermal insulation — Determination of steady-state thermal transmission properties — Calibrated and guarded hot box* [Online]. Geneva; 1994. Available from https://www.iso.org/standard/16519.html [Accessed 6 February 2026]

[31] European Committee for Standardization. *EN 1934:1998 Thermal performance of buildings — Determination of thermal resistance by hot box method using heat flow meter — Masonry* [Online]. Brussels; 1998. Available from https://knowledge.bsigroup.com/products/thermal-performance-of-buildings-determination-of-thermal-resistance-by-hot-box-method-using-heat-flow-meter-masonry [Accessed 6th February 2026]

[32] International Organization for Standardization. *EN ISO 12567-1:2010 Thermal performance of windows and doors — Determination of thermal transmittance by hot box method — Part 1: Complete windows and doors* [Online]. Geneva; 2010. Available from https://www.iso.org/standard/50327.html [Accessed 6 February 2026]

[33] International Organization for Standardization. *EN ISO 12567-2:2005 Thermal performance of windows and doors — Determination of thermal transmittance by hot box method — Part 2: Roof windows and other projecting windows* [Online]. Geneva; 2005. Available from https://www.iso.org/standard/40053.html [Accessed 6 February 2026]
[34] International Organization for Standardization. *ISO 8301:1991 Thermal insulation — Determination of steady-state thermal resistance and related properties — Heat flow meter apparatus* [Online]. Geneva; 1991. Available from https://www.iso.org/standard/15421.html [Accessed 6 February 2026]
[35] International Organization for Standardization. '*ISO 8302:1991 Thermal insulation — Determination of steady-state thermal resistance and related properties — Guarded hot plate apparatus*' [Online]. Geneva; 1991. Available from https://www.iso.org/standard/15422.html [Accessed 6 February 2026]
[36] European Committee for Standardization. *EN 12664:2001 Thermal performance of building materials and products — Determination of thermal resistance by means of guarded hot plate and heat flow meter methods — Dry and moist products of medium and low thermal resistance* [Online]. Brussels; 2001. Available from https://knowledge.bsigroup.com/products/thermal-performance-of-building-materials-and-products-determination-of-thermal-resistance-by-means-of-guarded-hot-plate-and-heat-flow-meter-methods-dry-and-moist-products-of-medium-and-low-thermal-resistance [Accessed 6 February 2026]
[37] European Committee for Standardization. *EN 12939:2001 Thermal performance of building materials and products — Determination of thermal resistance by means of guarded hot plate and heat flow meter methods — Thick products of high and medium thermal resistance* [Online]. Brussels; 2001. Available from https://knowledge.bsigroup.com/products/thermal-performance-of-building-materials-and-products-determination-of-thermal-resistance-by-means-of-guarded-hot-plate-and-heat-flow-meter-methods-thick-products-of-high-and-medium-thermal-resistance [Accessed 6 February 2026].
[38] European Committee for Standardization. *EN 673:2024 Glass in building — Determination of thermal transmittance (U value) — Calculation method* [Online]. Brussels; 2024. Available from https://knowledge.bsigroup.com/products/glass-in-building-determination-of-thermal-transmittance-u-value-calculation-method [Accessed 6 February 2026].
[39] European Committee for Standardization. *EN 674:2011 Glass in building — Determination of thermal transmittance (U value) — Guarded hot plate method* [Online]. Brussels; 2011. Available from https://knowledge.bsigroup.com/products/glass-in-building-determination-of-thermal-transmittance-u-value-guarded-hot-plate-method [Accessed 6 February 2026].
[40] European Committee for Standardization. *EN 675:2011 Glass in building — Determination of thermal transmittance (U value) — Heat flow meter method* [Online]. Brussels; 2011. Available from https://knowledge.

bsigroup.com/products/glass-in-building-determination-of-thermal-transmittance-u-value-heat-flow-meter-method [Accessed 6 February 2026].

[41] International Organization for Standardization. *ISO 10077-1:2017 Thermal performance of windows, doors and shutters — Calculation of thermal transmittance — Part 1: General* [Online]. Geneva; 2017. Available from https://www.iso.org/standard/67090.html [Accessed 6 February 2026].

[42] European Committee for Standardization. *EN 12114:2000 Thermal performance of buildings — Air permeability of building components and building elements — Laboratory test method* [Online]. Brussels; 2000. Available from https://knowledge.bsigroup.com/products/thermal-performance-of-buildings-air-permeability-of-building-components-and-building-elements-laboratory-test-methods [Accessed 6 February 2026].

[43] European Committee for Standardization. *EN 1026:2016 Windows and doors — Air permeability — Test method* [Online]. Brussels; 2016. Available from https://knowledge.bsigroup.com/products/windows-and-doors-air-permeability-test-method [Accessed 6 February 2026].

[44] European Committee for Standardization. *EN 12153:2023 Curtain walling — Air permeability — Test method* [Online]. Brussels; 2023. Available from https://knowledge.bsigroup.com/products/curtain-walling-air-permeability-test-method [Accessed 6 February 2026].

[45] Erkoreka A., Gorse C., Fletcher M., Martin K. *Logic and use of the decision tree for optimizing full scale dynamic testing* [Online]. Leuven, KU Leuven, 2016. Available from https://www.iea-ebc.org/Data/publications/EBC_Annex_58_Final_Report_ST2.pdf [Accessed 19 July 2026]

[46] University of Salford. *Thermal measurement laboratory* [Online]. Available from https://energyhouselabs.salford.ac.uk/labs/thermal-measurement-laboratory/ [Accessed 6 February 2026].

[47] Lucideon. *Hygrothermal testing* [Online]. Available from https://www.lucideon.com/uk/testing-characterisation/hygrothermal-testing [Accessed 19 July 2026]

[48] Fraunhofer Institute for Building Physics. *Thermal resistance and thermal transmittance testing* [Online]. Available from https://www.ibp.fraunhofer.de/en/building-physics-testing/thermal-resistance-thermal-transmittance.html [Accessed 6 February 2026].

[49] CSTB. *Thermal and energy testing laboratories* [Online]. Available from https://www.cstb.fr/en-us/our-offers/all-our-offers/thermal-and-energy-testing [Accessed 6 February 2026].

[50] ift Rosenheim. *Laboratory for thermal building physics* [Online]. Available from https://www.ift-rosenheim.de/en/laboratory-thermal-building-physics [Accessed 6 February 2026].

[51] FIW Munich. *Thermal insulation testing* [Online]. Available https://fiw-muenchen.de/en/technical-insulation [Accessed 6 February 2026].

[52] RISE Research Institutes of Sweden. *Thermal properties testing* [Online]. Available from https://www.ri.se/en/expertise-areas/thermal-properties [Accessed 6 February 2026].

[53] Peutz. *Façade performance testing* [Online]. Available from https://www.peutz.com/expertise/facades [Accessed 6 February 2026].

[54] Applus+ Laboratories. *Façade testing laboratories* [Online]. Available from https://www.appluslaboratories.com/global/en/what-we-do/service-sheet/building-enclosure–fenestration-testing [Accessed 6 February 2026].

[55] National Research Council Canada. *Hygrothermal performance testing facilities* [Online]. Available from https://nrc.canada.ca/en/research-development/nrc-facilities/hygrothermal-performance-buildings-testing-facilities [Accessed 6 February 2026]

[56] Oak Ridge National Laboratory. *Rotatable guarded hot box facility* [Online]. Available from https://www.ornl.gov/content/tools-apparatus [Accessed 6 February 2026]

[57] Intertek Group plc. *ASTM C1363 hot box testing* [Online]. Available from https://www.intertek.com/building/standards/astm-c1363 [Accessed 6 February 2026].

[58] UL Solutions. *Building envelope and fenestration testing* [Online]. Available from https://www.ul.com/services/building-envelope-window-and-door-fenestration-product-testing [Accessed 6 February 2026]

[59] CSIRO. *Thermal insulation testing services* [Online]. Available from https://www.csiro.au/en/work-with-us/services/testing-and-certification/thermal-properties [Accessed 6 February 2026].

[60] BRANZ. *Thermal performance testing* [Online]. Available from https://www.branz.co.nz/insulation-thermal-performance-testing/ [Accessed 6 February 2026]

[61] Masy G., *et al*. 'Lessons learned from heat balance analysis for Holzkirchen twin houses experiment'. *Energy Procedia*. 2015;78:3270–5

[62] Beizaee A., Allinson D., Lomas K. J., Foda E., Loveday D. L. 'Measuring the potential of zonal space heating controls to reduce energy use in UK homes: The case of un-furbished 1930s dwellings'. *Energy and Buildings*. 2015;92:29–44

[63] Gillott M., Rodrigues L. T., Spataru C. 'Low-carbon housing design informed by research'. *Proceedings of the Institution of Civil Engineers - Engineering Sustainability*. 2010;**163**(2):77–87

[64] Ritosa K., Saelens D., Roels S. 'Optimizing retrofit strategies for terraced dwellings: ConstrucThor case study'. *Proceedings of the 19th International Building Performance Simulation Association*; Brisbane, Aug 2025. Toronto: IBPSA; 2025

[65] Swan W., Fitton R., Brown P. 'A UK practitioner view of domestic energy performance measurement'. *Proceedings of the Institution of Civil Engineers - Engineering Sustainability*. 2015;**168**(3):140–7

[66] Chapman J., Lowe R., Everett R. *The Pennyland Project, no. 53, September 1985* [Online]. Available from https://oro.open.ac.uk/19860/ [Accessed 19 July 2026]

[67] Gupta R., Gregg M., Passmore S., Stevens G. 'Intent and outcomes from the Retrofit for the Future programme: Key lessons'. *Building Research & Information*. 2015;**43**(4):435–51

[68] Webborn E., Few J., McKenna E., *et al.* 'The SERL observatory dataset: Longitudinal smart meter electricity and gas data, survey, EPC and climate data for over 13,000 households in Great Britain'. *Energies*. 2021;**14**(21):6934

[69] Ji Y., Fitton R., Swan W., Webster P. 'Assessing overheating of the UK existing dwellings - A case study of replica Victorian end terrace house'. *Building and Environment*. 2017;77:1–11

[70] Farmer D., Gorse C., Swan W., *et al.* 'Measuring thermal performance in steady-state conditions at each stage of a full fabric retrofit to a solid wall dwelling'. *Energy & Buildings*. 2017;156:404–14

[71] Marshall A., Fitton R., Benjaber M., Swan W. 'Investigating the impact of renewing floor coverings on the energy performance of dwellings with suspended timber floors, tested under controlled conditions'. Presented at Sustainable Design of the Built Environment: Research in Practice; London, UK, 2018

[72] Pelsmakers S., Fitton R., Swan W., Marshall A. 'Heat-flow variability of suspended timber ground floors: In situ measurements in a full-scale test facility'. *Energy and Buildings*. 2017;138:396–405

[73] Tsang C., Farmer D., George B., Roberts B., Henshaw G. 'Evaluating the energy-saving potential of innovative window blinds in UK homes using simulation' Presented at International Retrofit Conference; Salford, UK, 2025

[74] Simpson A., Fitton R., Rattigan I. G., Marshall A., Parr G., Swan W. 'Thermal performance of thermal paint and surface coatings in buildings in heating dominated climates'. *Energy and Buildings*. 2019; 197:196–213

[75] Farmer D., Henshaw G., Roberts B., Fitton R., Swan W. *DEEP report 5. 02 Salford energy house heating systems testing* [Online]. London, UK: Department for Energy Security and Net Zero; 2024. Available from https://www.gov.uk/government/publications/demonstration-of-energy-efficiency-potential-deep [Accessed 19 July 2026]

[76] Johnston D., Farmer D., Miles-Shenton D. 'Quantifying the performance of a passive deaerator in a gas-fired closed loop domestic wet central heating system'. *Building Services Engineering Research and Technology*. 2016;**38**(3):2016:269–86

[77] Alzetto F., Pandraud G., Fitton R., Heusler I., Sinnesbichler H. 'QUB: A fast dynamic method for in-situ measurement of the whole building heat loss'. *Energy and Buildings*. 2018;174:124–33

[78] Henshaw G., Fitton R., Farmer D., Swan W. 'Using thermography to map U-values in homes: Validation of the Heat3D system in controlled conditions'. *International Journal of Building Pathology and Adaptation*. 2026;**44**(6):1709–24

[79] Gorse C., Jones Newport D., Booth C., *et al.* (eds.). 'Sustainable ecological engineering design' *Proceedings of the International Conference of Sustainable Ecological Design for Society (SEEDS) 2021*, Leeds, UK, 2021. Cham: Springer; 2025, pp. 595–605

[80] Lomas K. J., Oliveira S., Warren P., *et al.* 'Do domestic heating controls save energy? A review of the evidence'. *Renewable and Sustainable Energy Reviews*. 2018;93:52–75

Chapter 9

An overview of the IEA EBC Annex 71 project: measuring the fabric energy performance of homes using smart meters and onboard data

Richard Fitton[1], Staf Roels[2] and Grant Henshaw[1]

9.1 Introduction

In order to reduce the energy use of buildings, policymakers impose ever more stringent requirements for the energy performance of new and renovated buildings, as well as requiring the use of renewables, as in the upcoming Future Buildings Standard in the UK [1]. Many compliance checks and labelling of the energy performance of buildings are undertaken in the design phase by calculating the theoretical energy use, such as the Standard Assessment Procedure (SAP) model in the UK [2]. It is surprising that more attention is not paid to the measured energy consumption of buildings as, despite regulation and policy, monitoring of actual energy performances reveals, in many cases, a significant gap with theoretically designed targets [3–7].

Sources of deviation between the actual and expected performances can be attributed to *the design phase,* such as limitations, inaccuracies, and assumptions in the numerical models used to predict the energy performance; *the construction phase*, including quality of workmanship and differences between assumed and actual installed materials, components, and systems; and *the operational phase*, where energy consumption may be impacted by poor functioning of systems and/or no match between assumed and actual building usage. The observed performance gap, in combination with the increasing integration of innovative systems, such as intelligent elements, low-energy technologies, and active solar systems, accentuates the need to develop reliable methods and procedures that can be applied on site to assess the actual performance of buildings.

Two research projects within the framework of the International Energy Agency (IEA) are related to assessing the actual as-built thermal performance of

[1]Energy House Labs, University of Salford, Salford, UK
[2]Building Physics and Sustainable Design, KU Leuven, Belgium

buildings. The aim of the IEA, established in 1974 as an autonomous body within the framework of the Organisation for Economic Co-operation and Development (OECD), is to foster international co-operation in the area of energy policy. The IEA co-ordinates research activities through a comprehensive portfolio of Technology Collaboration Programmes, which includes the IEA Energy in Buildings and Communities (IEA EBC) Programme. The mission of the IEA EBC programme is to develop and facilitate the integration of technologies and processes for energy efficiency and conservation into healthy, low-emission, and sustainable buildings and communities through research and innovation. This research is undertaken through a series of research projects, or 'Annexes'. From September 2011 to June 2015, the IEA EBC Annex 58 project worked with international experts on the topic of 'Reliable building energy performance characterisation based on full scale dynamic measurements'.

The global objective of Annex 58 [8] was to develop the necessary knowledge, tools, and networks to achieve reliable on-site dynamic testing and data analysis methods that could be used to characterise the actual energy performance of building components and whole buildings. Within this project, a first step was taken to characterise the actual energy performance of buildings based on full-scale dynamic measurements. These on-site assessment methods mainly focus on the thermal performance of the building fabric by investigating the possibilities and limitations of black and grey box system identification models. Black-box modelling relies entirely on data with no knowledge of the system's internal workings, while grey-box modelling combines the data-driven methods with an understanding of the physics of the building to improve accuracy [9]. Guidelines have been developed on methods to assess the overall heat transfer coefficient of a building, starting from dynamic measured data instead of static coheating tests. The work identified several possible approaches, but all data was collected by making use of intrusive tests. These are generally large-scale tests that require a lot of equipment and distributed sensors and require artificially high internal temperatures in unoccupied buildings. Although Annex 58 showed that on-site quality checks are feasible, the project highlighted the need for non-intrusive methods, where not only the quality of the building fabric is assessed, but also the performance of systems and the impact of users in terms of their desired outcomes. The need for measurement systems that do not disrupt the occupants is crucial.

Currently, more buildings incorporate monitoring systems recording real-life data [10]. Smart meters, home automation systems, and wireless technologies for data management and communication have been developed in recent years. As a follow-up activity, the IEA EBC Annex 71 project built on the findings of IEA EBC Annex 58 but aimed to make a step towards monitoring buildings in use. Annex 71 focussed on the development of replicable methodologies to assess the actual energy performance of buildings based on onboard monitoring data. This means that buildings would be monitored based on data from existing sensors already located in the building as part of its systems, such as heating systems and controls. Characterisation and performance assessment of buildings can serve many purposes, such as identifying the air tightness of the building fabric, the efficiency of systems, or operational use by occupants.

The Annex 71 project focus was on two topics:

1. Building behaviour characterisation to predict heat demand/indoor temperature for a given time horizon.
2. Quality assessment to pinpoint the most relevant building performance indicators, such as the actual overall heat transfer coefficient of the building fabric.

Both approaches can be used for different applications. Within Annex 71 [11], building behaviour characterisation has been used for fault detection and to model predictive control. The main part of the project focussed on quality assessment by exploring and optimising identification techniques to determine the overall heat transfer coefficient of individual dwellings based on onboard-monitored data [12]. The global objective of the project was to develop a comparison system, or matrix, that compares the different applications of system identification and heat loss characterisation techniques in terms of cost, time scale, intrusiveness, and accuracy.

9.1.1 Understanding the performance gap

As noted, Annex 71 seeks to provide methods for estimating the energy performance of occupied dwellings using a metric known as the Heat Transfer Coefficient (HTC). This figure is at the heart of all energy performance certificates required by the Energy Performance of Buildings Directive (EPBD) [13]. In energy models, this figure is generally delivered using a very simple formula. In the UK, for example, U values of all the elements are multiplied by the area of the heat loss elements and combined with a figure relating to the airtightness of the building known as air permeability [14]. The values are used to make the calculation for heating and cooling demand. Importantly, this makes the HTC a value that can both be modelled and measured.

The problem that led to the creation of Annex 71 was that significant quantities of research indicated that there is often a gap between the calculated value and the value that is measured on site, also known as the As-Built In-Use (ABIU). This can contribute to an issue known as the Energy Performance Gap (EPG). This gap is a numeric representation, often presented in percentage terms, of the difference found between the modelled energy performance of a building compared to its ABIU value.

The EPG for dwellings has been researched globally for at least 20–30 years. EPG values of over +100% have been found in research on new-build properties by Leeds Beckett University in the UK [15]. The EPG is also well recorded in retrofit scenarios, with a recent study finding gaps of up to 71% [16]. Both of these studies specifically explored the HTC values between measured and modelled values, rather than annual running costs.

This is not a problem that is unique to the UK; other studies globally have found significant performance gaps between modelled energy performance and ABIU. The measured buildings are predominantly performing more poorly than the designed energy model suggests. Some examples of this are illustrated in Table 9.1.

As Table 9.1 demonstrates, the issue of the performance gap is not a localised one. The EPG issue is important for several reasons.

Firstly, energy models are used in a great number of countries to determine the energy performance of existing buildings and to assist in the design of new buildings. When these are used for building compliance, this can cause issues if performance gap factors are widespread, as suggested above. Therefore, when energy models are used to create scaled-up models of the performance of large collections of dwellings, stock models used for policy-making for example, these errors will be more pronounced. Secondly, the cost-effectiveness of a property for customer-facing records, such as energy certification, or for cost studies on retrofits or new builds, is often intrinsically linked to the HTC provided in the energy model. If this figure does not represent reality, then the running costs for the property will also vary. Thirdly, the HTC is a building physics metric that is used for the sizing of the heating and cooling equipment in a dwelling. If the HTC is calculated incorrectly, or does not reflect ABIU, then there is a chance that the building systems will not heat or cool as intended, which may lead to poor levels of thermal comfort or unexpected running costs.

The EPG is a critical issue, but it is not an unsolvable one. Where homes are constructed well and the models are accurate, then the EPG issue can be eliminated. There is often minimal EPG in Passivhaus dwellings, which have more stringent modelling and post-construction testing.

Johnston carried out a study of a sample of over 2000 Passivhaus standard homes. All of them met the energy performance to which they were designed, with all buildings performing better than the minimum prescribed standard [21]. A smaller study found an average EPG of only 8% [22]. However, Passivhaus construction only accounts for a small number of new build and existing houses globally, currently around 47000 homes [23]. We are still left with an issue; how do we reduce the EPG? Clearly, a component of this issue is how we can quickly and

Table 9.1 A review of global EPG studies with sample size

Country	Sample size (N)	Average performance gap
Canada [17]	1	74%
Germany [18]	3400	30%
United Kingdom [15]	25	50%
Switzerland [19]	50000	11%
Italy [20]	6	45%

cost-effectively diagnose the EPG. A process is needed; this process is known as 'Energy Pathology' [24]. This involves, as a starting point, the measurement of the HTC.

9.2 The current methods of measuring an HTC

It is important to differentiate between the two methods of determining the fabric performance:

(I) **Testing (cross-sectional):** Tests are a single point in time measurement usually carried out over hours or days, with a maximum of several weeks, the output being generally the HTC, U values, or air permeability.
(II) **Monitoring (longitudinal):** A monitoring campaign could last for weeks, months, or even years. The outcome of these campaigns can be multivariate, for example, thermal comfort, annual heating performance, and system performance.

The following section will only cover the testing, or cross-sectional, methods, for the HTC measurement process and will provide a review of existing methods that are well developed, as well as emerging techniques, some of which have been developed within the IEA Annex 71 group members.

The tests will be described in the order of maturity; we will judge maturity here by the number of buildings that have been tested by each method. Under this approach, the more tests that have been undertaken, the more mature the method is viewed to be. However, this should not be viewed as a metric of a method's robustness. The respondents were asked how many of each test had been carried out, and it is included next to the name in brackets.

9.2.1 Coheating (112 buildings tested)

Coheating is seen by some to be the 'gold standard' of building fabric testing for energy performance, and it has been in use in various forms since the 1970s [25]. The method has been developed through many iterations, the most recent being the method devised by Leeds Beckett University [26]. A review of this method was carried out by Bauwens [27]. This testing method approach has now had a European standard created to sit alongside it, known as EN17887 [28].

The method relies on simple physics principles. Temporary electrical heaters are added to the building and the temperature is elevated to 25 °C. This is generally undertaken for a period of two to three weeks, during the heating period. This ensures a temperature differential of at least 10 °C for the majority of the duration of the test. This has the effect of creating a monodirectional heat flow through the building fabric. Solar radiation and electricity to power the heating are considered as energy inputs. Using these data, as well as internal and external temperature measurements, allows analysis to ascertain how many watts it takes to raise the internal building temperature by one degree. This is expressed in watts per Kelvin (W/K).

9.2.2 QUB (105 buildings tested)

The QUB test derives its name from the term Quick U-value of Buildings and is an innovation developed by construction product manufacturer Saint-Gobain. The main feature of this method is the ability to measure an HTC in 48 hours or less. The procedure requires a temporary electrical heating system to be added to the dwelling, which must be vacated for the duration of the test. A duration of heating is followed by a cooldown period overnight to avoid solar radiation. This cooling period can continue through to the following day [29]. The method has a number of academic papers focussing on accuracy and repeatability and also has a large number of buildings tested [29–31].

9.2.3 ISABELE (20 buildings tested)

ISABELE (In Situ Assessment of the Building EnveLope pErformance) [32] is a test designed to measure the HTC and identify EPG in dwellings. This test has similarities to both the coheating method and QUB, in that a large amount of heat is injected into the building using a temporary electrical heating system, usually between 25 °C and 35 °C, for a duration of a day. This temperature is held at a steady state during this period. The building is then allowed to cool for a period of time (around 0.5 days). The heating and cooling periods may be elongated if the building has large amounts of thermal mass and/or is insulated externally.

9.2.4 CoDY method (10 buildings tested)

This is one of the newer HTC test methods. As the name suggests, this method uses the coheating method as its foundation but does not require steady-state conditions. This allows for the number of test days to be reduced from approximately 14 to 5 days [33]. This method, as with many others, relies on portable electric heaters being added to the property to inject heat into the building at a pre-programmed time. The test must be carried out in a vacant building during the heating season.

9.2.5 Loughborough In-Use Heat Balance (LIUHB) (6 buildings tested)

This method differs from other HTC methods in that it uses the heating system built into the home to provide the heat input. This might be the installed gas central heating system, for example. This negates the requirement for a separate heating system, reducing the cost for the test in terms of equipment and labour. The building can also be occupied during the measurement period [34]. The building is heated to typical thermal comfort temperatures, around 21 °C, and can be switched off if needed, as would be done in occupied properties. The test must be carried out in the heating season and requires temperature sensors to be placed in the dwelling. The test period should be at least three weeks.

9.2.6 CAM(B)BRIDGE (5 buildings tested)

CAM(B)BRIDGE (acronym for Calculation and Measurements in Buildings: Bridging the Gap) dates back to 2018, when it was developed in Belgium for the specific purpose of measuring the EPG [35]. This has similarities to other models in that portable electric heaters are used. However, it differs from all other methods in that a system of Lego-like blocks is used to provide a small, controllable environment adjacent to the facade; these blocks also contain elements of the monitoring equipment required, such as solar radiometers. As with other methods, the dwelling is heated to a high temperature during the test period, which is approximately 4 days.

9.2.7 Integrated coheating method (4 buildings tested)

This method has similarities to the LIUHB method in that the installed heating system is used to heat the building. It was developed by members of the research team responsible for the coheating method detailed above. The method uses the dwelling's heating system to provide the heat for the test. This should also be supplemented by a good-quality heat meter, which needs a tradesperson to install, internal temperature monitors in all significant rooms, and a weather station capable of measuring solar radiation [36]. The internal temperature should be elevated to 25 °C and should also be unoccupied for the duration of the test, which is 14–21 days. This test should be carried out during the heating season.

9.2.8 Average method (4 buildings tested)

This method relies on the calculation of the energy balance for the dwelling, such as internal heat from heating and appliances, and heat losses such as through the elements through conduction, convection, and radiation [37]. The method is designed to work with building automation systems (BAS) in homes and commercial buildings, using a technique that harvests data from the BAS to identify cloudy days when solar radiation does not affect the building significantly for at least three days. The method processes the energy inputs and temperatures to calculate an HTC figure over this period. The building can remain occupied, but the average temperature of the building should be similar at the beginning of the test to that at the start of the test. The test should be accurate after 3 days of appropriate weather conditions.

9.2.9 PSTAR (2 buildings tested)

This is an older method that was once a pioneering technique. However, it is software-based and is not freely available. It has not been updated since 2014, which can make the method difficult to use. The method itself was created in the 1970s and was used to test innovative low-energy buildings. It shares some characteristics with the coheating test, in that the building must be unoccupied, temporary heating systems are used, and it should be carried out during the heating season [38]. However, the measurement period is shorter, taking around 4 days, and also has a period where the building cools with no heat injection, similar to the QUB method.

9.2.10 Summary of methods

These research methods were either part of the Annex 71 group or were found in the literature review undertaken for the Annex. Only those methods supported by published peer-reviewed articles or papers were considered to be suitable for inclusion. Each research team was contacted and asked to complete a summary form on their method, collated in Table 9.2. The terms used in Table 9.2 are as follows:

(i) Typical duration of measurement (days): The amount of time to achieve a successful HTC measurement for a building is influenced by variables such as weather, stability of internal temperature, and the thermal capacity of the dwelling. This factor is common across all of the methods examined. This indicator should be treated with caution, as it represents, in most cases, a mean of the range of the values presented in the research papers or directly provided by research teams as a response to a questionnaire.

(ii) Typical expected errors: Teams presented error values in two main ways. The average of the range of the errors as stated in the papers quoted, or the range of errors provided by the research teams who developed the method. These are not fixed values and will likely change dynamically depending on variables for each test. These should be considered indicative values only.

(iii) Cost per test (€): If one house was observed in a standalone project, this would be the cost charged, not including any extras, such as travel or subsistence.

(iv) Equipment purchase cost (€): This figure reflects the cost of purchasing the total equipment to carry out a test of one dwelling, although much of this can be reused.

(v) Number of variables: This reflects the number of measurands being collected. These may include the following: temperature, energy consumption, and solar aperture. This also includes measurements that may be repeated at multiple points, such as temperature in rooms.

(vi) Number of buildings tested: Dwellings/buildings tested up until February 2021, when the data was collected.

Table 9.2 Summary of the main metrics of the current HTC test methods

Method	Typical duration of measurement (days)	Typical expected error %	Cost per test (€)	Equipment purchase cost (€)	Number of variables	Number of buildings tested to current	Level of complexity (1 easy 10 difficult)
Coheating	14	10.0	€1,400.00	€4,500.00	12	112	8
QUB	2	13.0	€1,000.00	€2,500.00	7	105	7
ISABEL	10	8.0	€1,000.00	€3,000.00	7	20	7
CoDY method	5	12.5	€1,000.00	€8,500.00	18	10	8
Dynamic Integrated method	60	4.0	€1,000.00	€500.00	5	8	6
Loughborough In-Use Heat Balance (LIUHB)	21	15.0	€1,000.00	€500.00	5	6	6
RC Dynamic method	30	4.0	€1,000.00	€500.00	5	6	6
CAM(B)RIDGE	7	3.0	€1,500.00	€16,098.75	12	5	8
Integrated Coheating method	17	6.0	€820.00	€4,000.00	10	4	6
Average method	5	8.0	€1,000.00	€500.00	5	4	6
PSTAR	3	30.0	€1,000.00	€4,500.00	10	2	6

(vii) Level of complexity: This is not a formal metric; however, it does provide an indication of how complex the measurement process and related data analysis are, with 1= Self working/practically no expertise required, and 10 representing only researchers in this area with advanced knowledge of the method could carry out this work.

The results found in Table 9.2 provide an overview of the current testing methods, and from this, we can draw the following conclusions.

HTC measurement is still an undeveloped area. Even considering the methods from 50 years ago, the total number of buildings to have been tested is just 282, with the assumption that each test was on a unique building, rather than the repeated tests on the same building.

These methods are generally expensive to apply for the following reasons. Firstly, there is no economy of scale. With a limited number of tests, it is hard to make savings by grouping tests or bulk purchasing equipment. Secondly, the up-front costs of the equipment are high and may be considered to be prohibitively expensive when we consider that the periods when one may carry out most of these tests are constrained by the heating season. This means that simultaneous testing will be required and, as such, several sets of equipment will be required to operate at scale. Thirdly, the disruptive nature of the tests, the timescales involved, the lack of automation, the requirement for skilled people, and the lengthy process of data analysis can all increase costs, with the average cost for an HTC measurement exercise being 1,065 (€) and the average equipment cost for one test being approximately 4,100 (€).

The number of variables required for some methods is high. This will clearly have a cost impact for both data collection, analysis, and the possible disruption to the home. Additionally, the timescales have a wide range, from two days to two months. It does not follow that a long test is more accurate, in terms of either initial outlay or cost of the test. In fact, there appears to be very little correlation between cost/accuracy, duration/accuracy, or cost/duration. Finally, accuracy is fundamental to establishing trust in HTC as a metric. The research has found a wide range in measurement error, ranging from 3% to 30%. We assume that this is due to the lack of formal development of some tests compared with others and the way that errors are recorded/predicted, which also differs from method to method. This potentially requires further investigation to ensure more transparency between methods.

9.3 Measurement and modelling gap

As the previous section mentioned, there is an area of concern in the measurement of an HTC due to uncertainty. This can be expressed in terms of percentages. A smaller percentage uncertainty means a more precise measurement, while a larger percentage uncertainty means a less precise measurement. It has been identified that some methods have uncertainties of up to 30%. Therefore, these measurements should be taken in the context of their uncertainty when considering how they reflect the dwelling's performance. Therefore, when we discuss the EPG, we should consider that it may be a measurement issue, a non-representative energy model of the dwelling, or a combination of both factors. Rather than an EPG, there may be a gap in both the way that we model, creating a modelling gap, and in measuring the building, leading to a measurement gap [39].

9.3.1 Measurement gap

The coheating test is often used as a referent against which other HTC methods are compared as a baseline. Although this method is the oldest and has the most tested buildings, its accuracy has been recently questioned. A report issued by the National House Building Council Foundation (NHBCF) used a referent house to assess the resultant HTC estimation by researchers using the coheating method. This was undertaken to identify how different research teams might measure input data and analyse these data differently. Six teams applied the method to the referent house and the range between the lowest and highest measured HTC was 20.4 W/K, or approximately 30% [40]. Some teams also used different methods to deal with solar gains, which are another source of debate and error among users of the method.

While the coheating test has been used on multiple buildings over a long period, issues still exist in terms of repeatability, especially when carried out by different researchers. Some fundamental issues also exist with the coheating test; these were recently the focus of work by Stamp, who presents 21 issues with the coheating test that may affect its uncertainty [41]. These cover variables such as wind speed, relative humidity in the dwelling, and the range of data analytics that can be applied to the data. A significant finding was that the coheating standard had scope for improvements, with standardisation being required. Work is currently progressing on a CEN standard [42]

9.3.2 Modelling gap

Another significant component of the EPG is the energy model, against which dwellings are compared to establish the EPG. This makes the energy model significant. When we consider energy models, an assumption can be made that they are an accurate representation of the building. However, several researchers have found significant issues that lead to errors in the output of models. Strachan led a case study approach to model validation. A referent building was measured and detailed in terms of building fabric and services. These details were sent out to 21 modelling teams internationally. Each team chose its own energy modelling software, with 12 different software packages being used. The inputs and outputs of these models from all teams were analysed by Strachan. He identified the issues that contributed to an erroneous model; the most significant one was user error, with the issue of misinterpretation and incorrect data input. However, the research also discovered that some of the fundamental physics assumptions behind the models were sometimes handled in different ways in each software package, not necessarily incorrectly, but in ways that caused variations in the outputs generated in the energy simulations. Examples of these varying parameters were the treatment of longwave radiation and differences in the way that some software modelled thermal bridges. Another significant difference was the approach taken by the different software to glazing transmission of heat and internal air convection treatment. Other researchers have also presented evidence of differing software generating varying outputs when the same input data are presented [43,44].

Addressing the components of the EPG and understanding that there may be more to this issue than a simple comparison of measurement to model, identifies that the issues may not solely lie in the poor construction of buildings and/or installation of services. It is possible that it is a combination of variations in the models and measurements, as well as the characteristics of the building itself.

This suggests a requirement for a standardised and validated approach to the measurement of buildings on site. This method should be cost-effective, time-efficient, and validated and tested on controlled buildings if it is to be effectively scaled up and applied by industry. This is the work of Sub Task 3 of the Annex 71 project and is discussed in the next section of this chapter.

9.4 Towards the development of a standard

The movement to on-site quality assessment is clearly visible from several regional and national initiatives, such as 'zero-on-the-meter' in the Netherlands [45], the SMETERS project in the UK to assess measured heat transfer coefficients [46], and energy performance measurements in France [47]. A survey undertaken as part of the IEA EBC Annex 71 project confirmed the interest of stakeholders to assess the actual energy performance of buildings based on on-site measured data. This survey examined the views of the retrofit, engineering, building performance sectors, as well as academics from the field [48].

To respond to this, the Annex 71 project evaluated and improved identification methods that can be used in residential buildings. These wider stakeholders will have different motivations to apply building performance assessments. A construction company may want to guarantee its quality, while a social housing company may seek the most effective retrofit strategy. An engineering office may need actual information on the thermal response of buildings to design or optimise a thermal network, to embed the building behaviour in model predictive control (MPC), or for fault detection and diagnostics (FDD) strategies. Energy Supply Companies (ESCO) and policymakers may want actual performance indicators to optimise their strategies to reduce carbon emissions. Each of these applications seeks different information on the actual building performance, such as the overall heat transfer coefficient of a building, the efficiency of the heat recovery system, and the capacity and response of the building to solar gains.

9.4.1 Data collection, costs and application

Over the recent decades, a wide range of identification methods have been developed that can be used for these purposes. They range from simple linear regression models to advanced dynamic system identification techniques. The methods not only differ in ease of use, some are straightforward while others demand advanced statistical expertise, but also require different input data in terms of quantity, duration, and frequency [27,49]. An important aspect in evaluating those methods is balancing the cost of the measurement method and the analysis against the achieved accuracy. The accuracy depends on the analysis technique, the measurement time, and the input data, including measurement error. To map all these interactions, an evaluation matrix was developed, linking the accuracy of the outcome needed for a specific application to the requirements of the measurements and the best-suited statistical data analysis technique. For instance, determining the HTC of a building from smart meter data will have a different cost and accuracy than performing a full coheating test. Where, for some applications, the accuracy of the smart meter approach will be appropriate, such as an estimate of the insulation installation quality for renovation purposes, a building quality assessment might require a higher accuracy and, hence, a more accurate test. Figure 9.1 presents the basic idea of the matrix. Starting from a level of accuracy (y-axis) as required by the application, a certain method and corresponding cost of data acquisition (x-axis) can be determined. Alternatively, knowing the data acquisition technique, the most applicable method and maximum attainable level of accuracy of the outcome can be determined. This allows stakeholders undertaking a characterisation or quality assessment of a building to make a well-informed choice depending on the application and the available budget.

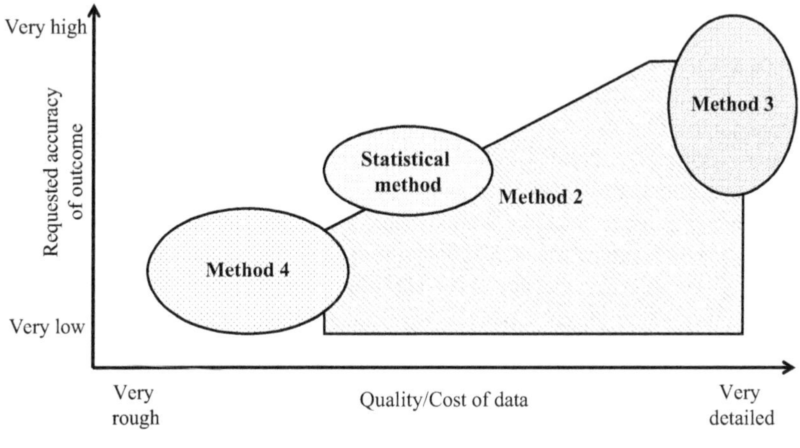

Figure 9.1 Outline decision matrix, linking the statistical methods to the quality of the input data and achievable accuracy of the outcome

Based on case studies, the necessary input data and obtained accuracy were evaluated for different data analysis methods and for different applications. Improvements of the identification techniques can be obtained by reducing the cost, moving to the left on the *x*-axis, or enhancing the accuracy, moving to the top on the *y*-axis. As such, the evaluation framework was refined during the project for specific applications. The project did not intend to optimise all possible methods for all possible applications but rather to develop the global framework and to explore it for selected specific applications as identified by the stakeholders. Subtask 2 of the Annex 71 project [11] explored the framework for model predictive control and fault detection. Subtask 3 [12] evaluated different identification methods that have been tested and optimised to assess the actual HTC of a building.

Mapping the interaction between potential application, corresponding performance indicator, expected accuracy, statistical method, and available input data is the main purpose of the decision matrix. Supported by the decision matrix, stakeholders should be able to select a specific application based on data to be monitored, how to analyse the data, and the corresponding accuracy that can be achieved. The link to cost, through the *x*-axis, of the data acquisition (see Figure 9.1) was considered essential by the stakeholders, but during the course of the Annex 71 project, it became clear that it is difficult to attribute a certain cost to an onboard measuring campaign. Firstly, some sensors will be available in the building as part of the systems, so it appears correct to only account for additional sensors needed for a certain application. Sensor costs were found to be highly dependent on factors such as type, accuracy, installation cost, and reusability. Furthermore, the cost of sensors is rapidly changing over time and is often geographically dependent. Therefore, a shift was made from specific costs of the data collection towards an *x*-axis showing possible sets of prescribed data packages for a specific application. The simplest and lowest cost data package for HTC-characterisation may be a smart meter and room thermostat, providing global energy use and the living room indoor temperature. Using only this data will require several assumptions, such as building use, actual averaged indoor temperature, and solar and internal heat gains, to estimate the HTC of the building. One can expect the uncertainty of the HTC estimate to be high, but it may be precise enough to get a first estimate of the renovation potential of the building. If a more precise assessment is needed, additional measurement points, such as additional room temperatures, flow meters to split energy use for space heating and domestic hot water, or a local weather station, can be added to define additional data packages that can reduce the uncertainty on the HTC estimate. Table 9.3 shows an example of possible data packages for HTC estimation.

In addition, the survey revealed time factors to be crucial for many stakeholders. Short measurement campaigns are generally preferred. Exploratory studies indicated a strong correlation between time span and frequency of the available data and accuracy. For example, a simple data package collecting data over a longer time span might result in the same accuracy as a monitoring campaign with a very extensive data package running over a few days. This allows us to clearly see that time issues, in a sense, are related to cost.

Table 9.3 Examples of data collection packages for HTC estimation

Examples of packages	Equipment	Variables available	Notes
Package 1	Existing home with smart meters and smart thermostat.	Gas consumption, aggregate electric consumption, internal temperature, occupancy patterns and heating setpoints.	External weather can be sourced from online sources.
Package 2	New build home with air-source heat pump and smart meters.	Aggregate electric consumption, ASHP consumption, internal temperature at several zones, external temperature and setpoint/occupancy pattern.	
Package 3	New build zero-carbon home with localised controls.	Aggregate electric consumption, ASHP consumption, internal temperature in each occupied room, external temperature, setpoint/occupancy pattern, ventilation system energy consumption and input from solar PV system.	PV system can help to analyse the solar radiation input into the dwelling.

To address these concerns, the x-axis of the decision matrix was modified by introducing a z-plane, which allows the selection between various data packages and the duration of the monitoring campaign. The underlying concept of this three-dimensional decision matrix, as illustrated in the diagram within the Annex 71 report [10], demonstrates how different data analysis methods can be assessed. If achieving a specific level of accuracy is essential for an application – such as ensuring the quality of the HTC in a newly constructed dwelling – a horizontal cut-off plane can be used to identify all statistical methods, together with the required data packages and monitoring durations, that meet the acceptable level of uncertainty. Conversely, where the main concern is the length of the monitoring period, a vertical intersection can be used to pinpoint all methods and data packages suitable for the specified timeframe, along with the expected accuracy that can be achieved.

It is important to note that different use cases, such as fault detection, HTC assessment, model predictive control, or estimations of solar aperture, will probably result in different decision matrices. While we have considered the analytical frameworks, the first part of this chapter has focussed on the practical application of the methods and their potential application to use cases. In the next section, we will discuss different analytical approaches in more detail.

9.5 Statistical methods

Different statistical methods can be used to determine the overall heat transfer coefficient of a building. Techniques using onboard data vary from simple stationary methods to advanced dynamic data analysis methods. Here, we provide short description of the most important characterisation methods is given together with their main benefits and limitations. With these methods, we will try to determine the HTC of a building starting from a single one-zone heat balance equation:

$$C_i \frac{\partial \theta_i}{\partial t} = \Phi_h + \Phi_{int} + \Phi_{sol} + \Phi_l + \Phi_{tr} + \Phi_v + \Phi_m \tag{9.1}$$

with C_i the effective heat capacity and θ_i a representative averaged indoor temperature of the dwelling. Φ_h corresponds to the heat input by the heating system, Φ_{int} the internal heat gains, Φ_{sol} solar heat gains through transparent fabric parts, Φ_l latent heat exchange due to moisture loading and unloading of interior objects and building parts, Φ_{tr} the transmission heat exchange towards neighbouring buildings and the external environment, Φ_v the ventilation heat exchange, and Φ_m the heat exchange when loading/unloading the thermal mass of interior objects and building parts. A detailed elaboration of this heat balance equation and the corresponding assumptions can be found in the Sub Task 3 report [12]

A simplified approach is taken when we use the existing methods such as coheating

$$\Phi_h = \Phi_{sol} + \Phi_{tr} \tag{9.2}$$

In many cases, the energy consumption data of buildings in use are only collected at low frequencies – on a weekly, monthly, or even yearly basis, generally for billing purposes. Although we might lose accuracy, when combined with other data, such as climatic boundary conditions and averaged indoor temperatures, these data might still yield sufficient information to assess the overall HTC. The dynamic behaviour, though, needs to be neglected and Eq. 3. further simplifies to

$$C_{i,eff} \frac{\partial \theta_i}{\partial t} = 0 = \Phi_h + \Phi_{int} + \Phi_{sol} + \tilde{\Phi}_{tr} + \Phi_{vent} \tag{9.3}$$

This simplification is applied in the static methods, such as the average method, linear regression and the energy signature method.

9.5.1 The average method

The average method is described in ISO 9869 [50] as a simple method to determine the U-value of homogenous building elements based on on-site heat flux and environmental temperature measurements. Given the analogy between the definition of the HTC and that of the U-value, the average method is also applicable on a whole building scale to determine the HTC. As demonstrated in Eq. 4, the method presumes that the division of the mean heat flow rate by the mean internal-external temperature difference asymptotically converges to the HTC if the average is taken over a sufficiently long period.

$$HTC = \frac{\sum_{j=1}^{n}\left(\Phi_{h,j} + \Phi_{int,j} + \Phi_{sol,j} + \Phi_{vent,j} + \Phi_{tr,j}^{adj}\right)}{\sum_{j=1}^{n}\left(\theta_{i,j} - \theta_{e,j}\right)} \tag{9.4}$$

where j corresponds to the individual measurement points. To ensure that the steady-state conditions are met, the heat content of the building fabric should, on average, remain constant throughout the experiment.

Linear regression techniques are also commonly used to determine the stationary thermal properties. Where in the averaging method, detailed, short-interval data can be used and the stationary values follow from the averaging technique, the linear regression method typically makes use of daily averaged values to cancel out short-term effects of thermal mass [27]. Applying multiple linear regression allows us to determine, not only the overall heat transfer coefficient, but also to gain some information on the solar transmittance.

9.5.2 Energy signature method

The energy signature method can be considered a special case of linear regression analysis in which internal and solar heat gains and intended ventilation are related to a so-called 'base temperature', the outdoor temperature for which the building is in thermal balance with its environment. An elaborated description of the method is covered by Danov [51].

Compared to the previous methods, dynamic methods allow the inclusion of the varying thermal response of the dwelling. The most commonly used are AR(MA)X and state space models.

9.5.2.1 ARX and ARMAX models

ARX and ARMAX models are standard methods for identifying so-called linear time-invariant systems. In the abbreviation AR stands for AutoRegressive: the current output is related to the previous values of the output; MA (Moving Average) refers to the noise model used, and X refers to the fact that eXternal inputs are used. The system relies not only on the current input value, but also on the history of the input. The most common ARX-model structure is the simple linear difference equation, which relates the current output at time, t, to a finite number of past outputs and inputs. A main problem when applying AR(MA)X models is, firstly, the selection and validation of the model and, secondly, the extraction of physical information from the model parameters. In an ARX model, each individual parameter lacks a direct physical meaning. However, steady-state physical parameters can be obtained by calculating the stationary gain of the transfer function from each input to the output. We refer to [49], in which the principles of ARX modelling are extensively described.

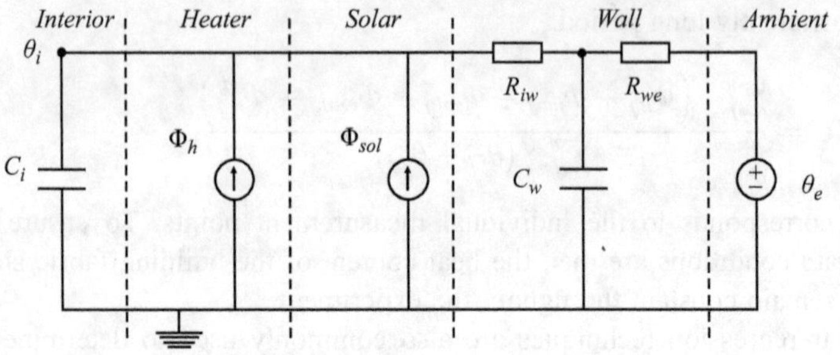

Figure 9.2 Example of a two-state grey box model applied to characterise the dynamic thermal response of a building

9.5.2.2 State space model

Another common method to characterise the HTC of a building is the use of state space models. State space models consist of simple resistance/capacitance schemes that can be used to model the thermal dynamics of a building as a linear time-invariant system. The parameters are readily estimated with an approach, such as the maximum likelihood technique, and most often a suitable model is identified with a forward model selection approach [49]. In this approach, the analysis starts with fitting a very simple model, which is then stepwise extended until the log-likelihood no longer increases significantly compared to the previous model and the residuals, identified as the difference between the measured and predicted output, correspond to white noise.

Both the initial model, and possible extensions thereof, simplify the building's thermal behaviour; their parameters have a direct physical interpretation. Defining them thus requires, in contrast to the ARMAX model, some prior physical knowledge. That is why state space models are often referred to as grey box models. Figure 9.2 shows, as an example, a two-state grey box model for a one-zone building, considering heat input by heater and solar radiation, capacity of the interior and walls, and (conductive) heat flow through the walls.

9.5.3 Building Energy Simulation

Finally, Building Energy Simulation (BES) models can also be used to determine the actual physical parameters of an existing building by calibrating the model using on-site measured data. However, although the models try to incorporate all physical phenomena as precisely as possible, they still simplify the full complexity of a building and its many independent interacting variables. Therefore, within the IEA EBC Annex 71 project, BES models were not used to deduce physical parameters from on-site measured data. Instead, the question was raised as to how reliable the models are in real-world building operations. To answer this question, a detailed empirical validation exercise was performed on a full-scale test building to verify the model predictive accuracy of BES models for real-life applications, including heating systems and (synthetic) users. All details on this validation exercise can be found in the report from Sub Task 2 [52].

9.6 Evolution towards quantification based on onboard data

To apply the different statistical methods presented above, we have to determine several input parameters starting from the limited on-site measured data. In real-life applications, several input parameters are not explicitly measured. Variables, such as the actual internal heat gains, the varying infiltration losses, and even the precise heat input by the heating systems or by solar gains, are not always clear. Therefore, the IEA EBC Annex 71 project investigated how the input for different terms of the simplified single-zone heat balance equation (Eq. 2) can be optimised starting from limited onboard monitored data in practical applications. For each of the terms, different approaches have been formulated, ranging from rough estimates, when limited data is available, to sophisticated modelling approaches or more extensive monitoring set-ups, for a more detailed and precise input.

As an example, Figure 9.3 shows the impact on the obtained HTC for different approaches to determine the dwelling-averaged indoor temperature for one of the case study dwellings within the project. Approach 1 assumes that no measured data is at hand. The indoor temperature can be assumed to be a constant value, as suggested in national standards or Energy Performance of Building Directive (EPBD) regulations. Alternatively, the time-weighted setpoint temperatures of the room thermostat could be used as global indoor temperature. If we have onboard measured data, often the indoor temperature is only measured by the room thermostat in the living room. In approach 2 of Figure 9.3, this indoor temperature is used as the input for the single-zone heat balance equation. However, as most dwellings typically maintain a significantly lower temperature in the night zone of the dwelling, a second indoor temperature can be synthetically created based on the assumption that the night zone of a dwelling is always 2 °C lower than the living room temperature (approach 2a). This value of 2 °C stems from the difference between the temperatures prescribed by the Belgian Annexes to standard EN 12831-1 for heat load calculations for a living room and bedroom. The measured and synthetic signal can then be weighted in different ways. In the current graph, a simple arithmetic mean is taken as global indoor temperature, but volume weighting, or even weighting according to the façade area, could be applied. In approach 3a, indoor temperatures are monitored in the living room and in the main sleeping room. Again, similar weighting methods can be used.

Similar approaches have been defined for heat input by the *heating system*, from gross energy use for space heating and domestic hot water to net heat input into the building zone. The assessment of *solar gains* can be assessed using a simplified constant solar aperture or advanced sun position-based spline models, separating between diffuse and direct solar radiation. *Internal heat gains* can be assessed using standardised occupancy profiles or presence determination based on

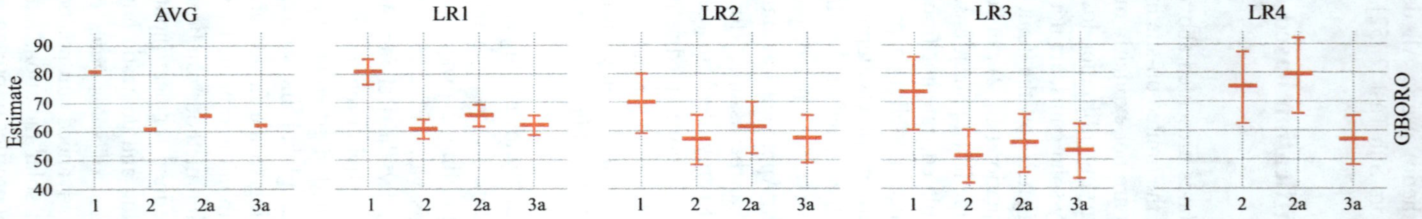

Figure 9.3 *Estimates of the HTC in the exploration of different indoor temperature inputs for one of the test cases in the IEA EBC Annex 71 project. Different statistical methods are applied. From left to right: average method (AVG), three types of linear regression.*

RH or CO_2 measurements. *Outdoor weather conditions* can be evaluated using on-site weather stations and the use of open-source climatic conditions.

For the different case study dwellings, the HTC has been calculated using the different approaches, while keeping the other input parameters simplified. This way, the impact of the approach on the HTC accuracy could be investigated. More details can be found in the Sub Task 3 report [12].

9.7 Investigating the accuracy and uncertainty of the different statistical methods

In the final step, the achievable accuracy and uncertainty of the different statistical methods have been investigated on five test cases with varying complexity. Some of the test cases were test facilities with well-controlled measurement campaigns, making use of synthetic users. Others were actual occupied buildings, including the typical disturbance one can expect in real occupied dwellings. Table 9.4 gives an overview of the test facilities and their main characteristics.

For each of the test cases, the optimal approach for the different terms in the building's heat balance (Equation 2) was applied starting from the available data. In this stage, 'optimal' means most precisely measured, or the most advanced method, according to our expert knowledge.

Figure 9.4 compares for each of the test cases the HTC estimate as predicted by the different statistical methods. Details on the selection of input data and modelling approaches can be found in the Sub Task 3 report [12].

Based on the knowledge gained while exploring the five test cases, 3D graphs as presented in the Annex 71 report were developed [10]. In addition, guidelines have been written to assist stakeholders in their choice of which input data to collect, i.e., which onboard monitoring to install, and which statistical method to choose to obtain the most accurate HTC estimate, depending on the available data.

Table 9.4 Overview of the main characteristics of the five test cases

Case study	**Type of house**	**Heating generator**	**Occupant**	**Weather data**	**Construction year**	**Renovated**	**Ref. HTC [W/K]**
TH N2	Detached single-family house	Electrical heaters	Synthetic	On location	1980s	Yes	110
TH O5	Detached single-family house	Air-source heat pump	Synthetic	On location	1980s	Yes	106
Uccle	Semi-detached house	Gas boiler	Real	Free database	2010	No	106
LBORO	Semi-detached house	Gas boiler	Synthetic	Station 8 km away	1930s	No	375
GBORO	Terraced house	Gas boiler	Real	On location	2012	No	40

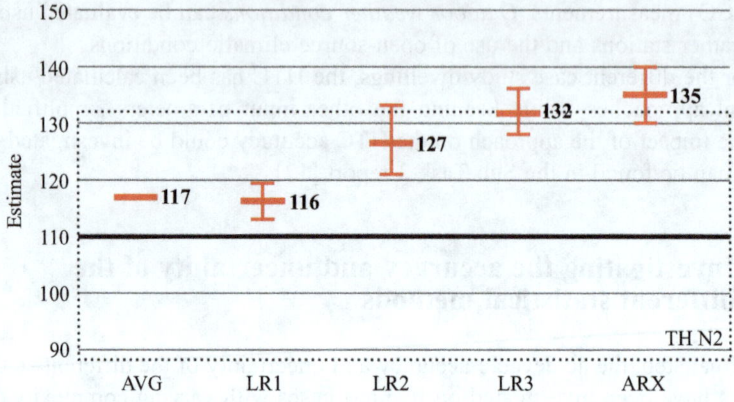

Figure 9.4 Example of figure for twin house test case TH N2

9.8 Understanding and application of HTC by stakeholders

We have spent much of this chapter discussing the HTC. It is clearly a useful, all-encompassing aggregated dataset. We also have learnings on its development and some of the inconsistencies. However, is it really a 'useful metric'? When we view the EPBD certification and the Energy Performance Certification that EU bodies have mandated, we find little reference to the HTC. Policymakers in many countries, as well as the general public, are more content with ratings, annual heating bills, and weighted metrics such as kWh/m^2. This means that we need to consider not only the validity of the HTC but also the manner in which it is presented. Next, we discuss some alternative approaches to the presentation of the HTC.

9.8.1 Current status of HTC use in a selection of countries

As part of the work by Sub Task 1 of Annex 71, a survey was sent to each member country. This was to gather information on the current method used by which the HTC is calculated, recorded, and viewed in the software for their national domestic energy model, as required under EPBD [13]. Under these regulations, all EU states have the requirement to model their housing. Non-EU states (UK, Switzerland, and Norway) also have models that meet these criteria. The way that they calculate and report the HTC varies across the states.

Data was gathered from the following countries: Austria, Belgium, Denmark, Estonia, France, Germany, Netherlands, Norway, Spain, Switzerland, and the UK.

The HTC is represented in all the models within these countries. It is seen as important, as this is the fundamental value that the models use to calculate the annual heating load and running costs for a building. However, the way that the value is calculated can differ across states, examples of which are outlined below.

Most states use a simple method *(where U= U-value of the element, A= area of the element and I = infiltration level of the building):*

$$HTC = (U.A) + I \tag{9.5}$$

Germany does not include infiltration in the HTC, opting for

$$HTC = (U.A) \tag{9.6}$$

France uses a normalised HTC, locally known as Ubat, or U Bâtiment, in English, U-value of the building, or whole building U-value, using the internal floor area of the building (m^2)

$$U\ Batiment = \frac{(U.A) + I}{Internal\ Floor\ Area\ m^2} \tag{9.7}$$

No country allows for a direct input of the HTC to replace the calculated HTC component with a measured value. Some of the responses explained ways the HTC could be changed with 'creative' styles of energy modelling, such as altering the U values or air permeability; however, this is not recommended by the guidance.

This research highlights some doubts about whether the HTC can be amended in an ABIU by insertion into the software. The survey also found a lack of visibility of HTC, with most respondents (58%) stating that the HTC could only be viewed by the assessor carrying out the work, with none of the tools making the HTC available for the public to see.

Although the HTC plays a crucial role in energy modelling, it cannot be altered in any way to represent the ABIU of a dwelling. This presents a quandary; Annex 71 and other projects globally have had, and are developing, cost-effective ways of measuring the HTC. However, when it comes to using these measured figures, it is not possible to apply these values for regulatory purposes for ABIU or completed dwellings. These need to be addressed by policymakers.

9.8.2 Creating a standard, understandable metric

Normalisation is the term that can be applied to a metric that may make it easier to understand; some may also refer to this as weighting.

9.8.2.1 Heat loss parameter

In property terms, the internal floor area of a building has many uses; sales details, spatial requirements for designers, and reporting of housing standards in many countries. As such, the idea of using this metric in line with the HTC can make sense; W/m^2K becomes the metric. This is useful in that, although the HTC is well understood by researchers, to the layperson, it may not be obvious as to its implications. For instance, a large home will generally have a larger HTC than a small home. This is not to say that the large house does not have good fabric energy efficiency; it is simply larger and will require more energy to heat. Using an area weighting helps with this understanding. This Heat Loss Parameter is used in some EPBD-based energy models but is not visible to the public.

9.8.2.2 HTC divided by dwelling volume

Another approach may be to use a volume-weighted measure, W/m^3K. This allows for features such as ceiling height to be considered. A large apartment block built in the 1930s will likely have higher ceilings than those found in an apartment block from the 1990s. If we were to normalise the HTC using only the floor area, then this important component would not be considered, and a false impression of fabric energy efficiency may be presented.

9.8.2.3 HTC/heat loss area

The HTC is a measure of fabric efficiency, and as such, relates directly to energy loss through the boundary between conditioned and unconditioned space. A sensible normalisation is to use the area of this surface. This allows for a comparison to be made on the actual thermal performance of the enclosing structure. This is important for homes that are terraced or apartment blocks, which may only have one external wall. This type of dwelling may provide a skewed HTC if weighted using the volume or area methods.

9.9 Use cases for HTC

A wider understanding of an accessible HTC metric has a number of benefits. These can be applied to specific use cases, which are outlined below.

9.9.1 Quality assurance of new-build properties

There is mounting global evidence to support the existence of an EPG. Energy certificates, such as the EPC in the UK, have been shown to contain substantial errors, with recent UK research finding errors in 27% of the UK EPCs within the national database [53]. Other research points to the EPC building measurement process leading to substantial mislabelling of the energy performance of buildings, 3–18% for dwellings in bands B–E [54]. This level of uncertainty has led some to believe that this is a contributing factor to the EPG. While these types of tools are needed as aids to design and rate buildings and to ensure regulatory compliance, they may not always be suitable for determining the actual ABIU performance of a dwelling. This may be particularly important when properties have a guaranteed energy performance or savings are guaranteed [55]. Where this is the case, the modelled energy performance may not be enough and further verification, such as a measured HTC, will be beneficial.

9.9.2 Evaluation of pre- and post-retrofit

Where dwellings are being retrofitted for energy improvements, it is often the case that, due to funding and time commitments, evaluation is often carried out on retrofitted buildings in their finished state, rather than measuring a baseline beforehand. It is considered to be good practice to take building performance measurements pre- and post-retrofit [56]. The measurement of the HTC will be useful information to evaluate the performance of the installation of retrofit measures. Knowing the original baseline HTC of the building will allow a more robust analysis of the savings made by the measures. Where older buildings are concerned, recent research has found that they may not perform as poorly as previously thought. An example of this was found in the UK, where solid wall buildings were modelled to have a U-value of 2.1 W/m^2K. However, following a large field trial examining solid walls ($N = 85$), the U values were measured and the median value was found to be 1.59 W/m^2K, a 32% difference from the value used in the UK energy model for over 10 years [57]. As such, measurements taken pre- and post-retrofit should be considered.

9.9.3 Building commissioning and system design

As buildings, both old and new, are coming under more pressure to decarbonise, many states are considering low-carbon heating solutions as a way of delivering their net-zero aspirations [58]. However, systems such as heat pumps and low-carbon technologies must be sized correctly to achieve efficient performance. Where units are either under- or oversized for the property, then the energy efficiency of the unit, the coefficient of performance (COP), may be significantly reduced [59], as can the thermal comfort within the property. It is for this reason that an effective solution to sizing the heating or cooling system for a dwelling may be to use an HTC. It is important to note that the HTC is a global measurement incorporating the whole dwelling, so local sizing of radiators will not be informed by the test. However, this could be an area of development for the additional supporting methods.

9.9.4 Energy pathology

Energy pathology is defined as 'the systematic investigation of the energy performance of buildings, with the aim of detecting any elements whose performance in energy terms is not as intended' [24]. This is often used, be it in a new build or retrofit, where a client or homeowner feels that the building is not performing to a satisfactory standard, usually in terms of energy performance and/or thermal comfort. This can involve a range of different methods, such as U-value measurement, air permeability, and thermography. There is a level of convenience here also, as the conditions set up by some whole-house in situ HTC tests can be conducive to work, such as U-value measurement and thermography, where a 'heat soaked' building is required. Hence, when we can carry out a whole-house test for HTC and couple these with other methods. A more informed outcome can be achieved with the benefit of the disaggregation of the energy flows in the building rather than the HTC metric alone. This will help to form a more robust opinion as to whether the building has underperformance issues, what the issues are, and how to remedy these issues.

9.10 Conclusion

We have discussed the EPG along with its constituent elements. The requirement to measure buildings to identify and resolve the EPG has also been explored. An overview of the current methods used to measure the HTC by industry and academia has been presented, identifying differences between the various methods in terms of accuracy, duration of measurement, cost, and complexity. The measurement gap component of the EPG and how this varies between methods was considered. There was an exploration of emerging innovative measurement of the HTC, using both in situ and onboard data; the initial findings from Annex 71 workstreams were provided alongside the results. This means that upon future development of these methods, HTC measurement could become accurate, cost-effective, and convenient. The HTC is being studied by many researchers in many different countries, and yet we find no agreement on the most effective way to represent this measurement in terms of fabric efficiency. We have several options around choice of metrics, with a view to be taken by policymakers, consultants, or engineers as to which one is the most effective for their individual context.

The Annex 71 work identifies several promising methods of determining the HTC, and we discussed the more established methods, some of which are working towards a standardised approach. However, when it comes to putting the HTC into use, the problems become more complex. EU states are under an obligation to produce energy modelling data for their housing stock, but ABIU data is not a requirement. Even the mechanisms of inserting a measured HTC into a model are a challenge in all of the states covered in this study. It is clear that scientific work on building measurement does not exist in a vacuum and that adoption may be strongly dependent on an individual country's regulatory and reporting frameworks. It should also be noted that there is considerable work going on outside of the research discussed in IEA Annex 71, particularly in the private sector. The risk is that these studies happen in silos, and the research findings are not coordinated or are carried out using methods that do not allow for comparison.

Future work for HTC measurement should ensure that the current and new methods are validated on buildings in the field, with each building typology, such as light and heavyweight construction, having characteristics that may be troublesome for some methods but not others. The data from the Annex 71 activities is publicly available and this information can be used by researchers and innovators to develop further methods for HTC measurement, with the advantage of an 'answer' being provided alongside for verification. It is also possible that the SMETERS [46] project will take the same approach, although the timeline is unknown.

In summary, we have advanced methods, both tests and monitoring, that can be used to measure HTC, with costs and timescales generally reducing with each innovation. However, now the difficult task lies ahead of standardising and verifying these methods, followed by the even more difficult task of seeking the backing of policymakers and regulators to include the HTC as a value for ABIU dwellings.

References

[1] Future Homes Hub. *Ready for Zero: Evidence to inform the 2025 Future Homes Standard* [online]. 2023. Available from https://irp.cdn-website.com/bdbb2d99/files/uploaded/Ready%20for%20Zero%20-%20Evidence%20to%20inform%20the%202025%20Future%20Homes%20Standard%20-Task%20Group%20Report%20FINAL-%20280223-%20MID%20RES.pdf [Accessed 22 June 2026]

[2] BRE. '*The Government's standard assessment procedure for energy rating of dwellings (SAP 10.2)*'. Watford: BRE; 2017

[3] Marchio D., Rabl A. 'Energy-efficient gas-heated housing in France: Predicted and observed performance'. *Energy and Buildings*. 1991;**17**(2): 131–9

[4] Branco G., Lachal B., Gallinelli P., Weber W. 'Predicted versus observed heat consumption of a low energy multifamily complex in Switzerland based on long-term experimental data'. *Energy and Buildings*. 2004; **36**(6):543–55

[5] Majcen D., Itard L., Visscher H. 'Actual and theoretical gas consumption in Dutch dwellings: What causes the differences?' *Energy Policy*. 2013; 61:460–71

[6] Lowe R. J., Wingfield J., Bell M., Bell J. M. 'Evidence for heat losses via party wall cavities in masonry construction'. *Building Services Engineering Research and Technology*. 2007;**28**(2):161–81

[7] Hens H., Janssens A., Depraetere W., Carmeliet J., Lecompte J. 'Brick cavity walls: A performance analysis based on measurements and simulations'. *Journal of Building Physics*. 2007; 31:95–124

[8] Roels S. *Reliable building energy performance characterisation based on full scale dynamic measurements (Annex 58): Project summary report* [online]. London, UK: AECOM Ltd; 2017. Available from https://www.iea-ebc.org/Data/publications/EBC_PSR_Annex_58.pdf [Accessed 22 June 2026]

[9] Janssens A, *IEA EBC Annex 58. Reliable building energy performance characterization based on full scale dynamic measurements. Report of sub-task 1b: Overview of methods to analyse dynamic data* [online]. IEA; 2016. Available from https://bwk.kuleuven.be/bwf/projects/annex58/data/A58_Final_Report_ST1b.pdf [Accessed 3 March 2021]

[10] Fitton R. *Building energy performance assessment based on in-situ measurements: Challenges and general framework* [online]. IEA EBC Annex 71, KU Leuven, Belgium, ST1 4, 2021. Available from https://iea-ebc.org/Data/publications/EBC_Annex71-ST1-4-Challenges_and_general_framework.pdf [Accessed 15 August 2025]

[11] Reynders G., Saelens D. IEA EBC *Annex 71. Building energy performance assessment based on in-situ measurements. Building behaviour identification* [online]. IEA, Belgium, 2021. Available from https://bwk.kuleuven.be/bwf/projects/annex71/ [Accessed 22 June 2026]

[12] Bauwens G., Ritosa K., Roels S. *IEA EBC Annex 71. Building energy performance assessment based on in-situ measurements. Physical parameter identification* [online]. KU Leuven, Belgium 2021. IEA, Belgium, 2021. Available from https://bwk.kuleuven.be/bwf/projects/annex71/ [Accessed 22 June 2026]

[13] European Commission. *Directive 2010/31/EU of the European Parliament and of the Council in the energy performance of buildings* Brussels: European Union; 2012

[14] BRE. *BRE: Standard Assessment Procedure (SAP 2012)* [online]. 2012. Available from http://www.bre.co.uk/sap2012/page.jsp?id=2759 [Accessed 20 March 2015]

[15] Johnston D., Miles-Shenton D., Farmer D. 'Quantifying the domestic building fabric "performance gap"'. *Building Services Engineering Research and Technology*. 2015; **36**(5):614–27

[16] Miles-Shenton D., Wingfield J., Sutton R., Bell M. *Energy efficient renovation of an existing dwelling: Evaluation of design & construction and measurement of fabric performance': Temple Avenue Project Part 2* [online]. Leeds: Leeds Metropolitan University; 2011. Available from https://eprints.leedsbeckett.ac.uk [Accessed 1 December 2015]

[17] Rouleau J., Gosselin L., Blanchet P. 'Understanding energy consumption in high-performance social housing buildings: A case study from Canada'. *Energy*. 2018;145:677–90

[18] Galvin R. 'Making the "rebound effect" more useful for performance evaluation of thermal retrofits of existing homes: Defining the "energy savings deficit" and the "energy performance gap"'. *Energy and Buildings*. 2014; 69:515–24

[19] Cozza S., Chambers J., Patel M. K. 'Measuring the thermal energy performance gap of labelled residential buildings in Switzerland'. *Energy Policy*. 2020; 137:111085

[20] Ballarini I., Corrado V. 'Application of energy rating methods to the existing building stock: Analysis of some residential buildings in Turin'. *Energy and Buildings*. 2009;**41**(7):790–800

[21] Johnston D., Siddall M., Ottinger O., Peper S., Feist W. 'Are the energy savings of the passive house standard reliable? A review of the as-built thermal and space heating performance of passive house dwellings from 1990 to 2018'. *Energy Efficiency*. 2020;13:1605–31

[22] Mitchell R., Natarajan S. 'UK Passivhaus and the energy performance gap'. *Energy and Buildings*. 2020;224:110240

[23] IPHA. *International Passive House Association | iPHA* [online]. Available from https://passivehouse-international.org/index.php?page_id=65 [Accessed 15 May 2025]

[24] Fitton R. 'Energy pathology: Measuring a dwelling's energy performance using smart meter and IoT data'. *Journal of Building Survey, Appraisal & Valuation*. 2020; **9**(2):181–90

[25] Sonderegger R. C., Madera M. P. '*Electric co-heating: A method for evaluating seasonal heating efficiencies and heat loss rates in dwellings*' [online]. Lawrence Berkeley Laboratory report LBL-8069. Berkeley, CA: Lawrence Berkeley Laboratory; 1979. Available from https://www.osti.gov/biblio/12246320 [Accessed 22 June 2026]

[26] Johnston D., Miles-Shenton D., Wingfield J., Farmer D., Bell M. *Whole House Heat Loss Test Method (Coheating)*. Leeds: CEBE; 2012

[27] Bauwens G., Roels S. 'Co-heating test: A state-of-the-art', *Energy and Buildings*, 2014;82:163–72

[28] BSI. *BS EN 17887-1:2024 Thermal performance of buildings. In situ testing of completed buildings. Data collection for aggregate heat loss test* [online]. Available from https://www.en-standard.eu/22-30437135-dc-bs-en-17887-1-thermal-performance-of-buildings-in-situ-testing-of-completed-buildings-part-1-data-collection-for-aggregate-heat-loss-test/ [Accessed 22 June 2026]

[29] Pandraud G., Fitton R. 'QUB: Validation of a Rapid Energy Diagnosis Method for Buildings', in International Energy Agency Annexe, vol. 58. Paris: International Energy Agency; 2014, pp. 1–6

[30] Alzetto F., Pandraud G., Fitton R., Heusler I., Sinnesbichler H. 'QUB: A fast dynamic method for in-situ measurement of the whole building heat loss'. *Energy and Buildings*. 2018;174:124–33

[31] Meulemans J., Alzetto F., Farmer D., Gorse C. 'QUB/e: A novel transient experimental method for in situ measurements of the thermal performance of building fabrics' in Dastbaz M., Gorse C., Moncaster A. (eds.) *Building Information Modelling, Building Performance, Design and Smart Construction*. Cham: Springer; 2017. pp. 115–26

[32] Thébault S., Bouchié R. 'Refinement of the ISABELE method regarding uncertainty quantification and thermal dynamics modelling'. *Energy and Buildings*. 2018;178:182–205

[33] Deltour J., Heijmans N., De Sloover K., 'Assessing the building envelope performance during occupancy'. 12th Nordic Symposium on Building Physics; Tallinn, Estonia, Sep 2020. Les Ulis: E3S Web of Conferences; 2020; 172:22004

[34] Jack R. *Building diagnostics: Practical measurement of the fabric thermal performance of houses* [online]. PhD thesis, Loughborough University; July 2015. Available from https://repository.lboro.ac.uk/articles/thesis/Building_diagnostics_practical_measurement_of_the_fabric_thermal_performance_of_houses/9453527 [Accessed 22 June 2026]

[35] Masy G., Hinque S. 'Assessment of the energy performance of building facades through different testing conditions and analysis techniques' Presented at SSB 2018 - System simulation in Buildings; Liege, Belgium, 2018. Available from https://www.researchgate.net/publication/331258785 [Accessed 21 April 2020]

[36] Farmer D., Johnston D., Miles-Shenton D. 'Obtaining the heat loss coefficient of a dwelling using its heating system (integrated coheating)'. *Energy and Buildings*. 2016;117:1–10

[37] Erkoreka A., Gorse C., Fletcher M., Martin K. *Reliable building energy performance characterisation based on full scale dynamic measurements Report of Subtask 2: Logic and use of the Decision Tree for optimizing full scale dynamic testing* [online]. IEA; 2016. Available from http://www.kuleuven.be/bwf/projects/annex58/index.html [Accessed 22 July 2016]

[38] Palmer J., Pane G., Bell M., Wingfield J. '*Comparing primary and secondary terms analysis and re-normalisation (PSTAR) test and co-heating test*

results'. *Final Report: BD2702*. London: Department for Communities and Local Government; 2011

[39] Marshall A., Fitton R., Swan W., *et al*. 'Domestic building fabric performance: Closing the gap between the in situ measured and modelled performance'. *Energy and Buildings*. 2017;150:307–17

[40] Butler D., Dengel A. '*Review of Co-heating Test Methodologies*'. NHBC Foundation report NF54. Milton Keynes: IHS BRE Press/NHBC Foundation; 2013

[41] Stamp S. F. *Assessing uncertainty in co-heating tests: Calibrating a whole building steady state heat loss measurement method*. London: University College London; 2016

[42] CEN. *CEN - Technical bodies - CEN/TC 89, CEN Technical Standards* [online]. Available from https://standards.cen.eu/dyn/www/f?p=204:7:0::::FSP_ORG_ID:6072&cs=1838C2B207A6B7D88562A05376928F458 [Accessed 11 February 2020]

[43] Roberts B. M., Allinson D., Diamond S., *et al*. 'Predictions of summertime overheating: Comparison of dynamic thermal models and measurements in synthetically occupied test houses'. *Building Services Engineering Research and Technology*. 2019; **40**(4):512–52

[44] Lomas K. J., Eppel H., Martin C. J., Bloomfield D. P. 'Empirical validation of building energy simulation programs'. *Energy and Buildings*. 1997; **26**(3): 253–75

[45] EFL topic group Energy Efficient housing. '*The Dutch approach for energy efficient renovations*'. Amsterdam: EFL Working Group; 2016

[46] BEIS. *Smart Meter Enabled Thermal Efficiency Ratings (SMETER) innovation programme* [online]. Available from https://www.gov.uk/guidance/smart-meter-enabled-thermal-efficiency-ratings-smeter-innovation-programme [Accessed 9 June 2026]

[47] SEREINE, *Sereine, un projet issu du programme PROFEEL* [online]. Available from https://programmeprofeel.fr/projets/sereine/ [Accessed 24 March 2021]

[48] Deb C., Gelder L. V., Spiekman M., Pandraud G., Jack R., Fitton R. 'Measuring the heat transfer coefficient (HTC) in buildings: A stakeholder's survey'. *Renewable and Sustainable Energy Reviews*. 2021; 144:111008

[49] Madsen H., Bacher P., Bauwens G., *et al*., *Thermal performance characterization using time series data - IEA EBC Annex 58 guidelines* [online]. IEA; 2015. Available from https://orbit.dtu.dk/en/publications/thermal-performance-characterization-using-time-series-data-iea-e [Accessed 3 March 2021]

[50] ISO. *BSI Standards Publication Thermal insulation — Building elements — In- situ measurement of thermal resistance and thermal transmittance Part 2: Infrared method for frame structure dwelling BS ISO 9869-2:2018*. London: BSI; 2018

[51] Danov S., Carbonell J., Cipriano J., Martí-Herrero J. 'Approaches to evaluate building energy performance from daily consumption data considering dynamic and solar gain effects'. *Energy and Buildings*. 2013;57:110–8
[52] Kersken M., Strachan P. *IEA EBC Annex 71. Building energy performance assessment based on in-situ measurements. Design, description and results of the validation of building energy performance simulation programs* [online]. Belgium: IEA; 2021. Available from https://bwk.kuleuven.be/bwf/projects/annex71 [Accessed 22 June 2026]
[53] Hardy A., Glew D. 'An analysis of errors in the Energy Performance certificate database'. *Energy Policy*. 2019;129:1168–78
[54] Crawley J., P. Biddulph, P., Northrop, P. J., Wingfield, J., Oreszczyn T., Elwell C., 'Quantifying the measurement error on England and Wales EPC ratings', *Energies*. 2019; **12**(18):3523
[55] Winther T., Gurigard K. 'Energy performance contracting (EPC): A suitable mechanism for achieving energy savings in housing cooperatives? Results from a Norwegian pilot project'. *Energy Efficiency*. 2017; **10** (3):577–96
[56] Fitton R. 'Energy monitoring in retrofit projects' in Swan W., Brown P. (eds.) *Retrofitting the Built Environment*. Chichester: Wiley; 2013, pp. 141–53
[57] Hulme J., Doran S. '*In-situ measurements of wall U-values in English housing*'. Watford: BRE; 2014
[58] Kavvadias K., Jiménez-Navarro J. P., Thomassen G., *Decarbonising the EU Heating Sector: Integration of the Power and Heating Sector*. EUR 29772 EN. Luxembourg: Publications Office of the European Union; 2019
[59] Domanski P. A., Henderson H. I., Payne W. V. '*Sensitivity analysis of installation faults on heat pump performance*'. Gaitherburg: NIST; 2014

Chapter 10

Building performance measurement: a framework for building consumer trust and market confidence

Luke Smith[1], Richard Jack[1] and John Littlewood[2]

10.1 Introduction: why measurement and trust matter

The global agenda for retrofitting existing buildings is enormous [1], driven by the urgent need to tackle the climate change emergency, energy security, and the cost-of-living crisis [2]. In 2021, buildings accounted for more than 30% of global energy use and 40% of energy-related CO_2 emissions [3], making accelerated investment in decarbonising the existing stock essential to meet national and international NZC goals by 2050. Yet despite clear environmental, economic, and social benefits, average annual retrofit rates remain low in most major markets, far short of what is needed [4].

One key reason is trust. Consumers, funders, and policymakers struggle to reconcile the *promise* of retrofit performance with what is *delivered* in practice [5]. The conventional retrofit journey often begins due to inadequate internal environmental conditions that negatively impact occupants, such as cold and damp, high energy bills leading to fuel poverty [6], or grant-driven opportunities for landlords [7]. These triggers typically lead to a visual inspection-based survey of buildings to be retrofitted, some professional judgement about the construction of unseen elements of the building, and the production and oversight of an Energy Performance Certificate (EPC), a simplified energy audit report that often fails to capture the as-built uniqueness of each property [8]. Such modelling and predicting of building performance rarely reflects reality and is widely referred to as the performance gap [5,9].

What is less commonly discussed, however, is the *trust gap* that is created alongside it. Householders, also known as residents or occupants, and investors

[1]Build Test Solutions, Lutterworth, UK
[2]Sustainable and Resilient Buildings, Cardiff Metropolitan University, UK

find it difficult to fully support investment decisions that are based on models and assessments that do not resonate with their lived experience [10]. Without robust insights into the true performance of the existing building stock, industry, policymakers, and investors are hampered in their ability to make informed investment decisions. Inspiring confidence and building trust with all stakeholders is difficult, not only at the consumer level, but throughout the entire retrofit supply chain.

Building Performance Evaluation (BPE) provides the evidence base to address this challenge. BPE can be defined as the systematic gathering and interpretation of quantitative and qualitative data to draw conclusions about how a building is actually performing, as defined by BS40101:2022 [11]. From a practitioner's perspective, BPE is essential: *what you don't measure, you can't manage*. By embedding BPE in retrofit processes, practitioners can:

- **Build trust** – giving confidence to funders, clients and occupants [12].
- **Better inform investment decisions** – targeting limited resources for maximum impact [10].
- **Close feedback loops** – learning from projects to improve processes [13].
- **Validate outcomes** – proving delivered performance matches investment claims [9].

Throughout this chapter, we distinguish between measurement, point-in-time testing, often carried out pre- or post-works, and monitoring, long-term data collection, often using sensors and smart meters. Both have value, but this discussion focuses principally on the application of *practical measurement techniques* that can be integrated into projects alongside existing building surveying practices.

The UK's commitment to achieving net zero by 2050 has often placed housing and building performance at the centre of national strategy [2]. The *Heat and Buildings Strategy* [2] set out ambitions for decarbonisation but recognises the limitations of EPC-driven decision-making. Meanwhile, PAS2035, the retrofit standard introduced in 2019, requires a 'whole-house, fabric-first' approach, underpinned by risk assessment and quality assurance [14]. Yet compliance with PAS2035 is still largely model-driven, dependent on assumptions that frequently fail to capture reality [15]. In Europe, similar challenges exist. The Energy Performance of Buildings Directive (EPBD) mandates EPCs, but research consistently shows limited correlation between EPC ratings and actual energy use [16]. Internationally, initiatives such as *EnerPHit* (Germany) [13] and *Energiesprong* (Netherlands) embed performance monitoring more directly into retrofit delivery, providing models for the UK to draw from [12].

Against this backdrop, BPE offers, not just a technical tool, but also a policy instrument: it allows performance-based standards, finance, and assurance to be built on evidence rather than projections. The strategic themes that measurement underpins may be summarised as follows:

1. **Quality and assurance** – Without independent measurement, retrofit quality is difficult to assure. BPE offers objective sign-off.
2. **Trust and communication** – Objective assessment of the unique performance of a given property is much more convincing than a basic, generic assessment. Outputs from certain measurements can also be visually engaging, such as thermography or air quality dashboards. All of this helps to build trust by making performance tangible to non-experts.
3. **Finance and investment** – Measured baseline performance helps to better direct and target investment, while measured outcomes enable new financial models, such as green mortgages and pay-for-performance contracts [17].
4. **Innovation and market growth** – Embedding BPE unlocks new procurement and finance models, engages a wider audience, and supports ongoing supply chain development by providing data and feedback on performance, what works well and not so well, to product manufacturers, system designers, and those ultimately involved in specifying and delivering retrofit programmes.
5. **Policy and accountability** – Policymakers can use BPE data to design outcome-based regulation and track progress against targets [18].

BPE unquestionably provides an opportunity to address some of the central problems in retrofit delivery. However, without a structured approach, we run the risk of generating data that will not deliver value to stakeholders. Here, we outline a proposed approach, providing structure for BPE projects to ensure effective gathering and use of data.

10.2 A practitioner framework for building performance evaluation

To move from principle to practice, practitioners require a clear framework. Here we outline a five-step framework, adapted from the Build Test Solutions *Thermal Performance Measurement Handbook* [19], that structures BPE as an iterative, cyclical process, rather than a one-off test.

1. **Survey** – Establish baseline assumptions and identify areas of risk. Surveying the building or, for a new building, using its design parameters, allows us to first determine the assumed performance of each element, as well as the building as a whole.
2. **Measure** – Apply testing techniques to quantify actual performance. Carrying out performance measurements allows us to ultimately assess whether the building is performing as predicted.
3. **Assess and validate** – Compare measured results with predictions or benchmarks. By comparing the predicted and actual performance, it is possible to identify if there are areas that do not align.

4. **Diagnose** – Investigate discrepancies and identify root causes. If the aspects of the building's performance do not match expectations, we might carry out further diagnostic testing to find out why.
5. **Act** – Apply insights to improve design, delivery, and assurance. Using the measurements for quality control, identifying remedial works, and measuring their effect, informing updated retrofit designs, and feeding back to improve build processes, or the assessment of specific installed building materials or products.

In practice, performance evaluation is often fragmented. A contractor may undertake an airtightness test, or a researcher may install heat flux plates, but these tests are rarely integrated into a wider cycle of learning. The framework presented here is designed to:

- Provide **structure**: linking surveys, testing, and corrective actions.
- Enable **consistency**: allowing results to be compared across projects.
- Support **decision-making**: showing where measurement adds value.
- Embed **learning loops**: ensuring feedback informs future work.

The framework is designed to bring structure to a disparate group of methods and allow them to be applied in a consistent way, with a clear understanding of objectives and outcomes.

The framework complements, but differs from, existing approaches such as Post-Occupancy Evaluation (POE) [20] and BSRIA's Soft Landings [21]. While POE is often undertaken after project completion, the five-step framework is designed for *integration throughout the delivery process*, from initial survey to post-installation validation. This makes it more applicable to practitioners who need evidence at multiple stages, not only to evaluate, but also to actively inform and steer projects.

More closely matched is the framework provided by BS40101:2022 [22]. This standard provides useful definitions, lists many building performance evaluation techniques, and sets out a tiered approach comprising 'preliminary evaluation', 'light BPE', 'standard BPE' and 'investigative BPE'. It is a helpful standard, but adoption has been limited due to its highly prescriptive nature, making it difficult to adhere to in full. As is the case with many formal standards, it is also difficult to reflect innovation and the evolution of industry best practices.

The framework outlined herein is intentionally pragmatic. It does not prescribe a fixed set of methods but aligns techniques to each stage of the process. For example, *Survey* might involve an EPC, Full Standard Assessment Procedure (SAP), or Passivhaus Planning Package (PHPP) modelling. *Measure* could involve SmartHTC whole-building heat loss measurement [23] or Pulse low-pressure airtightness testing [24]. *Diagnose* may draw on U-value measurements, thermography, or use of building pressurisation equipment alongside tracer smoke. This flexibility allows practitioners to adopt appropriate methods for different contexts, budgets, and risks while maintaining a coherent process.

10.3 Application of the BPE framework

The energy consumption in any building is driven by four key parameters, the thermal performance of the building itself, the efficiency of the heating and cooling systems, the actions of the occupants, and the local weather conditions. In this section, we describe a 5-step process that provides a framework for how measurement and monitoring can be embedded into practice. We intentionally focus on principally understanding the thermal performance of a building because, while the building services, occupants, and external environment regularly change, the underlying thermal performance of the physical asset does not. Ensuring good thermal performance is key to ensuring buildings are truly low-energy and efficient throughout their operational lifespan.

It is important that measurements are not carried out in isolation or taken out of context. They must instead be used to enhance the richness, credibility, and accuracy of our existing surveys, models, and investment decisions. We start with what is already widely practised; a **survey** of the building in order to establish what it is that we think we are dealing with. From here, we overlay some initial performance **measurements** to verify how well our survey observations reflect reality. Step 3 is to **assess and validate**, identifying if there are performance aspects that do not align. We subsequently **diagnose** the root cause of any misalignment between expectation and reality with further diagnostic testing methods as applicable. Step 5 is to ensure the measurements drive **action**. This could be to feed into a quality control process, trigger remedial action, inform an updated retrofit design or specification, or serve as formal feedback to improve build processes, materials, or products.

10.3.1 Step 1: survey

In order to consider measurements in context, so that it is possible to assess whether results are good or bad, the first step is to survey and model the building. This is a well-established process applicable to both new builds and retrofits, using techniques and tools such as SAP and Reduced Data Standard Assessment Procedure (RdSAP) [25], Energy Performance Certificate assessment procedures, the PHPP [26], or for non-residential buildings, the Simplified Building Energy Model (SBEM) [27], or other dynamic thermal simulation modelling tools. As surveying is so commonplace and typically required for compliance purposes, it is very possible that this stage has already been carried out, so that no new work is required.

In order to compare with measurements, the following building performance metrics must be included in the modelling. These may be based on design intent and associated manufacturer-supplied specifications or may be inferred based on the type and age of the building:

- Elemental performance, with U-values for key elements such as floors, walls, ceilings and windows.
- Airtightness.

- Ventilation provision.
- Whole-building heat loss rate, or Heat Transfer Coefficient (HTC).

All of these metrics are essential inputs and outputs to all modelling packages, including, but not limited to, SAP, RdSAP, PHPP, SBEM, and Dynamic Simulation tools, such as Integrated Environmental Solutions (IES) or EnergyPlus. They are the parameters that ultimately define the fabric performance of a building, which is used to determine space heating and cooling demand, the largest energy load in the majority of buildings.

10.3.2 Step 2: measure

The central challenge in BPE is obtaining reliable, cost-effective, and actionable data about how a building actually performs in operation. Methods have evolved from highly technical research tools into practical techniques that can be deployed at scale by contractors, surveyors, and consultants.

During the Measure stage, three key measurements are recommended to allow an overall assessment of whether the building is operating as designed or assumed by an as-built assessment. These are measurements of the overall thermal performance, the airtightness, and the ventilation provision.

10.3.2.1 Whole-building thermal performance

The Heat Transfer Coefficient (HTC) is a fundamental measure of a building's thermal performance [28]. HTC quantifies the total rate of heat transfer (W/K), combining both fabric transmission and ventilation losses. HTC is a great barometer of the overall fabric efficiency of a building, and the single best indicator of how much energy is needed to maintain indoor comfort. In temperate climates, it is more common to use HTC to quantify heat loss during winter and, hence, space heating demand. It can equally be used to assess the rate of heat gain when considering space cooling demand [29]. The HTC of a dwelling is directly correlated with its space heating and cooling requirements [30]. One of the main uses for HTC measurements is to assess whether the as-designed, or as-calculated, thermal performance of the building is correct. HTC measurement is ideal for this use, as the measured HTC can be compared with design predictions from energy models such as SAP, RdSAP, and PHPP.

The HTC of a building is measured using two different methods. *In-use measurements* are carried out while a building is occupied and used as normal. As such, they include all sources of heat transfer: through the fabric of the building and by infiltration and by ventilation, including window opening. They have the advantage of being less disruptive and can be less expensive, as they generally use less measurement equipment. However, great care is required to accurately account for the effects of occupants and the period of the year during which measurements are possible, which is generally shorter than tests in vacant buildings. These tests are sometimes referred to as 'SMETERs' after the UK Government's Smart Meter Enabled

Thermal Efficiency Ratings technologies project, which demonstrated that in-use measurements could be carried out accurately and reliably [23]. *Tests in vacant buildings* are carried out while access to the building is strictly limited. Many of these methods deliberately remove some aspects of ventilation during the measurement by turning off mechanical systems and temporarily closing and/or sealing passive vents. As the buildings are vacant, heat transfer due to window opening is not included. This is done to create a more controlled environment during the test. As a result, these sources of heat transfer must be accounted for when comparing the rest of an HTC *test* with a calculated HTC. Tests in vacant buildings can have shorter test durations than those in occupied buildings, as the conditions can be more closely controlled. Tests can also be applied for a longer period of the year, as it is possible to heat the building to elevated levels without disrupting residents.

Both methods are based on measuring a rate of heat transfer, and to do so accurately requires there to be a significant amount of heat transfer during the measurement. To this point, all HTC measurement methods currently work by measuring the rate of heat input to a building; this is because it is easier to measure accurately than the rate of cooling input. Practically, testing is typically limited to periods of cooler weather outside, e.g., in the northern hemisphere, the heating season of October-March, inclusive.

In-use measurements can be undertaken while a building is occupied and operational, although these measurements can also be used in vacant buildings. They generally involve measuring the heat input to a building over a period of time and then inferring the rate of heat loss per degree of temperature difference between inside and outside. To provide an accurate HTC measurement, all significant heat inputs to the building must be accounted for, including the primary space heating, any secondary heating, solar gains, and internal gains from electricity consumption and other uses.

Most methods use internal temperature data collected within a building, though some can provide less precise HTC measurements without internal temperature. Data can be collected from energy service meters, plus either internal sensors temporarily installed specifically for the HTC measurement or data from installed systems like smart thermostats. The measurement equipment required is limited and inexpensive.

There are commercially available HTC measurement methods, for example, SmartHTC offered by Build Test Solutions and Knauf Energy Solutions, and in-use HTC measurements have now been carried out on thousands of houses [23].

Taking an example to illustrate the application of such a method, *SmartHTC* estimates HTC using temperature and energy-use data from a property over a period of 21 days. It requires only smart meter data and a few indoor/outdoor sensors, making it minimally intrusive. Its strength lies in scalability. As it can be applied without major disruption, SmartHTC has been trialled across hundreds of homes (Figure 10.1) [31]. A limitation is that testing requires a positive internal-external temperature difference, which limits use to heated buildings during colder weather.

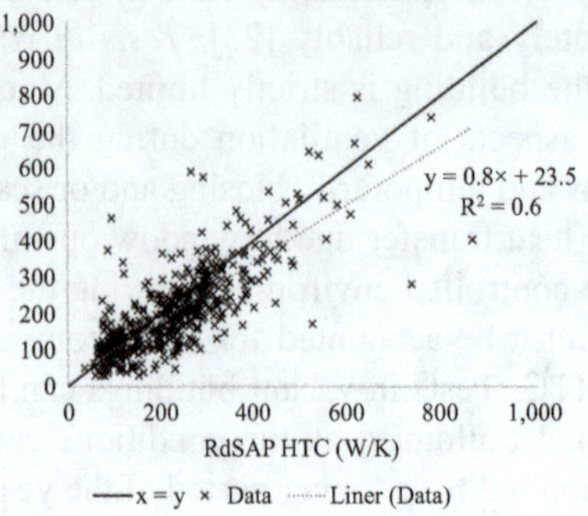

Comparison between rdSAP predicted and SmartHTC measured heat loss rate (HTC) across 503 homes, showing a weak correlation, but with many significant outliers and a tendency to overestimate heat loss.

SmartHTC allows for the measurement of whole building heat loss using simple low cost temperature sensors and smart meter energy data.

Figure 10.1 Smart HTC data and equipment for measurement in use

Tests in vacant buildings were the first HTC measurement methods to be developed and have provided most of the evidence for the performance gap in building thermal performance [9,32]. The 'co-heating test' was the first HTC measurement method to be used at any scale. It was developed at Leeds Beckett University [33], and the method has since been standardised and renamed Aggregate Heat Loss testing. An early example of the insights that HTC measurement can provide is the identification of party wall thermal bypass as a significant additional source of internal-external heat transfer, which had not previously been characterised [34].

Other tests are applied under dynamic conditions and can be carried out in a shorter period of a few days. This includes the PSTAR method developed in the USA in the 1980s [35], Saint-Gobain's QUB (Quick U-value of Buildings) developed in the 2010s and now in commercial application [36], and the Sereine test developed in the 2010s by a French consortium [37].

For example, the *QUB* (Quick U-value of Buildings) method works by applying short, controlled heating pulses over 12–24 hours. These methods offer faster results than SmartHTC and other SMETER-style techniques, which makes them more attractive for use cases where a quicker result is required or the building is not fully occupied. They are, however, more disruptive, requiring the building to be temporarily vacated, and are generally more expensive.

HTC measurements in vacant and occupied buildings are different routes to the same destination, both capable of arriving at an accurate measurement of the actual thermal performance of a building. The two measurement types are best suited for different applications. If a result is required quickly, then a test in a vacant building is likely to be more suitable, but it comes with additional cost and disruption. For these reasons, testing in vacant buildings may be more attractive for new build testing. Where a measurement is not required as quickly, or vacating the building is difficult, then an in-use measurement is likely to be more suitable, making it well-suited for existing buildings and retrofit projects.

All whole-building heat loss measurement methods provide an overall check on performance but are not as strong at identifying the causes of performance gaps. For example, if the predicted HTC is 150 W/K but measured HTC is 250 W/K, it is clear that something significant is wrong, but it may not be clear what the problem is. Using a medical analogy, HTC measurements are therefore a great tool for triaging buildings into those with high, medium, or low performance but are not a diagnostic tool capable of identifying the cause of the problem.

10.3.2.2 Airtightness testing

Airtightness has long been recognised as a major factor in thermal efficiency and indoor comfort [38]. Excessive air leakage leads to uncontrolled heat loss, while over-sealing without ventilation risks poor indoor air quality [39]. The blower door test is a steady-state test method used to depressurise and/or pressurise a building to measure building air leakage [40] with results typically reported at a pressure difference of 50 Pascals (Pa) between the inside and outside of the building. The test is carried out by temporarily installing a fan in an opening and measuring the fan flow rate and the resulting building pressure. Blower door tests are widely used in both compliance regimes, such as Approved Document L testing in new build construction, and retrofit quality assurance [5]. An additional benefit of the test is that the elevated pressure difference can be used to find air leakage paths, either by simply feeling the air flow, by using neutrally buoyant smoke to visually trace leakage paths, or thermography to show where external air penetration changes the surface temperature of building elements. This type of leakage diagnostics is extremely valuable, but largely qualitative. Disadvantages of blower door tests include disruption, reliance on a non-representative pressure differential, and the need for trained technicians [41]. It can also be practically difficult to mount the fan in an airtight manner, and the opening that is used to house the fan is not included in the measurement.

Alternatively, the low-pressure Pulse method uses compressed air as a means of measuring the air leakage rate of a building or space within it [24]. By releasing a short pulse of air and measuring the pressure response, it dynamically measures airtightness directly at 4 Pa, much closer to the pressures that buildings experience under normal day-to-day conditions [42]. It is non-intrusive, can be carried out in occupied homes, and does not require any penetrations across the envelope, thus testing the full enclosure. The Pulse technique is described in full in CIBSE TM23:2022, meaning that it is approved for use for new build compliance (Part L of the Building Regulations), as well as for retrofits (PAS2035) in the UK.

The Pulse method can also be used to measure the airtightness of spaces where it is practically difficult to mount a blower door fan. Examples include cleanroom facilities, data centres, and vehicles. Under the Insulation Assurance Authority's (IAA) Background Ventilation Testing Scheme, Pulse has also been used to measure the airtightness of individual rooms to test the distribution of infiltration around a building. These results are not simple to interpret, as a significant amount of the air movement in any internal room will be to an adjacent space rather than to outside, but they provide a useful quantitative guide to whether an individual room is particularly airtight, which can be useful when classifying risk of under-ventilation [43]. Pulse tests take a few seconds and do not exert a constant elevated pressure difference across a building. The test is therefore not useful for qualitative leakage diagnostics in the same way that blower door testing is.

10.3.2.3 Ventilation and Indoor Air Quality (IAQ)

Retrofit often improves airtightness, which heightens the importance of verifying that ventilation systems perform as intended. In the UK, Approved Document F of the Building Regulations (England and Wales) mandates the testing and commissioning of all installed mechanical ventilation systems [44]. Here we outline some of the main methods.

Flow hoods can be placed over supply and extract grilles to measure volumetric airflow, providing direct verification of whether designed ventilation rates are being achieved. Ventilation types for testing could include simple through-the-wall fans (dMEV), through to centralised ducted systems, such as mechanical extract ventilation (MEV) and mechanical ventilation heat recovery (MVHR). The available test methods are generally classified across three categories.

Method 1, the unconditional method, uses a power flow hood that eliminates back pressure and places no additional restrictions on the fans being tested. This method is the most accurate and reliable, especially when measuring lower supply and extract rates or wishing to carefully balance a ducted ventilation system with numerous supply and extract points.

Method 2, the conditional method, covers a wide variety of testing equipment to be used, including rotating vane anemometers, air capture hoods, thermal anemometers, thermal capture hoods, differential pressure anemometers, and pressure pans. These offer useful spot measurements that are quick and simple to carry out, but it is important that the readings are adjusted using a conversion factor specific to the fan make and model and the type of equipment being used. In the UK, the conversion factors must be provided by the fan manufacturer and are based on third-party testing by a UKAS-accredited body.

Method 3, the minimum benchmark method, uses the same hardware as method 2 but, in place of conversion factors, simply relies on a minimum benchmark level to be met, which takes into account the uncertainty associated with the equipment. If the fan performance exceeds the minimum benchmark value indicated, it would be reasonable for an inspector to assume that the fan is performing to the required level and accept the test results as showing compliance with Part F of the Building Regulations.

For more detailed studies, *tracer gas decay* methods, such as CO_2 release and monitoring, can also be used as a means of estimating air change rates. While accurate, these methods are less practical for everyday retrofit projects due to time and equipment needs. The results can be highly variable from day-to-day, which reflects a real variation in infiltration driven by varying internal and external conditions, particularly temperatures, wind speed and direction, and sheltering. This variation can be very insightful, but the results can also be more difficult to interpret and use when adhering to general ventilation guidance, which tends to cite a single air permeability value.

10.3.2.4 Performance monitoring technologies

In addition to building performance measurements, which tend to be discrete in time, there are also many types of performance monitoring that can be carried out for periods of differing length, or even continuously. When working with this type of data, there is a risk of collecting large quantities of data points that are difficult to interpret and act upon. It is crucial to monitor with purpose and to design and use clear, actionable metrics.

With the smart meter rollout now well advanced across the UK, *gas and electricity consumption data* can be readily accessed at half-hourly intervals for most buildings. Treated carefully, this data can be useful to understand building performance [45], though great care must be taken to understand the effects of weather and occupants. Linked with indoor/outdoor temperature sensors, this enables in-use HTC measurement methods such as SmartHTC [23].

Low-cost Internet of Things (IoT) sensors provide continuous data on temperature, humidity, CO_2, and other airborne pollutants. These support longitudinal understanding of comfort and air quality, as well as identifying risks such as mould, condensation, or overheating risk or occurrences [46]. Over time, this type of monitoring can be useful to understand in-use ventilation performance [47]. The main challenges are data quality, connectivity, and long-term maintenance, along with the need to balance data richness with simplicity of deployment.

10.3.2.5 Digital twins

Digital twins allow us to integrate both measurement and monitoring data into virtual building models. These allow predictive simulations, scenario testing, and real-time optimisation. While currently used mainly in commercial or new-build contexts, the potential for retrofitted housing stock also exists.

This is increasingly being enabled through advances in building scanning technology. These specialised cameras combine high-resolution photography with depth-sensing technology, such as structured light or Light Detection and Ranging (LiDAR) (Figure 10.2). This allows the devices to capture not only images but also

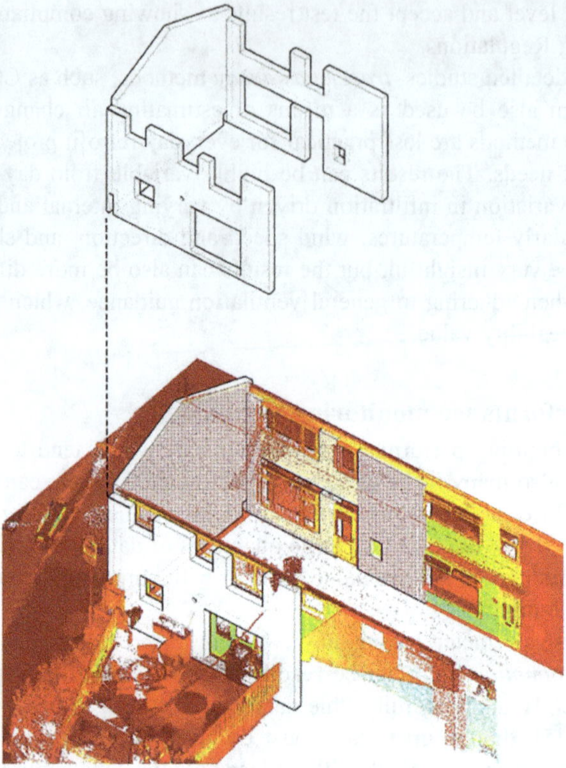

Figure 10.2 3D survey data from Lidar scan of an existing dwelling to be retrofitted

precise spatial data, including measurements and the layout of a space. This can be a significant improvement to current building survey and modelling practices, as well as providing a digital representation of a building where supporting measurements can be stored for later use.

10.3.2.6 Other considerations for practitioners

Even at this earliest measurement stage, practitioners must balance accuracy, cost, and disruption. In occupied housing, methods that require decanting may be impractical, while long-term monitoring may be deemed too invasive or resource-intensive. Moreover, measurement should be aligned with decision-making needs. For strategic planning, the use of low-cost, non-invasive, scalable techniques that help to build householder and/or investor trust, while simultaneously helping to guide the investment strategy. Quality assurance and detailed retrofit design require methods that support the understanding of the pre- and post-performance of retrofits that are part of the works programme. Measurements can also inform the detailed design of the retrofit, while also allowing for checking and validating post-installation outcomes. Finally, for contractual purposes or dispute resolution, it is important to have techniques that can be mobilised by energy assessors, surveyors, and consultants to help reveal and rectify defects.

10.3.3 Step 3: assess and validate

Gathering measurement data alone is rarely enough. It must be assessed, validated, and reviewed in context. Once both the predicted and measured headline performance are known, it is possible to compare them and check that the building is operating as intended.

- Compare the measured and predicted HTCs to see if the overall thermal performance is as expected or if there is a 'performance gap', indicating a higher heat loss rate than expected. When performing this comparison, consider the confidence interval of the measurement, which is typically around $\pm 15\%$; if the performance gap is less than the confidence interval, then the building can be viewed as performing as expected.
- Compare the predicted and measured airtightness and ventilation levels to ensure that the building has an adequate supply of fresh air in accordance with Part L and Part F of the Building Regulations and Guidance in PAS2035:2023.
- Break the total heat transfer down into fabric, airtightness, and ventilation components and compare the measured and predicted performance for each. It is possible that the overall thermal performance could match, but for the wrong reasons. For instance, if the building is more airtight than it is designed to be, there would be less heat loss but at the cost of insufficient ventilation. An example of this type of comparison is shown in Figure 10.3 below. In this case, the measured HTC is lower than the prediction, but this is due to insufficient ventilation compared to the design values.

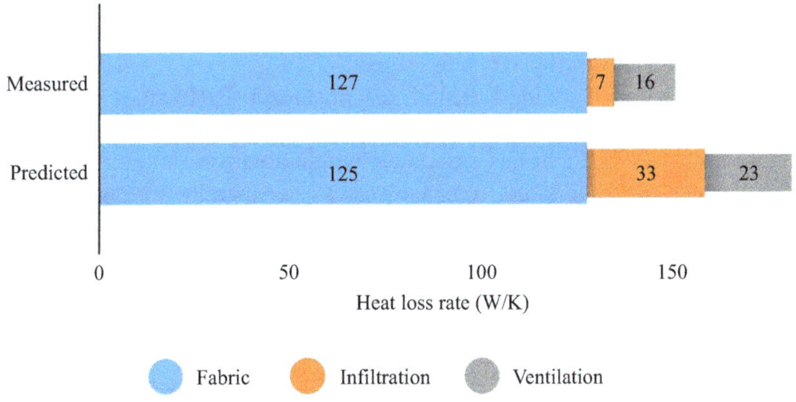

Figure 10.3 A breakdown of predicted total heat loss versus measured total heat loss, compiled through the on-site measurement of HTC and airtightness

Practitioners and policymakers alike need confidence that results are robust, interpretable, and comparable.

10.3.3.1 Statistical confidence and reproducibility

When measuring or calculating heat loss from buildings, it is important to recognise that all such assessments come with a degree of *uncertainty*. Uncertainty is a statistical concept known as the confidence interval. Understanding confidence intervals can help improve the design and efficiency of heating systems by providing clarity on the reliability of heat loss measurements and calculations. A confidence interval defines a boundary within which we can be reasonably certain the true value lies. In most cases, a 95% confidence interval is used, meaning that we can be 95% confident that the correct answer is within this range. This statistical approach acknowledges the inherent variability and limitations in data collection and analysis. Within the given confidence interval, the reported result is the most likely correct answer. The size of a measurement confidence interval is influenced by both the equipment used, such as temperature or pressure sensors, and contextual factors, like the internal and external conditions during the measurement.

Repeatability is the ability to obtain consistent results when measuring the same building under similar conditions and this tends to be much better than the given confidence interval. This is because several sources of uncertainty remain constant within a single building. As a result, even if the confidence interval is broad, it reflects broader uncertainties rather than inconsistencies in measurement.

Despite the importance of understanding uncertainty, it is almost unheard of for confidence intervals to be calculated and reported for heat loss calculations. However, that uncertainty does exist. For heat loss calculations, the uncertainty is introduced as the calculations are informed by a visual survey and assumed or calculated values for key inputs. Metrics like airtightness, whether there is insulation in a cavity or if it is well installed, or how well the insulation under the floor and in the walls joins up, are virtually impossible to determine by a visual survey. However, measurements show that even within a particular construction type, in situ performance can vary by 100% or more.

For example, in a Building Research Establishment (BRE) study [48] of more than 300 solid walls, the average U-value was 1.6 W/m2K, but the range of results was more than 2 W/m2K. In this case, even within a narrow definition of a particular wall type, the range in performance was greater than the average. In simple terms, this means that the performance of seemingly similar walls, floors, and roofs can vary widely, such that it is difficult to accurately represent them with a single figure for an archetype. In other words, there is significant uncertainty in the input, which introduces uncertainty into the outputs.

When you use an assumption based on an archetype as an input to a heat loss calculation, such as RdSAP or BS12831, you introduce that possible error, or uncertainty, in that input to the result. The uncertainty in the reliability of assessing input values applies to the U-values, airtightness, ventilation, and thermal bridging, which are not measured in situ. It would be possible to sum these uncertainties according to a standard method [49] and calculate and report the total uncertainty for a calculation. If this method were followed, then it would demonstrate that the confidence interval of heat loss calculations is very high, with the amount varying for each calculation depending on the details of the building and how detailed the knowledge of the inputs is. To indicate the size of the uncertainty in calculation results, errors of 100% or more have been found regularly [9,32]. Understanding this context is critical to understanding the value of a measurement with a given confidence interval.

Reproducibility is equally critical. If one operator measures a wall U-value with heat flux plates [50] and another with Heat3D, a quantitative thermography approach used to measure U-values [51], are the results comparable? Calibration, protocols and accreditation schemes are needed to ensure that results can be trusted across projects and providers.

10.3.3.2 Integration with design tools and benchmarking

With the original building performance calculation from Step 1 and measurements from Step 2 in hand, we can not only make direct comparisons but also feed the measurement values directly back into design tools such as SAP, PHPP, or dynamic simulation models. This essentially offers recalibration of the model, allowing for improved scenario modelling or appraisal of improvement options onwards. In most calculation tools, it is possible to directly override values, such as airtightness, ventilation rates, U-values, and even the heat transfer coefficient. This subsequently impacts the remainder of the calculation, improving overall confidence in the updated outputs. In situ measurement is not as commonplace today as we argue it should be, there should be greater emphasis on the highly predictive and uncertain nature of building performance calculations and that the integration of measurements serves to greatly enhance understanding of the building(s) in question. In a similar vein, it is seldom the case that building performance measurements would stand alone without onwards integration with an energy calculation.

What can also be helpful are benchmarks. The Chartered Institute of Building Services Engineers (CIBSE) has generally led the way in this regard, with documents such as TM46 [52] and its Guide A: Environmental Design handbook [53]. As approaches to in situ measurement develop and become more commonplace, there will be good scope to further update such benchmarks, particularly in relation to HTC, building airtightness, ventilation rates, and elemental U-values. This will help both practitioners and policymakers to better contextualise BPE results. Such updated benchmarks may, in turn, feed back into the reference tables used by models, creating an iterative feedback loop whereby measurements in the field support continual improvement of tools used for building design and policy.

10.3.4 Step 4: diagnose

If the building's performance is not as expected, the next step is to diagnose the problem and identify root causes. If the building is performing as expected, or better, this step could be skipped. It is sensible to follow a staged approach for diagnostics, starting with the simplest and most inexpensive tests on the most likely failure points, building up in complexity and expense until the cause of a discrepancy between predicted and actual performance is found. It is this pragmatism that is currently missing on many BPE projects, with practitioners more typically opting 'to go all out', conducting an exhaustive set of measurements and tests on one-off buildings rather than triaging and only introducing additional tests and checks on certain buildings where it is deemed necessary. The information from the assessment step should always be used to inform the diagnosis and guide whether the unexpected performance is linked to air movement or fabric heat loss. There are a diverse range of tools available for the purpose of diagnostics, many of which are discussed in this section.

10.3.4.1 Elemental methods

While whole-building HTC tests provide a big-picture view from a thermal performance perspective, elemental methods help pinpoint weaknesses in specific components such as walls, windows, and roofs.

Elemental U-values help us understand the thermal performance of specific elements. A common method is to use *heat flux plates* attached to surfaces over several days or weeks, which can be combined with temperature sensors to derive U-values. They provide robust, standardised measurements but require careful installation and extended monitoring periods [50]. Their use is, therefore, more common in research or forensic evaluation than in routine practice.

A more recent innovation, *Heat3D*, uses thermal imaging combined with calibrated heating to rapidly derive U-values for walls and floors. Measurements can be obtained in approximately one hour, with visual outputs that are easy for clients to interpret [54]. While accuracy depends on conditions, Heat3D is promising as a scalable, practitioner-friendly tool and the approach conforms to ISO 9869-2:2018 [51]; see Figure 10.4.

Thermography is widely recognised for its ability to reveal insulation defects, thermal bridging, and air leakage. Properly executed, under a sufficient temperature differential and with trained operators, it provides powerful visual evidence of a building's fabric heat loss, as shown in Figure 10.5. Limitations include sensitivity to environmental conditions and the risk of misinterpretation by untrained users. Nonetheless, as a communication tool for engaging clients and funders, thermography can be very effective.

In addition to measuring the overall air leakage rate from a building, from a diagnostics perspective, it may also be desirable to track, identify, and address *air leakage pathways*. The combination of a leak checker, a blower door fan providing depressurisation with tracer smoke, [57] or infrared thermography [58] all serve as effective methods for identifying leakage paths. Although only qualitative and indicative, these methods are visually engaging: a photograph of cold air or smoke infiltrating through a poorly sealed loft hatch often carries more weight with

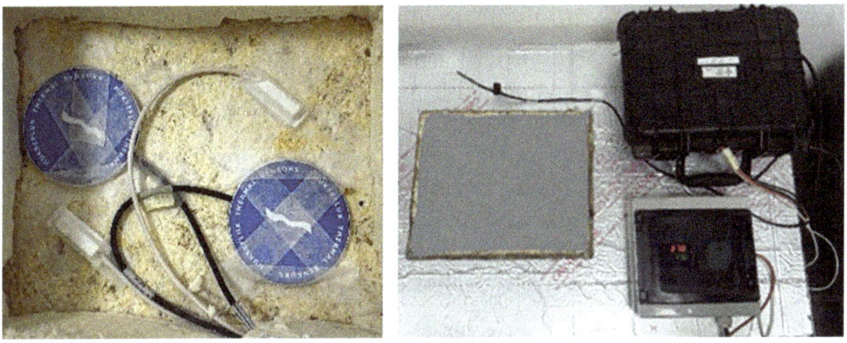

Figure 10.4 Two heat flux and two surface temperature sensors installed on the top of the material sample (left). Build Test Solutions datalogger and thermostat controlling internal temperature installed on top of the hotbox. [55]

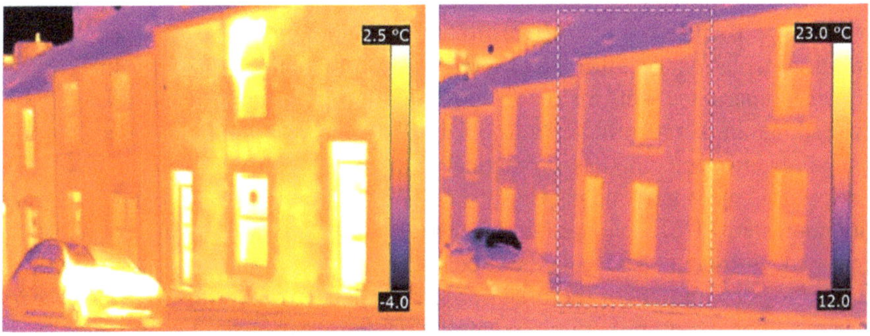

Figure 10.5 Thermography test undertaken prior to retrofit of exterior wall insulation (left) and following retrofit of EWI to existing pre-1919 terraced dwellings [56]

contractors and clients than a numerical leakage rate, which may be more difficult to understand.

In more recent years, there have also been advances in ultrasound-based leakage path detection [59]. Unlike smoke or infrared, these techniques are more precise and, in many cases, can quantify the size of the aperture or defect. An additional benefit is that these techniques do not require the whole building envelope to be under pressure, allowing for spot testing around windows, doors, and service penetrations before the primary air barrier is even complete. One practical limitation is the need for an emitter and receiver to be positioned on either side of the facade under test, requiring access and, in some cases, two operatives.

10.3.5 Step 5: act – applying insights to improve practice

The ultimate purpose of BPE is to take action to improve a building's performance. Measurement, assessment, and diagnosis must feed into tangible improvements in design, delivery, assurance, and policy.

Post-retrofit testing can provide objective *sign-off*, moving beyond paper-based compliance. For example, requiring a Pulse or blower door test before completion, ensures that workmanship-related defects are identified and addressed before handover [60,61]. This is particularly valuable for social landlords commissioning retrofit at scale, with assurance of quality reducing long-term maintenance liabilities.

Funders and investors increasingly seek performance-based assurance. The UK government's *SMETER* trials (Smart Meter Enabled Thermal Efficiency Ratings) demonstrated that in-use HTC measurement can underpin financial instruments, such as green mortgages or pay-for-performance retrofit contracts [12]. By tying funding to measured outcomes, risk is shared more fairly between funders, contractors, and occupants.

Policy is also expected to increasingly embed *measurement as a requirement* rather than an option. For example, PAS2035 [61] could evolve to require not only design-stage risk assessment but also post-installation verification through in situ testing. Internationally, the Dutch *Energiesprong* model mandates that retrofit providers guarantee performance, with monitoring to ensure compliance [12]. Embedding similar requirements in UK retrofit schemes would accelerate cultural change towards outcome-based practice.

Finally, measured insights and the arising plan of action must be clearly communicated to occupants and is central to how the cost-benefit of retrofit is presented. Although a largely non-technical audience, measurements presented alongside benchmarks or before/after comparisons, along with thermographic images and/or monitored environmental conditions, can make performance tangible and relatable. Engaging occupants builds trust and buy-in but also behavioural alignment. Occupants are more likely to ventilate appropriately or use heating efficiently when they understand the home and the motives behind the interventions.

10.4 Stakeholder perspectives and opportunities

The value of Building Performance Evaluation (BPE) extends across the retrofit ecosystem, but each stakeholder group engages with measurement differently, shaped by their priorities, risks, and responsibilities.

For households, the immediate concern is comfort, health, quality of life, [62,63] and affordability. Retrofitting promises warmer homes and lower bills, yet without measurement, these outcomes remain uncertain. Providing occupants with accessible evidence – before/after HTC comparisons, IAQ data, or thermographic images – bridges the *trust gap* and strengthens acceptance of the retrofit. In social housing, evidence is also critical for tenant engagement, demonstrating fairness, and accountability.

For builders, surveyors, and retrofit coordinators, measurement provides practical risk management. Defects identified early reduce costly call-backs and reputational damage. Evidence of airtightness or thermal improvements supports compliance with PAS2035 [61] and enhances differentiation in a competitive market. The key challenge is ensuring that tools are affordable, accessible, and integrated into standard workflows, rather than perceived as optional extras.

Investors in retrofit, whether public or private, require assurance that money spent delivers measurable outcomes. Energy efficiency loans, green mortgages, and pay-for-performance contracts all depend on robust evidence. Without BPE, financial flows remain constrained. Measurement therefore, acts as a 'currency of trust', enabling innovative finance products and unlocking capital for large-scale retrofit [64].

For government and regulators, measurement supports accountability and policy design. It enables outcomes-based regulation: instead of assuming energy savings, schemes can require proof. Aggregated datasets also inform housing stock models, ensuring national carbon budgets rest on evidence. The challenge is to design a policy that balances rigour with practicality, embedding measurement without creating excessive burden.

Professional bodies, accreditation schemes and training providers are all critical enablers. For BPE to scale, practitioners must be trained, accredited, and insured. This requires curricula that integrate performance measurement into surveying and retrofit training, as well as CPD routes for existing professionals.

10.5 Future outlook and research agenda

Looking forward, several trends are likely to shape the role of measurement in retrofit. The convergence of smart meters, IoT sensors, AI, and cloud-based analytics enables richer, lower-cost data collection and analysis [65]. Digital twins – dynamic models of buildings fed by real-time data – are emerging in commercial contexts and are increasingly being extended to housing projects [66]. These technologies promise continuous assurance of performance rather than one-off tests and siloed activity. Meanwhile, machine learning algorithms can identify patterns in energy use, detect anomalies, and predict failures. When combined with measured HTC or airtightness data, AI could further automate diagnosis, guiding contractors towards likely defects while perhaps reducing the need for some of the more costly expert diagnostics work. National retrofit strategies and underlying policy mechanisms require evidence across millions of homes. Scalable methods, such as SmartHTC [23], offer a route to gathering stock-level data without prohibitive cost. Linking measured data with housing databases, such as EPC registers, could transform national models from assumption-driven to evidence-based. The future of retrofit finance is likely to be performance-linked. Green mortgages, pay-as-you-save loans, and insurance products all depend on robust measurement [64].

As financial institutions adopt sustainability criteria, measured outcomes will become essential for unlocking capital.

Perhaps the most urgent requirement is human capacity. For BPE to move from niche to mainstream, training is needed, not just of practitioners, but also to educate clients and the wider supply chain as to the limitations of current models and calculations through to the application and benefits of measurement. This requires investment not only in technical expertise but also in communication, enabling practitioners to translate complex results into accessible insights for households, landlords, and funders.

10.6 Conclusions

The performance gap is no longer a matter of academic debate but a lived reality undermining the retrofit agenda. Less widely recognised is the *trust gap*: the loss of confidence that occurs when occupants, funders, and policymakers see promises unfulfilled. Both gaps can be addressed by embedding Building Performance Evaluation into practice.

This chapter has presented a five-step practitioner framework – **Survey, Measure, Assess and Validate and Diagnose and Act** – as a means of structuring BPE. It has reviewed key methods, from SmartHTC and Pulse to Heat3D and IAQ sensors, and demonstrated their application. It has also shown how measurement provides value for stakeholders, from households to investors, and outlined future opportunities in digitalisation, AI, stock-level evaluation, and finance.

For practitioners, the message is clear: measurement is not an optional add-on but a core component of quality practice. For funders and policymakers, it provides the assurance needed to unlock investment and drive large-scale retrofit. For occupants, it builds trust that promises will be delivered. Ultimately, what you do not measure, you cannot manage. Without robust, scalable measurement of building performance, retrofit risks fall short of their potential. With it, we gain not only data but also confidence – the foundation for a housing transformation capable of delivering comfort, affordability, and decarbonisation at scale.

References

[1] Salim S.S., Luxembourg S.L., Longa F.D., van der Zwaan B. 'Decarbonisation scenarios for the European residential sector'. *Building and Environment*. 2025;269:112408

[2] HM Government. *Heat and Buildings Strategy*. London: Business, Energy and Industrial Strategy; 2021

[3] Fenner A.E., Kibert C.J., Woo J., *et al.* 'The carbon footprint of buildings: A review of methodologies and applications'. *Renewable and Sustainable Energy Reviews*. 2018;94:1142–52

[4] Lai Y., Papadopoulos S., Fuerst F., Pivo G., Sagi J., Kontokosta C.E. 'Building retrofit hurdle rates and risk aversion in energy efficiency investments'. *Applied Energy*. 2022;306:118048

[5] Glew D., Smith M.B., Miles-Shenton D., Gorse C. 'Assessing the quality of retrofits in solid wall dwellings'. *International Journal of Building Pathology and Adaptation*. 2017;**35**(5):501–18

[6] Tjørring L., Gausset Q. 'Drivers for retrofit: A sociocultural approach to houses and inhabitants'. *Building Research & Information*. 2019;**47**(4):394–403

[7] Swan W., Fitton R., Smith L., Abbott C. 'Adoption of sustainable retrofit in UK social housing 2010-2015'. *International Journal of Building Pathology and Adaptation*. 2017;**35**(5):456–69

[8] Rye C. 'Are traditional buildings really carbon villains? Why it matters when insulating solid walls'. *Journal of Building Survey, Appraisal & Valuation*. 2015;4(2):119–26

[9] Building Performance Network. *State of the Nation Review: Performance evaluation of new homes* [online]. 2020. Available from: https://building-performance.network/wp-content/uploads/2020/06/State-of-the-nation-report-June-release-FINAL-UPDATED-1.pdf [Accessed 23 April 2025]

[10] Miller N., Bajjada T., Greaves C., Oliver D. *Evaluation of Arbed 3 Final Evaluation Report* [online]. Cardiff: Welsh Government; 2022. GSR report number 15/2022. Available from: https://gov.wales/evaluation-arbed-3. [Accessed 23 April 2025]

[11] Adeyeye K. 'From product to service–strategies for upscaling smart home performance monitoring'. *Building Research & Information*. 2024;**52**(1–2): 107–28

[12] Energiesprong UK. *Performance Guaranteed Retrofit: Learning from the Netherlands* London: Energiesprong; 2020

[13] Passivhaus Institut. *EnerPHit – The Retrofit Standard* Darmstadt: PHI; 2016

[14] Edwards J. 'Retrofitting traditional buildings for improved energy performance in the context of BSI PAS 2035'. *Journal of Building Survey, Appraisal & Valuation*. 2021;**9**(4):294–304

[15] Ministry of Housing, Communities and Local Government. *Reforms to the energy performance of buildings regime* [online]. 2025. Available from: https://www.gov.uk/government/consultations/reforms-to-the-energy-performance-of-buildings-regime/reforms-to-the-energy-performance-of-buildings-regime [Accessed 24 April 2025]

[16] Hardy A., Glew D., Gorse C., Fletcher M. 'Validating solid wall insulation retrofits with in-use data'. *Energy and Buildings*. 2018;165:200–5

[17] Brown D., Sorrell S., Kivimaa P. 'Worth the risk? An evaluation of alternative finance mechanisms for residential retrofit'. *Energy Policy*. 2019; 128:418–30

[18] Swan W., Fitton R., Brown P. 'A UK practitioner view of domestic energy performance measurement'. *Proceedings of the Institution of Civil Engineers-Engineering Sustainability*. 2015;**168**(3):140–7

[19] Build Test Solutions. *Thermal Performance Measurement Handbook*. [online]. 2025. Available from: https://www.buildtestsolutions.com/files/29b97df62c573ca94a53e0076e4f7f550148a620.pdf [Accessed 15 June 2025]

[20] Elsayed M., Pelsmakers S., Pistore L., Castaño-Rosa R., Romagnoni P. 'Post-occupancy evaluation in residential buildings: A systematic literature review of current practices in the EU'. *Building and Environment*. 2023; 236:110307

[21] BSRIA. *Soft landings framework* [online]. 2025. Available from: https://www.bsria.com/uk/consultancy/project-improvement/soft-landings/?srsltid=AfmBOoqpoYB8DzTpVk-ImvE1Taw_n85ZI9PBpENLcVj20DDMCoQY0B0U [Accessed 13 April 2025]

[22] British Standards. *BS 40101:2022 Building Performance Evaluation of Occupied and Operational Buildings (Using Data Gathered from Tests, Measurement, Observation and User Experience) Specification*. London: BSI; 2022

[23] Allinson D., Crawley J., Elwell C., *et al. Technical Evaluation of SMETER Technologies Project, Phase 2 Data: Heat Transfer Coefficients, In-Use Temperature and Energy Data, Local Weather, Dwelling Information, and Occupancy Surveys From 30 English Dwellings, 2019-2021*. Colchester: UK Data Services; 2024

[24] Hsu Y.S., Zheng X., Cooper E., Gillott M., Wood C.J. 'Evaluation of the indoor pressure distribution during building airtightness tests using the pulse and blower door methods'. *Building and Environment*. 2021;195:107742

[25] Kelly S., Pollitt M.G., Crawford-Brown D. Building performance evaluation and certification in the UK: Is SAP fit for purpose? *Renewable and Sustainable Energy Reviews*. 2012; **16**(9), pp. 6861–78

[26] Feist W., Pfluger R., Kaufmann B., Schnieders J., Kah O. *Passive House Planning Package 2007. Specifications for Quality Approved Passive Houses, Technical Information PHI-2007/1 (E)*. Darmstadt: PHI; 2007

[27] Building Research Establishment. *National calculation method, iSBEM for England* [online]. 2025. Available from: https://www.uk-ncm.org.uk/page.jsp?id=35 [Accessed 15 February 2025]

[28] Nepomuceno M.C., Martins A.M., Pinto H.A. 'A comparison between on-site measured and estimated based adjustment factor values used to calculate heat losses to unconditioned spaces in dwellings'. *Buildings*. 2022;**12**(2):146

[29] Eastwood M., Allinson D., Roberts B.M. 'Measuring the in-use performance of dwellings; improving the precision by accounting for variability'. *Journal of Building Physics* **49** (4): pp. 530–55

[30] Eastwood M. *Variability in the Heat Transfer Coefficient of dwellings*. Doctoral dissertation. Loughborough: Loughborough University; 2025

[31] Build Test Solutions. *Smart HTC validation report* [online]. 2024. Available from: https://www.buildtestsolutions.com/files/de368d4cb6a9f0ab62b0d95c82542d01540ece16.pdf [Accessed 12 May 2024]

[32] Johnston D., Miles-Shenton D., Farmer D. 'Quantifying the domestic building fabric "performance gap"'. *Building Services Engineering Research and Technology*. 2015;**36**(5):614–27

[33] Johnston D., Miles-Shenton D., Farmer D., Wingfield J. *Whole House Heat Loss Test Method (Coheating)*. Leeds: Leeds Metropolitan University; 2013

[34] Lowe R.J., Wingfield J., Bell M., Bell J.M. 'Evidence for heat losses via party wall cavities in masonry construction'. *Building Services Engineering Research and Technology*. 2007;**28**(2):161–81

[35] Subbarao K., Burch J., Hancock C.E., Balcomb J.D. *Short Term Energy Monitoring: Application of the PSTAR Method to a Residence in Fredericksburg, Virgina*. Solar Energy Research Institute Report, SERI TR-3356, 1988

[36] Alzetto F., Pandraud G., Fitton R., Heusler I., Sinnesbichler H. 'QUB: A fast dynamic method for in-situ measurement of the whole building heat loss'. *Energy and Buildings*. 2018;174:124–33

[37] Juricic S., Rabouille M., Challansonnex A., *et al.* 'The Sereine test: Advances towards short and reproducible measurements of a whole building heat transfer coefficient'. *Energy and Buildings*. 2023;299:113585

[38] Parry C., Strang M. 'Airtightness and thermal bridging in buildings'. *Environment*. 2021;(3):1–18

[39] Gorse C.A., Thomas F., Miles-Shenton D., Farmer D., Glew D., Johnston D. 'Understanding Building Performance: Implications of Heat Loss and Air Permeability on Building Control'. *Zero Carbon Building*; 2016;4:36–49

[40] Luther M.B. 'Air leakage in buildings–Review of international literature and standards'. *Environment Design Guide*. 2007;1:1–10

[41] Norton, P., Earle, L., Sparn, B., *et al. Field Test Best Practices*. Golden: National Renewable Energy Laboratory; 2023

[42] Solcher O., Rolfsmeier S., Simons P. *Natural pressure differential–infiltration through wind results of a long-term measurement* [online]. 2017. Available from: https://www.aivc.org/sites/default/files/68_0.pdf [Accessed 20 August 2025]

[43] Insulation Assurance Authority. *Background ventilation assessment of existing buildings* [online]. 2025. Available from: https://www.theiaa.co.uk/files/uploads/The%20IAA%20Ventilation%20Assessment%20of%20Existing%20Buildings%20-%20Best%20Practice%20Guide.pdf [Accessed 25 August 2025]

[44] HM Government. *Approved Document F: F1 Means of Ventilation*. London: MHCLG; 2010

[45] Few, J., Manouseli, D., McKenna, E., *et al.* 'The over-prediction of energy use by EPCs in Great Britain: A comparison of EPC-modelled and metered primary energy use intensity'. *Energy and Buildings*. 2023;288:113024

[46] Kempton L., Kokogiannakis G., Cooper P. 'Mould risk evaluations in residential buildings via site audits and longitudinal monitoring'. *Building and Environment*. 2021;191:107584

[47] Few J., Elwell C. *Measuring the ventilation rate in occupied buildings and adapting the CO2 tracer gas technique* [online]. 2019. Available from: https://www.aivc.org/resource/measuring-ventilation-rate-occupied-buildings-and-adapting-co2-tracer-gas-technique [Accessed 14 May 2024]

[48] Hulme J., Doran S. *In-Situ Measurements of Wall U-Values in English Housing*. BRE Output (290-102). London: DECC; 2014

[49] International Standards Organisation. *Guide to the Expression of Uncertainty in Measurement*. Geneva: ISO Technical Advisory Group, Working Group 3; 1993

[50] International Standards Organization. *ISO 9869-1:2014, Thermal Insulation - Building Elements - In-Situ Measurement of Thermal Resistance and Thermal Transmittance - Part 1: Heat Flow Meter Method*. Geneva: ISO; 2014

[51] International Standards Organisation. *ISO 9869 – 2:2018, Thermal Insulation — Building Elements — In-Situ Measurement of Thermal Resistance and Thermal Transmittance — Part 2: Infrared Method for Frame Structure Dwelling*. Geneva: ISO; 2018

[52] Chartered Institution of Building Services Engineers. *CIBSE TM46*. London: CIBSE; 2008

[53] Chartered Institution of Building Services. *Environmental Design, CIBSE Guide A*. London: CIBSE; 2015

[54] Build Test Solutions. *Heat 3D Validation Report*. Northampton: BTS; 2022

[55] Littlewood J., Howlett R.J., Lakhmi C.J. (eds.) Sustainability in Energy and Buildings; Madeira, Portugal, Sep 2024. Singapore: Springer; 2024

[56] Hopper J., Littlewood J.R., Taylor T., *et al.* 'Assessing retrofitted external wall insulation using infrared thermography'. *Structural Survey*. 2012;**30**(3):245–66

[57] Littlewood J.R., Alam M., Goodhew S. 'A new methodology for the selective measurement of building performance and safety'. *Energy Procedia*. 2017;111:338–46

[58] Pearson C.C. *Thermal Imaging of Building Fabric*. Bracknell, UK: BSRIA; 2011

[59] Dobrowolski D., Johnston D., Hodgkiss A. 'Benefits and accuracy when using ultrasound for acoustic leak detection'. Presented at 14th International BUILDAIR Symposium. Hannover, Germany, 2025

[60] Chartered Institute of Building Services Engineers. *TM23 Test Buildings for Air Leakage*. London: CIBSE; 2025

[61] Fylan F., Glew D. 'Barriers to domestic retrofit quality: Are failures in retrofit standards a failure of retrofit standards?' *Indoor and Built Environment*. 2022;**31**(3):710–18

[62] Jahic D., Littlewood J.R., Karani G., Atkinson J. 'Evaluating occupant wellbeing in retrofitted dwellings with the short form 36 questionnaire'. *International Conference on Sustainability in Energy and Buildings*; Gold Coast, Australia, Jun 2018. Cham: Springer International Publishing; 2018, pp. 359–68

[63] Littlewood J.R., Zhang X., Karani G. 'A pilot study to evaluate occupant quality of life in optimised retrofit dwellings in Wales, UK' *International Conference on Sustainability in Energy and Buildings*; Split, Croatia, Sep 2022. Singapore: Springer Nature Singapore; 2022, pp. 337–46

[64] Britton J., Minas A.M., Marques A.C., Pourmirza Z. 'Exploring the potential of heat as a service in decarbonization: Evidence needs and research gaps'. *Energy Sources, Part B: Economics, Planning, and Policy*. 2021;**16**(11–12): 999–1015

[65] Kelly S.D., Suryadevara N.K., Mukhopadhyay S.C. 'Towards the implementation of IoT for environmental condition monitoring in homes'. *IEEE Sensors Journal*. 2013;**13**(10):3846–53

[66] Bortolini R., Rodrigues R., Alavi H., Vecchia L.F., Forcada N. 'Digital twins' applications for building energy efficiency: A review'. *Energies*. 2022;**15**(19):7002

Chapter 11

Post-occupancy evaluation of residential building retrofits

Marianne Touchie[1,2] *and Yuan Cao*[2]

Occupant perceptions and resulting behaviours have a tremendous impact on building performance and must be understood to ensure the success of retrofits in the residential building sector. This chapter describes how Post-Occupancy Evaluations (POEs), including surveys, interviews, and monitoring, are used to assess performance and keep the occupant at the centre of the retrofit implementation process.

11.1 Introduction to Post-Occupancy Evaluation (POE)

Post-Occupancy Evaluations (POEs) have been used for decades to assess the performance of existing buildings and have become increasingly popular in recent years due to the proliferation of green building certification programmes [1], as well as the increasing complexity of buildings. The definition of a POE was formulated by Preiser, a POE pioneer, as a "process of systematically comparing actual building performance . . . with explicitly stated performance criteria . . . [and this] comparison constitutes the evaluation in terms of both positive and negative performance aspects" [2]. Further, the evaluation must happen after the building is occupied and may involve "measure[ment] and monitor[ing] . . . of [the] built environment using data gathered from behavioural, technical, and functional observation" [3]. Note that the selected POE methodology, of which there are many [4], will depend on the objective of the evaluation, which can be based on technical factors such as energy performance or carbon emissions and/or human factors, such as resident experience, perception, and behaviour.

As the POE process is intended to compare actual performance to design intent, performance gaps are usually revealed and can vary in nature depending on

[1]Department of Mechanical and Industrial Engineering, University of Toronto, Canada
[2]Department of Civil and Mineral Engineering, University of Toronto, Canada

the data collection methods [5]. Documenting and sharing findings from the POE process is essential for the building industry to learn how to improve outcomes and how to avoid making the same mistakes over and over again, thereby negatively impacting decades of building performance. However, these evaluations are often conducted by private firms for internal use only [6], which is problematic as the lessons learned are not available to the broader building community, including code officials. In a way, this lack of disclosure is understandable given that the evaluation often points to shortcomings in the design, construction, or commissioning processes, which can lead to litigation. That said, many researchers have undertaken POEs, or reviews of POEs, from various building sectors to synthesise lessons learned and share them with a broader audience [3]. Fortunately, residential buildings are often the subject of POEs [1,3], despite some of the challenges of doing research in this sector.

POEs are often conducted upon completion of the construction of a new building or when a performance issue has been identified, but POEs offer a promising opportunity in the context of building retrofits. Here, where the building is already occupied, we can conduct a "Pre-OE" prior to retrofit activities and then compare these baseline data with a POE following the retrofit to determine the efficacy of the measures employed. This Pre-OE approach can be used to identify specific performance issues that may influence utility consumption but not be visible through a utility bill analysis or a calibrated energy modelling exercise. A Pre-OE also allows the design team to identify synergies between environmental/financial objectives (e.g., reduced energy use) and retrofit design objectives related to an improved occupant experience (e.g., better thermal comfort and indoor air quality).

The remainder of this chapter will delve deeper into the approaches used to conduct occupancy-based evaluations (pre- or post) and provide examples of these evaluations in multifamily buildings, where extensive POEs are most common in the residential sector. Finally, the chapter will conclude with a summary of recommendations for the future of POEs in the process of retrofitting the residential building sector.

11.2 General POE considerations

There are various methods to conduct POEs in residential buildings, with the most common being energy consumption assessments, surveys, interviews or focus group meetings, and physical monitoring [7]. Most of these evaluation approaches require some level of interaction with building residents, as one of the most essential goals of POE is to evaluate the performance of buildings from the user's perspective [8]. When conducting studies with human subjects, one must consider several important aspects in the planning and design phase to ensure ethical standards are met and the participants are protected [9]. These considerations, including ethics review, participant recruitment strategies, confidential data collection, and resident privacy protections, are described further in this section.

11.2.1 Ethics review

POEs that involve collecting data from building residents require review by a research ethics board. This process consists of completing and submitting an application for the ethics review board to review the submission [10]. Ethics review boards are established to provide oversight in human studies to ensure compliance with applicable laws and regulations and safeguard the rights, safety, and benefits of the participants [11]. Depending on the region, scale, and organizational structure of the group administering the study, different level(s) of ethics review boards could be involved in the ethics review procedure. For instance, most research institutions have an internal ethics review board for domestic research, while collaborative studies with organisations from different countries might be subject to the approval of international ethics review committees.

While the required information for the ethics applications differs between various ethics review boards, some common sections can be expected. For instance, the ethics review documents established by the Government of Canada and the European Commission have similar provisions on research fairness, informed consent of participants, vulnerable population involvement, personal data collection and protection, and risks and benefits declaration, just to mention a few [12,13].

In terms of the review process, there can be different levels of reviews based on the project types, including exempt, expedited, and full board reviews [11]. Exempt reviews apply when the participants will be exposed to extremely low to no risk from the study, and for some review boards, the review and approval process will be optional [14]. An example of a study that might qualify for an exempt review could be an anonymous questionnaire without collecting any personal or identifiable information from the participants. Expedited reviews apply to applications with minimal risks or minor amendments to previously approved applications [15]. Such reviews are generally completed in a relatively short period of time, as applications are often accepted on a rolling basis and reviewed by a number of experienced reviewers. Finally, a full board review would be required for a study associated with a higher risk level. This type of review is the most comprehensive and time-consuming, as it must be approved by everyone in the review board. Fortunately, the majority of POEs are not subject to full board review as they pose minimal risks to the participants. However, POEs that involve vulnerable populations, such as children or individuals with cognitive impairments, may be required to undergo a full board review [11].

Generally, POE studies should be designed to minimize risk to the participants and maintain anonymity, wherever possible. Those individuals and groups conducting the POE should collect only the minimum amount of data required to answer the specific research questions that are driving the study.

11.2.2 Participant recruitment

The recruitment of participants can often be a challenging part of the POE process, and inappropriate selection of these participants could introduce significant bias to the study, so this stage is also critically important. Two major aspects are usually considered when recruiting participants [15]. First, the selection of human subjects should be equitable and fair. This requires that studies demonstrate inclusiveness when choosing the target population so that all racial, ethnic, gender, and age groups are fairly considered in the study participation. Moreover, when a study involves vulnerable populations, special attention and additional protective measures should be included to safeguard their wellness [16]. The second consideration is that no coercion, whether intentionally or unintentionally, of the participants by the researchers should occur. In most evaluation studies, there is a considerable power difference between the researcher and participants, and therefore, participants' decisions and feedback can be easily influenced by the researchers' actions [17]. Therefore, indirect and nonphysical recruitment of participants (e.g., via emails or posters) is generally preferred and recommended.

Informed consent is another important requirement during and after the process of participant recruitment. Sufficient information on the project purpose, methodology, risks, and benefits must be provided and disclosed to the potential participants to help them make a rational decision on whether to agree to participate. Informed consent is not a one-off event; rather, participants' consent should be checked and confirmed on a regular basis throughout the study [18].

11.2.3 Privacy and confidentiality protection

The protection of participants' privacy and confidentiality is a critical component of studies involving human subjects. It is important to distinguish between participants' rights to privacy and confidentiality. Privacy refers to the extent of control that the participants have over the shared information, and confidentiality implies that the shared information would not be accessed by, or disclosed, to a third party (i.e., other than the research team) without permission being granted by the participants [18,19].

To ensure the privacy of participants, numerous steps can be taken [11]. If participants' personal information is required, the researcher should provide sufficient justification for the collection of these data from the participant, such as name and contact information required to contact the participant for a later study phase. Further, any personal information should be removed from the data to be analysed and individuals with access to this sensitive information should be kept to an absolute minimum. To allow for matching, different data types for a given participant, monitored data, or survey data, for example, each participant can be assigned a special identifier code containing no personally identifiable information with which the datasets can be labelled. In terms of result reporting, only trends or average values should be reported whenever possible, but when analysis on an individual level is needed, no identifiable information on participants should be disclosed.

Similarly, confidential information can be protected in many ways [15]. For example, the use of hard-copy forms and questionnaires should be avoided whenever possible and destroyed completely once data is transferred to a digital format. Digital information and data should be encrypted once collected and stored in a password-protected system to ensure that no one outside the research team can gain access, and unencrypted data should never be shared via email.

Following this summary of general considerations required for any study involving data collection from residents, the next section outlines the various methods that can be employed when conducting a Pre-OE before a retrofit or a POE after a retrofit.

11.3 Summary of POE methods

The methods employed in POEs vary widely depending on the objectives of the study. These methods to conduct POEs in residential buildings can range from a simple building walkthrough and utility bill assessment to more involved methods, including building envelope performance evaluation, such as thermal performance or air tightness. Energy and environmental performance may be collected through energy submetering, indoor environmental quality spot measurements, or monitoring. Resident involvement can include surveys, interviews, or focus groups on comfort and satisfaction and, occasionally, resident behaviour or interaction monitoring, identifying interactions, such as window and blind operation or thermostat use [1,20]. This section provides further detail on these various approaches.

11.3.1 Surveys

The collection of residents' opinions through surveys is one of the most commonly adopted methods for POE studies [1]. Surveys can be used to capture and quantify the subjective insights from residents through various question types, including multiple choice, rating scales, ranking questions, and free-form responses, the latter of which are often the most fruitful but harder to analyse. Depending on the goals and scale of the POE study, a survey can be used as a one-off exercise following a retrofit or it can be conducted before and after a retrofit to capture changes in resident feedback [21].

11.3.1.1 Design and application

Surveys are usually conducted as one of the first steps in a POE study, as they can be easily distributed to a large sample size through various means. Due to this potentially broad reach, surveys can also be used as a participant recruitment approach for any subsequent investigations, such as interviews and monitoring. The most effective and comprehensive surveys are designed to address residents' perception and satisfaction with the full range of environmental factors including thermal comfort, air quality, acoustics, and lighting [22].

A well-designed survey should promote consistency between participants' interpretations of the questions, accurate responses to all questions, and a willingness to complete the survey [23]. To facilitate participants' interpretation of the survey, the formulation and wording of questions must be conducted carefully to avoid any ambiguities that might lead to misunderstandings and unintended responses by residents. One strategy to reduce the ambiguity in questions is to avoid technical jargon whenever possible and replace it with layman's terms to facilitate participants' understanding. In the case where technical terminology is unavoidable, sufficient supplementary notes and explanations for the terminology should be provided as part of the question. The scale and length of the survey can potentially influence participants' willingness to complete the questionnaire. Although it is intuitive that shorter and simpler surveys with a minimal number of questions and choices are more favourable from the participants' perspective, such surveys might not be able to provide sufficient information for the research team to discover significant issues and reach a valuable conclusion. However, having a longer and more comprehensive survey may not be ideal either, as participants might experience survey fatigue, leading to incomplete or inaccurate answers. Therefore, the number of questions and the level of detail of each question should be designed to align with the objectives of the POE study to ensure the efficiency and effectiveness of the survey. Once the survey questions are developed, the process of internal validation should be conducted by piloting the questions within the research team and with survey experts to assess the appropriateness of the questions and identify any ambiguities. Well-validated survey questions can be used repetitively in future POE studies and data collected from similar studies can then be compared and benchmarked against one another.

When designing surveys for a retrofit context, it is important to have an idea of the specific building characteristics that are of interest. For example, one might be interested in identifying operational issues with the building (e.g., uneven temperature or air distribution) prior to the retrofit so that these issues can be addressed during the design and retrofit construction process. For a single-family home, this information could be gleaned from a simple conversation with the resident but a survey would be a more effective tool for documenting the prevalence of particular issues in a multifamily building or a collection of houses undergoing similar retrofits. By employing the same survey tool after the retrofit, the efficacy of the retrofit strategy in addressing the issue can be evaluated.

11.3.1.2 Data collection and analysis

In terms of collecting survey responses, there are two main approaches that can be considered, namely, paper-based surveys and web-based surveys. Paper-based surveys generally have a higher response rate, but unless a mail-in method is used, they need to be administered in person, which can be labour-intensive and costly. In addition, as questions are answered on paper, the responses need to be manually converted to a digital format for storage and data analysis purposes, which can be a potential source of error. Web-based surveys can be administered remotely, and questions can be displayed in dynamic branching. For instance, different sub-questions with greater detail can be displayed based on responses to the main question. Although the platform used for conducting web-based surveys might be different from study to study, participants' responses can be easily exported and categorised for data analysis. When collecting survey data, a sufficient sample size needs to be achieved to represent the overall population of the target building [23]. Generally speaking, a 40% response rate of all occupants in the subject building would be considered sufficient to gather occupants' insight for POE studies [22], but the specific sample size will depend on the study objectives.

11.3.1.3 Common tools

Although many survey tools have been developed for POEs, there have not been standard POE surveys developed for the residential building context specifically [24]. In a published review of POE studies, two standard survey tools were identified and the authors recommended that these be customised and employed in a residential context [25]. One is the Occupant Indoor Environmental Quality (IEQ) Survey designed by the Center for the Built Environment (CBE) at the University of California, Berkeley [26], which evaluates building performance and indoor environmental quality. The CBE survey covers questions on occupants' comfort and satisfaction with acoustic, thermal, lighting, and air quality characteristics, as well as building layout, cleanliness, maintenance, and furnishings. The second standard survey is the Building Use Studies (BUS) questionnaire developed in the UK [27]. The BUS questionnaire collects data on occupants' demographics; comfort level in terms of air quality, thermal, acoustic, lighting, sick building symptoms, and general satisfaction with the building. However, both of these surveys are currently designed for commercial buildings.

11.3.1.4 Advantages and disadvantages

The use of surveys in POE studies has advantages and disadvantages. It is a highly cost-effective approach, as it can reach a large resident population within a relatively short period of time [11]. Surveys are also highly flexible, as different study teams can develop their own customised questions and data collection platform, either paper-based or web-based, depending on the characteristics of the target population and objectives of the study. From a cost perspective, surveys are relatively inexpensive, as no instrumentation is required for the data collection, and potential problems with building performance can be identified directly without any physical testing or monitoring [1]. When a standard survey format and questions are used in a POE study, the results from pre-retrofit and post-retrofit POEs can be compared and potentially benchmarked with a similar population, which would provide valuable insight and indication on the performance of the target building.

However, surveys might not be useful for all POE studies. For instance, this approach is less suitable for longitudinal studies with frequent sampling to monitor changes to building performance after a series of retrofits, as it requires considerable work for the residents and may lead to survey fatigue. For longitudinal studies, an alternative or supplementary approach to surveys is the use of ecological momentary assessments (EMAs) collected through wearables (e.g., smartwatches), which ask respondents very short questions but frequently over a period of time [28]. Compared to other POE methods, such as observation and measurement, results from surveys are highly subjective and can be potentially biased due to errors in survey design and data collection, such as inappropriate sampling and nonresponse from participants. In addition, asking residents about past events can be challenging and can result in inaccurate responses due to recall bias.

11.3.2 Interviews

Interviews can be conducted in place of surveys or be a complementary approach to survey-based data collection when used sequentially or concurrently. Interviews can help the research team gain valuable insights into residents' satisfaction and comfort and uncover more in-depth rationales related to residents' behaviour.

11.3.2.1 Design and application

Similar to survey questions, interview questions should also be piloted before data collection to ensure the questions are clear, suitable, and meaningful for the purpose of the study [29]. An interview protocol should also be developed before starting data collection. This protocol should outline the procedure, instruments, and instructions to ensure the consistency between interviews and reliability of findings [30]. Interviews can be more flexible than surveys and they can be conducted via different platforms, such as in-person, online, or via phone. Another convenient approach is via email, which can be an effective way to ask structured questions and save time in transcribing the interview responses [11]. The interviews can be formatted as structured, semi-structured, or open-ended.

Compared to individual interviews, which focus on individual perspectives, focus group interviews can be used to gather collective perspectives in a single interview, which saves time and other resources. Also, focus group interviews have the potential to discover new collective insights from group discussion, as residents are encouraged to express their insights and build upon each other's ideas. However, for focus group interviews, it is important to ensure that all participants have a chance to participate in the discussion and avoid some participants dominating the conversation [31].

11.3.2.2 Data collection and analysis

The interviewer should conduct the interview in a relaxed and comfortable environment and ask the questions in a clear and objective way without influencing the participants' responses, particularly attempting to avoid leading questions. Interviewers should be properly trained in the required techniques prior to conducting the interviews, such as using appropriate body language and effective note-taking. With participants' consent, interviews can also be recorded so they are available for transcription later. Privacy and confidentiality protection measures should be implemented. The transcription of the recordings to textual data can be a source of error; therefore, special considerations and best practice rules for interview transcription should be followed to eliminate errors and ensure accuracy [32,33]. A basic analysis method for interview data is to compare the responses from all interviews and identify similar opinions, or common themes, which could help identify specific building performance issues. More advanced methods include using computational methods to code the interview responses automatically [34–36].

11.3.2.3 Common formats

Depending on the objectives of the POE study, structured, semi-structured, or open-ended interview formats could be selected. In fully structured interviews, a fixed set of questions should be predetermined and followed strictly for all interviews, which would make the direct comparisons of results easier and more effective. On the contrary, in open-ended interviews without structured questions prepared beforehand, participants are encouraged to express their viewpoints on building performance. Compared to structured interviews, open-ended interviews have the potential to discover more in-depth or novel findings, but it can be challenging to make comparisons between responses as each interview might involve different discussion points. Semi-structured interviews combine the merits of both open-ended and structured interviews, where consistent questions are asked in all interviews and open-ended discussion is allowed for more flexibility.

One could also mix these formats before and after a retrofit. For example, an open-ended interview prior to the retrofit may identify building performance issues previously unknown to the research team. Then, the use of a structured or semi-structured interview after the retrofit can focus the interview on providing feedback on specific retrofit aspects.

11.3.2.4 Advantages and disadvantages

Compared to surveys, interviews have the potential to collect more in-depth and detailed resident feedback and discover new and, perhaps, more complex building performance issues than can be captured by a survey. However, the interview recruiting process can be more challenging compared to surveys due to the longer time commitment and higher level of involvement required by the participants. For the administrators of the study, the processes of collecting, transcribing, and analysing interview data are also time-consuming, labour-intensive, and error-prone. Moreover, responses can be biased, as some interviews are face-to-face and non-anonymous, and participants might be reluctant to express their true opinions. Interview results also tend to be less generalisable due to the limited sample size in most studies.

11.3.3 Indoor Environmental Quality (IEQ) monitoring

In addition to, or in place of, subjective residents' feedback collected by surveys and interviews, IEQ monitoring is sometimes used to capture the actual conditions of the thermal, visual, acoustic environment, and indoor air quality. Data collected from the monitoring can be benchmarked with national or international IEQ standards to determine compliance, or the level of performance, of the target building relative to standards. IEQ measurements can also be compared with the survey and interview responses to identify performance issues and the source of residents' dissatisfaction. Some commonly measured IEQ parameters in POEs include temperature, relative humidity, air movement, light intensity, noise levels, and indoor air quality in terms of particulate matter, gases (CO and CO_2), and volatile organic compound (VOC) concentrations [37].

11.3.3.1 Design and application

A number of factors need to be taken into consideration during the design of an IEQ monitoring campaign. Based on the objectives of the study, targeted IEQ parameters should be selected as the first step. Subsequently, different sensor options should be considered based on the desired measurement accuracy and sensitivity, suite conditions, layout and features, and monitoring duration and frequency. Prior to sensor selection, the researchers should conduct a physical walkthrough of the target monitoring space; after obtaining residents' consent, review the drawings of the buildings and talk with the facility management team to determine the space layout and any existing sensors. It is possible that some of the needed IEQ data might already be logged via a Building Automation System (BAS), or even smart thermostats, for instance, air temperature and relative humidity. However, care should be taken to verify that the BAS measurements agree with spot checks by research-grade equipment.

With this information, suitable sensors and data loggers can be determined based on the available budget for the study. Ideally, a detailed and comprehensive monitoring protocol should be developed prior to participant recruitment, which outlines the monitored parameters; required resources, including labour, instrumentation, hardware, software, and other infrastructure; sensor calibration approaches; installation and removal plan; and measurement frequency and duration, as well as any other special considerations during monitoring. Measurement frequency needs to be carefully considered to capture the fluctuation of various IEQ parameters and monitoring duration needs to be long enough to collect sufficient data for analysis of any seasonal trends. In addition to the monitoring protocol, estimated timing of visits to residents' homes, instructions on handling of sensors, and a privacy and data protection plan should be established and compensation for participation should be disclosed to the potential participants during the recruitment process.

11.3.3.2 Data collection and analysis

During data collection, sensor location, and position need to be selected carefully to ensure accurate measurement but also to minimise disruption to participants.

Whenever possible, measurements should be taken during different times of the year, the cooling, heating, and shoulder seasons, to capture the variations in indoor conditions, ideally both before and after the retrofit. Sensors should be calibrated prior to each deployment and co-located before and after all deployments to ensure the functionality and accuracy of the instrumentation. In addition to the sensor measurements, activity log sheets can be used during the monitoring period to document supplementary information, such as occupancy schedules and window operation.

The collected IEQ data could be benchmarked with industry standards such as ASHRAE Standard 55 for thermal comfort, ASHRAE Standard 62.2 for indoor air quality, and the CIBSE Code for Interior Lighting standards for different tasks. However, the suitability and validity of applying these standards to the POEs of residential buildings is open to discussion. For instance, ASHRAE 55 specifies the thresholds for various thermal parameters, but thermal comfort is ultimately defined by residents' satisfaction with the environment [38]. Therefore, the monitoring data should be analysed and interpreted together with survey and interview results to evaluate whether the indoor conditions of the target building are satisfactory to the residents.

11.3.3.3 Advantages and disadvantages

Unlike the collection of subjective data using surveys and interviews, IEQ monitoring can provide valuable objective data that reflects the actual indoor environment that residents are exposed to on a daily basis.

However, IEQ monitoring is often a significant undertaking. First, it is often more challenging to recruit participants for monitoring compared to surveys and interviews, due to the relatively long duration, potential risks associated with the measurement (real or perceived), and disruption of their daily lives. Second, the monitoring process can be costly, time-consuming, and labour-intensive. In addition, it is highly unlikely for the monitoring to capture the spatial variations of different IEQ parameters due to the limited coverage of sensors, and there are risks of equipment tampering by participants (intentionally or unintentionally). Instrument inaccuracy and improper calibration can lead to inaccurate data and errors in data analysis.

11.3.4 Combined approach

The relationship between buildings and residents is complex. Therefore, it is critical to conduct POEs from a multidimensional perspective that captures residents' feedback and the actual indoor conditions to which the residents are exposed [37]. As discussed, surveys, interviews, and IEQ monitoring are some of the most commonly adopted approaches in POE studies. While each approach involves different data collection strategies, they are complementary to each other and, when integrated, can provide a more complete picture of building performance prior to and after a retrofit.

11.4 Residential POE examples

As discussed, POEs are undertaken for a number of different reasons and the resulting data are used in different ways. This section describes some examples of POEs conducted in the multifamily residential building sector. An overview of the selected approach is provided, followed by a summary of the results and lessons learned. The purpose of this section is to demonstrate how POE approaches have been applied specifically in a retrofit context. There are three multifamily project POEs showcased here: one each in North America, South America, and Europe.

11.4.1 North American example

A Canadian social housing provider undertook extensive energy retrofits in seven of the buildings that it owned. As part of a planned energy retrofit, they wanted to ensure that the quality of the indoor environment for residents would not degrade and ideally improve. Thus, the housing provider worked with a local university and nonprofit agency to develop and implement an extensive in-suite monitoring program and survey protocol to assess in-suite conditions and resident perceptions. This study was conducted for one year prior to the retrofit, during the construction period, and for one year following the retrofit between 2015 and 2018. In addition to in-suite monitoring and surveys, energy submetering was also conducted on space and domestic hot water heating systems [39].

11.4.1.1 Method

The study buildings included low-, mid-, and high-rise buildings with varying resident types including seniors, families, and single occupancy. Prior to the retrofit, all seven buildings were heated through perimeter hydronic baseboards, without any in-suite thermostats or controls. There was no central cooling or ceiling-mounted fans in the suites, but almost half of the monitored suites had free-standing or window-mounted air conditioning (A/C) units. Outdoor air was provided through a pressurised corridor ventilation system [40].

Pre-retrofit surveys were conducted in early 2015 and the post-retrofit surveys were conducted in early 2018. The survey captured information about basic demographics, the number and age of suite residents, and behaviours, such as smoking, cooking, bathing, and cleaning. Residents were asked to rate their thermal comfort during the summer and winter seasons as well as indicate the existence of odours or draughts and excessive moisture or dryness in the air. They were also asked about any measures they had taken to alter their interior environment such as use of heaters, air conditioners, fans, odour control, humidifiers/dehumidifiers, operation of doors and windows, and use of weather stripping [41].

Sensor packages were deployed in a subset of the suites to assess various IEQ parameters including air temperature (±0.35 °C), relative humidity (±2.5%), a proxy for mean radiant temperature (±0.35 °C), and various IAQ parameters.

Measurements were recorded once every 15 minutes and data were collected from the loggers in each suite every three months [40].

All buildings, except one, had their boilers replaced and buildings A, B, F, and G received new air handling units. Buildings F and G had their radiators cleaned to improve heat transfer, as well as the installation of a reflective panel between the exterior wall and the radiator. Smart thermostats and thermostatic control valves were installed in Buildings A, B, F, and G to allow in-suite temperature control [41].

11.4.1.2 Results and lessons learned

Using the pre- and post-retrofit monitoring data, the percentage of time each suite could be considered thermally comfortable was determined based on the graphical method from ASHRAE Standard 55. The pre- and post-retrofit survey data were also compared between the buildings. Overall, there were 321 survey responses (180 pre- and 141 post-retrofit) and 30 suites with both pre- and post-retrofit monitoring data. For buildings A, B, C, and E, which all had high pre-retrofit modelled time comfortable, modelled time comfortable changed in opposite directions post-retrofit: two increased and two decreased. In Buildings A and B, where wintertime overheating was reduced, residents reported no change to thermal comfort on average in Building A and a decrease in Building B due to an increase in participants reporting they were too cold. For the high-rise buildings F and G, there was no significant change from their high modelled pre-retrofit time comfortable, but perceived comfort improved in both buildings, albeit from a very low level of reported comfort pre-retrofit [41].

One finding from this study overall was that similar retrofits applied in similar buildings can yield different results, though in the case of these buildings, those differences were small. The mechanical system retrofits, which were designed to improve energy performance (10–33% savings achieved), did not necessarily improve thermal comfort by a significant degree or at all [41]. Also, there was clearly disagreement between modelled and reported thermal comfort, both pre- and post-retrofit, so modelling approaches specific to the residential context need further development [41]. This study highlights the importance of collecting both survey and monitoring data to understand the actual changes in the indoor environment driven by the energy retrofits and how these differed from resident perceptions.

11.4.2 South American example

A POE approach was employed in a 400-suite building located in Chile and constructed in 2013. This study is detailed in Gonzalez-Caceres *et al.* [42] but is summarised here. Seventy-six of these apartments were impacted by various performance issues including odours, condensation, mould growth, and high moisture loads. However, there was no apparent pattern to the occurrence of these issues after considering a variety of factors including occupancy level, space use, and heating system type. So, the purpose of the investigation was to determine the cause (s) of these issues so that they could be corrected through retrofits.

11.4.2.1 Method

The researchers, along with an interdisciplinary team of project stakeholders, employed a three-stage approach to the POE, with each stage getting progressively more detailed. The first "indicative" stage involved a building walkthrough, which included evaluating any maintenance issues, as well as conducting resident surveys to evaluate demographics and schedules. The second "investigative" stage involved developing an energy model of the building and conducting in-suite monitoring of temperature, relative humidity, and carbon dioxide concentrations. The final "diagnostic" stage involved field and lab measurements, including air tightness testing, determining the thermal transmittance and rain penetration of the envelope, and assessing the thermal and vapour permeability of various components and materials.

11.4.2.2 Results and lessons learned

From the "indicative" stage, the researchers determined that there was no clear connection between the resident density or behaviour in a particular suite that could be attributed to the observed moisture issues and the second "investigative" stage did not provide any further insight into the causes. The final "diagnostic" stage revealed that the envelope had higher thermal conductance and moisture permeability than designed, as well as highly variable air tightness and ventilation issues. These findings are not unexpected given the variability inherent in on-site construction, but the authors did not conduct any sort of statistical analysis to establish a link between these performance features and the issues experienced in the building, so it is hard to draw any specific conclusions from this study.

One novel aspect of this study was the authors' use of BIM to share the collected data between stakeholders. The proposed framework could serve as a model for data management in future POE work and, by employing the same framework across a particular portfolio or jurisdiction, might allow for a more direct comparison on POEs between different buildings or between the same buildings before and after a retrofit to extract lessons learned.

11.4.3 European example

This study examined the performance of two similar six-story multifamily buildings located in Linköping, Sweden. This study is detailed in Liu *et al.* [43] but is summarised here. One of the buildings had been retrofitted and one had not. The retrofitted building was overclad with new exterior insulation, had double-glazed windows replaced with triple-glazed, and had insulation added to the roof. The ventilation system was upgraded from an exhaust-only system to one with heat recovery. Space heating was provided by a district heating system. The POE was undertaken because the local housing provider was using this first retrofitted building as a demonstration for future projects and wanted to fully understand the implications of the retrofit measures on residents.

11.4.3.1 Method

The researchers used in-suite environmental measurements, energy submetering, and questionnaires in their data collection. Three suites were selected for measurements including two from the retrofitted building and one from the non-retrofitted building. Measurements of temperature, RH, and various parameters over a one-week period in each suite were used to estimate the Predicted Mean Vote (PMV) and the Predicted Percent Dissatisfied (PPD). These measurements were used to develop a calibrated building simulation in the IDA Indoor Climate and Energy software package to examine the impact of the specific retrofit measures on energy use and indoor environmental conditions.

A standardised survey was also used to collect data on resident perceptions in the two study buildings, as well as the other four similar buildings in a nearby cluster.

11.4.3.2 Results and lessons learned

As a result of the envelope retrofits, the researchers determined that the retrofitted building's balance point temperature was reduced to 16.5 °C from 18 °C for the non-retrofitted building, indicating an increased resilience to changes in exterior conditions. A total of 42 surveys were also collected (11 from the retrofitted building and 31 from the other non-retrofitted buildings in the cluster). Generally, the residents in the retrofitted buildings were more satisfied with the indoor environmental quality, specifically moisture levels and odour transmission, than those living in the non-retrofitted buildings when considering only participants who had mechanical ventilation in their suites. The residents in the retrofitted building were also more satisfied with their thermal comfort, specifically related to the uniform temperatures and the ability to control in-suite temperatures. Residents in the non-retrofitted buildings reported poor thermal comfort conditions, experiencing draft sensations and cold temperatures during the winter months. However, residents in both building types reported summertime overheating. Interestingly, the model, which was calibrated with monitored in-suite temperatures, showed no significant differences between the buildings in PMV and PPD during the winter and improved PMV/PPD values in the retrofitted building during the summer, so without the resident perceptions collected via surveys, this issue would not have been uncovered.

In terms of acoustic comfort, about 20% of residents in the retrofitted building complained about noise from the new in-suite ventilation system, while the residents in the non-retrofitted building were more disturbed by outdoor noise and noise from pipes and other residents. This difference in perception of outside noise is likely related to the envelope retrofits. From an energy use perspective, in-suite electricity use increased in the retrofitted building because washing machines and dryers were added to the suites. Also, as more residents lived in the retrofitted buildings, there was a higher domestic hot water load; however, based on the energy simulation, the space heating energy loads are projected to be about 40% less than in the non-retrofitted buildings.

This study highlights how important POEs can be in identifying unintended consequences of building retrofits, including increased energy use due to changes in how spaces are used and new forms of discomfort for residents. By more regularly collecting these data following retrofits, we can ensure that these issues are considered at the design stage and ideally avoided post-construction.

11.5 Conclusions

Based on the discussion above and further review of the literature, several recommendations are made for the future POEs of residential building retrofits. First, a large portion of the existing POEs were carried out in an academic context for research and demonstration purposes. POEs for residential building retrofits are not widely adopted. This could be for reasons including cost, complexity, and lack of understanding of the benefits of a POE. When POEs are conducted, they are often done because of the high-profile nature of the project or to diagnose a performance issue. An effective way to promote the broader implementation of POEs is to make it compulsory in one form or another [6]. For instance, a top-down approach can be adopted where POEs become a part of the building regulatory requirement, where certain types of building projects, such as public housing, are required to conduct POEs pre- and post-retrofit to secure public funding. In terms of the private sector, an alternative approach would be the establishment of initiatives, or certification programmes, to encourage more building owners to conduct POEs as a part of their building management practices.

Second, there is a lack of standardised POE methodology designed for residential buildings that addresses various aspects of building performance and occupant feedback. Past POEs in residential retrofits have adopted distinct approaches due to their varying building contexts and project objectives, which makes the comparison and benchmarking of POE results from isolated studies extremely challenging. Although some standardised POE methodologies have been proposed in the literature, the majority of them were targeted to commercial buildings and, hence, cannot be applied directly to residential buildings [44]. Therefore, a reliable and standardised POE method for residential buildings to assess building performance and occupant satisfaction is needed. We suggest that this protocol be designed based on modifications from existing methods where lessons learned about prior use of the tool can be integrated. This new POE protocol should also be piloted extensively before widespread implementation.

However, a one-size-fits-all solution should not be the development goal of a standardised POE for residential buildings. Rather, a modular and staged approach with varying levels of detail would be more effective and flexible for implementation by building in different lifecycle stages or for building owners with different needs, newly constructed, pre- or post-retrofitted, or rehabilitated, for example [42,44]. In the modular and staged POE approach, for instance, less expensive and intrusive data collection, such as occupant surveys and building energy consumption, could be carried out in the initial stage of the POE to obtain a general understanding of building performance and identify any performance issues. Findings from this initial stage could then be used to decide which building performance areas, or issues, should be further investigated in subsequent stages by using more sophisticated data collection methods, such as short- or long-term monitoring and building performance testing [1].

Finally, the delivery and communication of results from POEs to the building retrofit community is crucial. In addition to the findings from the POE, any comparisons of building performance with design targets, industry benchmarks, successes, challenges, and lessons learned from the POE and recommendations for future POEs could be documented and communicated to the stakeholders. In addition, to maximise the benefits of POE results and findings, it is important to identify any additional end users of the POE prior to data collection and determine the best way and suitable level of detail for communication of results. For instance, building occupants could be informed on the key findings and educated on how to better operate and interact with the building through various activities, such as informative workshops, promotional videos, and brochures, or even exhibitions.

References

[1] Li P., Froese T. M., Brager G. 'Post-occupancy evaluation: State-of-the-art analysis and state-of-the-practice review'. *Building and Environment*. 2018; 133:187–202

[2] Preiser W. F. E. 'Post-occupancy evaluation: How to make buildings work better'. *Facilities*. 1995;**13**(11):19–28

[3] Kalantari S., Shepley M. 'Psychological and social impacts of high-rise buildings: A review of the post-occupancy evaluation literature'. *Housing Studies*. 2020;**36**(8):1147–76

[4] Göçer Ö., Hua Y., Göçer K. 'Completing the missing link in building design process: Enhancing post-occupancy evaluation method for effective feedback for building performance'. *Building and Environment*. 2015;89:14–27

[5] Coleman S., Touchie M. F., Robinson J. B., Peters T. 'Rethinking performance gaps: A regenerative sustainability approach to built environment performance assessment'. *Sustainability*. 2018;**10**(12):4829

[6] Stevenson F. 'Post-occupancy evaluation and sustainability: A review'. *Proceedings of the Institution of Civil Engineers: Urban Design and Planning*. 2009;**162**(3):123–130

[7] Sanni-Anibire M. O., Hassanain M. A., Al-Hammad A.-M. 'Post-occupancy evaluation of housing facilities: Overview and summary of methods'. *Journal of Performance of Constructed Facilities*. 2016;**32**(5):04016009

[8] Federal Facilities Council. *A State-of-the-Practice Summary of Post-Occupancy Evaluation*. Washington, DC: The National Academies Press; 2002

[9] Preiser W. F. E., White E., Rabinowitz H. *Post-Occupancy Evaluation*. New York: Routledge; 2015

[10] Head G. 'Ethics in educational research: Review boards, ethical issues and researcher development'. *European Educational Research Journal*. 2020;**19**(1):72–83

[11] Wagner A., O'Brien W., Dong B. (eds.). *Exploring Occupant Behavior in Buildings*. Cham: Springer International Publishing; 2018. doi:10.1007/978-3-319-61464-9

[12] Canadian Institutes of Health Research. *Tri-council policy statement: Ethical conduct for research involving humans* [online]. 2018. Available from https://www.nserc-crsng.gc.ca [Accessed 9 June 2026]

[13] European Commission. *Horizon 2020 - How to complete your ethics self-assessment* [online]. 2014. Available from H2020-Guidance-How-to-complete-your-ethics-self-assessment.pdf [Accessed 9 June 2026]

[14] Millum J., Menikoff J. 'Streamlining ethical review'. *Annals of Internal Medicine*. 2010;**153**(10):655–7

[15] Oakes J. M. 'Risks and wrongs in social science research: An evaluator's guide to the IRB'. *Evaluation Review*. 2002;**26**(5):443–79

[16] Kipnis K. 'Vulnerability in research subjects: A bioethical taxonomy' *in Ethical and Policy Issues in Research Involving Human Participants, Volume II: Commissioned Papers*. Rockville, MD: National Bioethics Advisory Commission; 2001

[17] Sieber J. E. *The Ethics of Social Research: Surveys and Experiments*. New York: Springer Science & Business Media; 2012

[18] Penslar R. L. *Protecting Human Research Subjects: Institutional Review Board Guidebook*. Washington, DC: US Government Printing Office; 1993.

[19] Beauchamp T. L., Childress J. F. *Principles of Biomedical Ethics*. New York: Oxford University Press; 2001

[20] Gupta R., Chandiwala S. 'Understanding occupants: Feedback techniques for large-scale low-carbon domestic refurbishments'. *Building Research & Information*. 2010;**38**(5):530–48

[21] Fink A. *How to Conduct Surveys: A Step-by-Step Guide*. Los Angeles: Sage Publications; 2015

[22] ASHRAE. *Performance Measurement Protocols for Commercial Buildings*. Atlanta, GA: ASHRAE; 2010

[23] Dillman D. A., Smyth J. D., Christian L. M. *Internet, Phone, Mail, and Mixed-Mode Surveys: The Tailored Design Method*. Hoboken, NJ: John Wiley & Sons; 2014

[24] Brown C. 'The power of qualitative data in post-occupancy evaluations of residential high-rise buildings'. *Journal of Housing and the Built Environment*. 2016;**31**(4):605–20

[25] Dykes C., Baird G. 'A review of questionnaire-based methods used for assessing and benchmarking indoor environmental quality'. *Intelligent Buildings International*. 2013;**5**(3):135–49

[26] Zagreus L., Huizenga C., Lehrer 'Listening to the occupants: a Web-based indoor environmental quality survey'. *Indoor Air*. 2004;**14**(8):65–74

[27] Leaman A. *BUS Occupant Survey Method: Details for Licensees*. 2010. Available from https://www.usablebuildings.co.uk/BUSMethodology.pdf [Accessed 9 June 2026]

[28] Jayathissa P., Quintana M., Abdelrahman M., Miller C. 'Humans-as-a-sensor for buildings—Intensive longitudinal indoor comfort models'. *Buildings*. 2020;**10**(10):174

[29] Griffee D. T. 'Research tips: Interview data collection'. *Journal of Developmental Education*. 2005;**28**(3):36–7

[30] Boyce C., Neale P. *Conducting in-Depth Interviews: A Guide for Designing and Conducting in-Depth Interviews for Evaluation Input*. Watertown, MA: Pathfinder International; 2006

[31] Liamputtong P. *Focus Group Methodology: Principle and Practice*. London: Sage Publications; 2011

[32] McLellan E., MacQueen K. M., Neidig J. L. 'Beyond the qualitative interview: Data preparation and transcription'. *Field Methods*. 2003;**15**(1):63–84.

[33] Widodo H. P. 'Methodological considerations in interview data transcription'. *International Journal of Innovation in English Language Teaching and Research*. 2014;**3**(1):101–7

[34] Dam G., Kaufmann S. 'Computer assessment of interview data using latent semantic analysis'. *Behavior Research Methods*. 2008;**40**(1):8–20

[35] Seale C., Rivas C. 'Using software to analyze qualitative interviews' in Gubrium J. F., Holstein J. A., Marvasti A. B., McKinney K. D. (eds.). *The SAGE Handbook of Interview Research: The Complexity of the Craft*. Thousand Oaks, CA: Sage; 2012. pp. 427–40

[36] Weston C., Gandell T., Beauchamp J., McAlpine L., Wiseman C., Beauchamp C. 'Analyzing interview data: The development and evolution of a coding system'. *Qualitative Sociology*. 2001;**24**(3):381–400

[37] Meir I. A., Garb Y., Jiao D., Cicelsky A. 'Post-occupancy evaluation: An inevitable step toward sustainability'. *Advances in Building Energy Research*. 2009;**3**(1):189–219

[38] ASHRAE. *ASHRAE Standard 55: Thermal Environmental Conditions for Human Occupancy*. Atlanta, GA: ASHRAE; 2004

[39] The Atmospheric Fund. '*Retrofitting Arleta manor: A TowerWise case study*' [online]. 2019. Available from https://taf.ca/wp-content/uploads/2019/09/Retrofitting-Arleta-Manor-September-2019.pdf [Accessed 31 March 2021]

[40] Touchie M.F., Tzekova E.S., Siegel J.A., Purcell B., Morier J. 'Evaluating summertime overheating in multi-unit residential buildings using surveys and in-suite monitoring'. *Proceedings of the 13th International Conference on Thermal Performance of the Exterior Envelopes of Whole Buildings XIII*; Clearwater, Florida, 5–8 December 2016. Oak Ridge: Oak Ridge National Laboratory; 2016. pp. 135–151

[41] Vakalis D., Diaz Lozano Patino E., Opher T., *et al.* 'Quantifying thermal comfort and carbon savings from energy-retrofits in social housing'. *Energy and Buildings*. 2021;235:110950

[42] Gonzalez-Caceres A., Bobadilla A., Karlshøj J. 'Implementing post-occupancy evaluation in social housing complemented with BIM: A case study in Chile'. *Building and Environment*. 2019;158:260–80
[43] Liu L., Rohdin P., Moshfegh B. 'Evaluating indoor environment of a retrofitted multi-family building with improved energy performance in Sweden'. *Energy and Buildings*. 2015;102:32–44
[44] Teasdale-St-Hilaire A. 'Post-occupancy evaluation framework for multi-unit residential buildings'. *Proceedings of the 12th International Conference on Thermal Performance of the Exterior Envelopes of Whole Buildings*. Clearwater, USA, Dec 2013. Oak Ridge: Oak Ridge National Laboratory; 2013

Chapter 12

Living with retrofit

Graeme Sherriff[1] *and Michael Lomas*[1]

12.1 Introduction

Retrofitting is an intrinsically socio-technical process, an intervention in the home with a profound impact on the lives of occupants. It is, therefore, essential to unpack the sociological and psychological aspects involved. In this chapter, we draw on over ten years of our own research with occupants at various stages of a retrofit journey. This research has involved conversations with occupants who have been part of retrofit schemes or received other forms of energy-related advice and support and has been conducted in the UK, Spain, and Australia.

Far from having a passive role, occupants are often the decision-makers, financiers, and instigators of retrofits. Whether they own their home or rent it privately or socially, they will have some stake in changes being made to their home. Meaningful engagement with them will likely mean that they have a better understanding of the changes being made and are able to achieve greater benefits. For them, this engagement potentially translates into a more comfortable home, cost savings, a better quality of life, and improved health. For policymakers and asset managers, who may be overseeing retrofit programmes as part of low-carbon transitions, this means that reductions in energy consumption and associated emissions are more likely to be achieved, and the performance gap associated with resident behaviours is reduced.

In the UK, most homes are privately owned. Homeowners make decisions about priorities for their homes and manage their household budgets. Rolling out a retrofit programme is therefore not only a technical exercise but also one of communication, facilitation, and incentivisation. Owner-occupiers vary greatly in their ability to pay for capital improvements on their homes and in their level of interest in retrofits. Cost and affordability are important factors, but the primary challenges are not purely financial: it is necessary to understand how to garner interest in retrofit interventions and how to translate this into take-up.

[1]School of Health and Society, University of Salford, UK

Some occupants, such as those in privately or socially rented housing, have comparatively less control over their living situation. For them, energy solutions involve a negotiation with their landlord or local housing standards office. Some in socially rented properties have benefitted from progressive policies and forward-looking asset managers, but they are sometimes left with new technologies for which insufficient explanation and training have been given. Irrespective of tenure, what people do in their homes will continue to affect energy consumption: an ongoing dialogue with occupants is therefore important if health, cost, and carbon goals are to be achieved.

Energy poverty – also referred to as fuel poverty, especially in the UK – is a condition whereby people struggle to keep their homes at a comfortable temperature. It is understood to stem from energy-inefficient housing, high energy costs, low incomes, and poor health [1,2]. It is occupants who suffer the impacts of poor homes, through ill health, financial stress, and wider associated implications, such as poor diet and educational attainment. Many of those who suffer the most are those least able to afford to make improvements to their homes or have minimal control over their living spaces. Yet they could benefit the most. It is, therefore, important to consider the focus of our research and policy: poor housing is a social issue and an emphasis on those who are able and willing to pay for retrofit could exclude those who are most in need of it.

Much has been written about the limitations of a focus on behaviour change [3] and it is certainly true that simply 'doing things differently' is not a suitable approach to fuel poverty: there are only so many extra layers you can put on in a cold, uninsulated home over which one has little control. It is likely that a combination of retrofit and changes in practice will have the greatest effect on carbon emissions [4]. Our position is that, whilst we should not rely on behaviour change, we also should not ignore occupant practices because it is clear that the success of retrofit – in both human and carbon terms – depends on proactive and sensitive engagement with occupants.

Working through the chronology of a retrofit intervention, this chapter explores how these factors play a role at various stages of the process. We start with some terminology and cross-cutting themes. We then look at issues that are likely to prompt retrofit, such as cold homes and financial stress. We then dive into decision-making on retrofit and energy efficiency, moving through installation and implementation and, finally, consider life in the retrofitted home.

12.2 Starting points

12.2.1 Retrofit, occupants and measures

Retrofit is a socio-technical process that involves building fabric, systems, and occupants. Occupants will have particular expectations and needs in relation to their home, will be involved in decision-making about it, and will likely be exposed to a degree of disruption as the project is implemented. What they do and how they live in the retrofitted home will be key factors in determining the extent of actual, rather than predicted, energy savings and this has implications for meeting carbon reduction and fuel poverty alleviation goals. Retrofit measures vary greatly: there is a substantial difference between a boiler replacement and a whole-house deep retrofit in terms of, not only the financial commitment, but also the level of disruption and upheaval to which occupants could expect to be subjected to.

To oversimplify the nature of retrofit is to risk limiting the potential for the relationship between occupants, technology, and fabric to shape the outcome of interventions, as well as discount the ways in which changes to a building will impact the health and well-being of occupants. As Karvonen [5] argues, we cannot reduce the story of retrofit to one of a building: 'Domestic retrofit is not [only] an activity of changing a house from one steady state to another – from poor energy performance to exceptional energy performance – but an intervention into the rhythms of domestic habitation'.

Talking about 'the occupant' may imply a level of universality, but just as retrofitting is not a uniform activity, householders are far from a homogenous group, reflecting multiple interpersonal and circumstantial differences that relate to location, living arrangements, life stage, housing tenure, financial status, and level of vulnerability. In particular, they vary in their ability to pay for retrofit and the extent to which they are interested in retrofit and prioritise it. Those experiencing fuel poverty are likely to need home improvements the most but are least likely to be able to afford them. With the recent energy price rises, those who are financially secure and have some savings are more likely to be incentivised to invest in improvements that could bring down energy costs.

12.2.2 Tenure, places and pathways

As with any green initiative, the principal challenge with retrofitting is to increase public buy-in [6,7]. Recent research conducted by the Green Finance Institute, completed in partnership with stakeholders from industry, academia, government, and members of the public, explored different strategies for using financial mechanisms to incentivise householders to engage with retrofit. In doing so, they segmented the housing sector into the three commonly used categories of tenure: owner-occupied, privately rented, and socially rented [8].

By dividing the housing sector in this manner, it becomes possible to identify challenges faced at various stages in a manner that takes account of the circumstances of the occupant. The report notes, however, that these broad meta-categories are insufficient for understanding the experiences and situations of residents in their entirety.

A person close to retirement and having paid off their mortgage, for example, will see the option of retrofitting differently to someone starting out in home ownership. Occupancy and ownership patterns are relevant and may make retrofitting more challenging: seeking agreement amongst multiple occupants in an apartment block to invest in property will be extremely challenging. Owner-occupiers may be subject to leasehold conditions that limit what they can change, particularly in relation to the outside of the building. Location is also a consideration: urban areas are often better able to benefit from economies of scale, such as street-by-street approaches, whilst rural and remote residents face extra expense in arranging retrofits [9].

It is useful to think of the experience of housing as a pathway. Butler and Sherriff [10] spoke to young people who described the different stages of 'parental home', 'temporary home' and 'independent home'. For some, rented accommodation may seem like a stepping stone to be endured in the short term, whereas for others, it will be the main housing throughout their life and the impact of poor conditions, therefore, more severe and long-lasting.

12.2.3 Health, vulnerability and disability

In considering the impact of retrofit upon residents, the process is as important as the outcome. Given the potential for a retrofit process to be disruptive and stressful, it is important to find ways to mitigate this so that occupants feel involved and respected. Residents have reported disruption from requirements, such as loft and room clearance, and stress from concerns about rent increases and the operation of new technologies [11]. Occupants vary in their experiences of retrofit and may have vulnerabilities that can compound the effect of the disruption in relation to health conditions, such as dementia and respiratory illnesses.

The redesign of homes can have significant implications for disabled people in particular, who may experience difficulties coping with changes in surface levels and restricted circulation space in rooms and passages; fixtures and fittings may be difficult to reach and new windows and doors challenging to open. Conditions such as dementia, sight loss, and arthritis can add to difficulties in understanding and operating new heating systems [12,13]. It is important to take into account patterns of behaviour and the ability of occupants to cope physically and mentally with the disruption of the process and the usability of the retrofitted home. In our research in the Outer Hebrides, an island group in Scotland, 'Moving Together' [9], tackling cold homes in close collaboration with health services, we found that the need for unfamiliar contractors to come into the homes of vulnerable people was a likely barrier to retrofitting [9].

12.3 Living without retrofit

The primary well-being issues being addressed by retrofit are thermal comfort and fuel poverty. The latter describes the inability to afford to adequately heat or cool homes as a result of a combination of financial stress, ineffective technologies, and/or poorly performing building stock [1], which can be accentuated by health issues, household composition, and other vulnerabilities. Even when individuals can pay their bills, this may be at the expense of other household budgets, such as food, school, and social activity. Fuel poverty is therefore a 'specific form of material deprivation that occurs when a person is unable to attain a socially- and materially-required level of energy services in the home' [14]. Alongside financial support, tariff reform, and energy advice, retrofit is a potential response to fuel poverty. For a range of reasons, not all of them financial, people experiencing fuel poverty are likely to be the least able to retrofit their home, whilst also being the ones who could benefit most from energy efficiency enhancements.

Fuel poverty is not the only reason to retrofit a home. The case for energy efficiency measures in achieving decarbonisation has been made elsewhere in this book. The reason for the focus on fuel poverty in this section is that it highlights the ways a poorly performing home can impact health and well-being.

12.3.1 Health and well-being

One of the main ways of understanding well-being is the capacity to live well, satisfy one's needs, and reach one's full potential [15]. The most established theory of well-being is Maslow's hierarchy of needs [16]. This conceptualises well-being as a series of human needs that must be satisfied sequentially before an individual can reach optimal well-being: safety, love and belonging, esteem and, finally, self-actualisation.

Each of these needs is in some way facilitated by the home and thus stands to be influenced by retrofit measures and there is consequently a well-established evidence base that connects thermally inefficient and poor-quality homes with physical and mental health issues, including respiratory, circulatory, and cardiovascular diseases, as well as increased levels of anxiety and depression [17–22].

A recent study of Better Housing Better Health [23], a programme of support in Oxfordshire, highlighted some of the ways in which cold homes can aggravate health conditions. People working in support organisations describe the challenges faced their clients:

> *'. . . it is because if someone's disabled or if they're not able to get out at all to do any exercise, if they're confined to the house, then . . . As you probably know yourself, if you're sitting around in the house, you need to have the heating on in the winter. You're not creating that energy of your own to create warmth, so they're having to keep the heating on a lot, and that's so expensive. That is a massive concern for them. The last thing you want to do is make the person you're caring for even more unwell by them being cold. Yes, really difficult.'* [22]

As part of the same study, a householder described the effect of their cold home on their physical health conditions and their associated mental health:

> *'I suffer with really bad arthritis. When you feel cold, that arthritis pain is really up to a next level. I also have Crohn's disease, and when you get a bad infection or you get a flare it lowers your body temperature, and it makes you feel rubbish anyway, but when you're cold there's nothing worse. The last 12 to 18 months I've struggled not only physically but [also] mentally because of it . . . '* [22]

The psychological impact of fuel poverty relates, not only to poor thermal comfort, but also to the stress of managing household budgets and the potential stigma associated with being unable to cope. Research examining the lived experience of fuel poverty [23] has enhanced this understanding, drawing attention to less tangible but critically important impacts upon well-being and quality of life connected to, for example, educational attainment and child development, stigma and self-worth, and family and social life. The use of the term *'energy vulnerability'* alongside *fuel poverty* reflects a broadening focus beyond narrow technical specifications of fuel poverty towards a conceptualisation that places value on subjective experiences [24].

In a study in the English Midlands with mental health service users [20,25], the linkages between mental health and thermal comfort were clearly illustrated. Mental health service users spoke of the effect of being cold, not only in terms of temperature, but also in relation to the way this limited how they experienced their home. Where they were unable to achieve comfortable temperatures, largely out of concern about high energy bills, interviewees opted to remain in bed for longer or to limit the number of rooms they used – their homes and their indoor horizons becoming severely limited. Interviewees spoke of the stress of managing fuel bills as part of strained household finances and the impact of this on their anxiety levels, with the implication that they did not seek out reductions, such as the Warm Homes Discount, to which they were entitled. In some cases, service users had particular conditions, such aş fibromyalgia, that meant that physical pain was caused by cold conditions in the home. Householders in our study in the Outer Hebrides highlighted this close association between cold homes, mental health, and general well-being:

> *'... there comes a point with my mental illness, I've noticed that I don't even notice that I'm also cold... It's like anybody, if you're cold you're not comfortable and everything is harder.'* [9]

12.3.2 Diverse experiences

Research has highlighted the commonality and divergence of different groups in terms of their experience of and exposure to fuel poverty, including mental health service users [25], young adults [10], private rented tenants [26], and people in remote and rural locations (14). The potential risks appear to be particularly significant amongst children and young people, with living in such conditions being associated with greater instances of respiratory ailments [18,27,28]. Living in cold homes has also been shown to lead to multiple mental health risks in children in comparison to those living in homes with affordable warmth, manifesting as issues including school truancy, expulsion or exclusion, as well as issues with the police [27].

12.3.3 Home life

The notion of 'spatial shrink' [29] describes the practice of limiting the use of the house to avoid cold and damp areas and/or heating only a single space. This can impact household life, leading to overcrowding, as well as reduced privacy and personal space. A particular impact of this is on the provision of quiet study space for children.

In this example, from Better Housing Better Health in Oxfordshire, the parent could not afford to heat the whole house, so they chose to heat the child's room. Their use of the word 'obviously' highlights the instinctive nature of this:

> *'I could not put my heating on in my living room. I have a child who is 12. I had to heat her room, obviously. We would argue as a parent that their welfare is more important than the adult. It's not, of course, but that's our logic. So I'd heat her room but not heat the living room.'*

In our research in the Outer Hebrides [9], we conducted interviews with vulnerable people, often older and living in isolated locations in quite poor-quality housing. In this case, they needed to avoid the bedroom because the damp affected them even if they attempted to use the heating:

> *'We used to have that room as a bedroom, but we can't have it any more because it's so damp. It's too dangerous. I slept there last winter as well, and it [affected] my chest… I was getting migraines and everything, and that's with the heater on pretty high up'* [9].

12.3.4 A changing context

Although fuel poverty research tended to be dominated by studies in the UK and Ireland, there has in recent years been a growing interest in understanding how the issue is experienced elsewhere in the world and how the concept can be developed to include a diversity of climates, people, and building types [28,30,31]. An increasing tendency to refer to *energy poverty and energy justice*, rather than *fuel poverty*, reflects a recognition of the importance of cooling technologies and electric and electronic systems. Research in Australia, for example, has highlighted the importance of approaches to staying cool in hot weather, including air conditioning and architectural techniques as well as adaptive practices [32] and also drawn attention to the social gradient in access to suitable housing [33]. With heatwaves becoming more severe [34], and likely to continue to do so in the coming decades [35], this is becoming a more urgent issue in all settings.

12.4 Deciding on retrofit

12.4.1 Homeowners

A central challenge for scaling up retrofits relates to motivating homeowners to invest in such initiatives and, depending on their tenure, to seek out opportunities for improvements to be made or to liaise with their housing provider to request improvements. It is therefore important to understand how people make decisions about energy efficiency and what barriers and challenges they face in progressing it. As we have established, tenure is instrumental here. For some, retrofit is a personal or household decision; for others, it is something for their private or social landlord to determine.

For local authorities and housing providers, such decisions are likely to relate to policy targets, stock maintenance, and budget allocation. For homeowners, the decision-making process is likely to be more personal and relational and less purely economic, a reason why rational choice models have tended not to adequately explain why people engage with energy efficiency interventions [36]. People understand their energy use in relation to their personal circumstances and retrofit interventions should be promoted with these in mind [37].

In terms of the delivery of retrofit programmes, working with the owner-occupied sector is challenging. A study of the Green Deal Communities programme in Greater Manchester [38] highlighted the stark difference when compared to social housing, with which the large housing contractors tended to work. Rather than dealing with social housing businesses with an asset management strategy and managers overseeing the programme, working with homeowners involves a large number of individual entities:

> '*... as an organisation we don't traditionally deliver solely in private housing. So when you're working with a social housing provider you know the number of properties. You get given the addresses and all of the details, where[as] this became a B to C [business to consumer] role. So we were engaging with individual private householders, cold calling, door-knocking...* ' [38]

Contractors commented on the challenges this presents in terms of engaging with non-professionals, for whom the home is a place to live rather than an asset to be managed:

> *... the general public are difficult to deal with. Swapping and changing their mind, not understanding what they're getting, whereas when you're dealing with a social landlord you're actually dealing with a social landlord, not individuals...* ' [38]

The problem here is not the preferences of homeowners, of course, but the way schemes are designed without taking into account the need to understand the circumstances and requirements of individuals.

Research conducted by the Green Finance Institute examined the main barriers to investment in retrofit within each of the tenure categories [8]. Owner-occupiers range from those who are highly mortgaged to those who own their homes outright. Consequently, within this category, they found significant differences in terms of socio-economic status and access to expendable capital. Some common motivators do exist, however, including lower energy costs, owning a more environmentally friendly home, increasing comfort, health, and aesthetics, as well as increasing property value. Whilst it may be the case that homeowners are concerned about energy costs, this is not always the main driver. Common barriers identified within this group are the upfront costs associated with retrofit investment, a lack of access to capital, beliefs in potential long-term financial savings, and impact on property value.

12.4.2 Finances

The cost of retrofit is clearly a barrier to homeowners [7], but the ways in which householders think about the finances of retrofit are far from straightforward. In part, this relates to debt and attitudes towards it. From our discussions with householders in Greater Manchester [39], a low-interest loan was not enough to convince them to borrow substantial sums of money.

The payback period of the retrofit was considered by these householders, but this was also far from simple. A two-year payback might be seen as an easy, quick win that they might just go ahead with, but anything longer prompted consideration of their medium-term plans. If the payback period was longer than the time they planned to stay in the home, then it might be perceived as a poor investment. One householder commented, for example, that, rather than spend large sums retro-fitting their current property, they might as well save up so that their next house purchase could be a more energy-efficient property. Generally speaking, though, these discussions did not suggest a *calculative* approach, but rather a general sense of what it was appropriate to spend on the house and what levels of debt it was appropriate to be in. The source of the money can also be important: Middlemiss *et al.* [36] found, for example, that getting in debt from commercial lenders was not seen to be the 'proper way', whilst using money from an inheritance for renovations was seen as a good 'fit'.

In some cases, householders do not expect to recoup costs and may be happy to pay for improved comfort. They may also be interested in making an initial outlay in order to keep bills down in the future: a particular example being a couple who were investing in their home at the start of retirement, not expecting to necessarily recoup their costs per se, but looking forward to having manageable energy bills over their retirement. This is something we also saw amongst those building their own low-energy homes in Australia [32].

These financial motivations and barriers need to be understood in context. They are unlikely to be sufficient on their own: the right financial offer might nudge someone to progress their retrofit plans, but they are unlikely to create the desire or aspiration from nothing [40]. This means retrofit needs to be desirable in itself. This relates to less tangible factors, such as social norms and aspirations. We need to understand how retrofitting needs to become as sought after as a new kitchen or a conservatory.

An effective approach could be to connect retrofit to other home improvements and make clearer connections between retrofit and other aspects of the house. KFW in Germany offered funds to householders to make more general home improvements, providing they make energy efficiency interventions at the same time and the UK-based Ecology Building Society offers finance tied to energy performance [40].

12.4.3 Disruption

The disruptive nature of the implementation of retrofit can also be a barrier to take-up [41]. With deep retrofit in particular, it is likely that rooms – or the whole house – will be unusable for substantial periods of time and that there will be dust and noise. It may also be likely that occupants would have to move to temporary accommodation for some of the time and this adds to the disruption and cost. When speaking to householders in Greater Manchester, one referred to his partner's long-term health condition, ME, which would make it very difficult to cope with the disruption. In another local study [42], an occupant told us that various health issues meant that being in her social housing home whilst works were carried out was extremely difficult – the dust and noise, in particular, affected her.

12.4.4 Trust and access to contractors

Trust in contractors and tradespeople is very important [43] and the importance of intermediaries in helping homeowners identify reliable contractors to build trust in retrofit solutions has been acknowledged [44]. When speaking to householders in Greater Manchester [39], they shared concerns about being able to find contractors who would be able to deliver on time with high quality. In many cases, these concerns stemmed from previous experiences. They mentioned particular occasions where contractors demonstrated an inadequate understanding of what was needed to attain high standards of energy efficiency. Stakeholder interviewees in West Yorkshire also shared these frustrations and saw it as a barrier to progress on retrofit in general [40]. Programmes to build and equip the supply chain are therefore important and could include training and accreditation focused on energy efficiency. These reflections are from a householder in Greater Manchester; they relate to their difficulties in convincing the contractor to add insulation above the current regulated minimum:

> *The biggest problem we had was finding contractors who actually understood the retrofit process... who could do it and could do it to the right standard. We never did find them, or at least on the odd occasions we did they were booked up literally years in advance... our constant experience of all trades and all builders is that they will do what the building specs say they have to do, and they'll do that, and they'll do that religiously. Trying to move them one inch beyond that is all but impossible.*
>
> *I can't remember what the regulations were at that point. Maybe it was 150 millimetres of Kingspan or something. They would religiously put in 150 millimetres of Kingspan. If you said to them, 'Well, 250 millimetres will fit in there. Why don't we put in 250 millimetres?' 'Oh, there's no point in doing that. It's a waste of time.' You just could not get them to do it, to the extent that I used to come around in the evening sometimes and put in extra insulation.* [45]

Trust is a particularly pertinent issue when considering vulnerable people. In the Outer Hebrides, conversations with older people living in isolated conditions highlighted how difficult it can be to allow people to come into your home, inspect it, and work on it, even in a small way. An important role of the Moving Together programme was building trust, something that involved patient work with vulnerable householders. This relates to trust, not simply in the sense of knowing the contractors will install insulation correctly, but rather in a broader sense of being comfortable with strangers coming into and working on their home:

> *That's a good way of describing it, trust building; it is all about trust because we can't – we don't have enough time, or we don't feel we've got enough time, to gain their trust. Sometimes it can take months: when we get a phone call, we arrange a survey, and people don't always want that. They're not always sure who's coming round with us. When someone gains their trust it makes a big, big difference.* [9]

12.4.5 Private renting

Owners of property in the private rented sector range from landlords with one or two properties to professional and corporate landlords of multiple properties. Common barriers to retrofit investment include high uptake costs (increasing exponentially per the number of properties owned); access to capital (exacerbated by the illiquidity of property assets); investment not further contributing to rental income; leasehold restrictions; lack of awareness and understanding; regulatory uncertainty; and the duration, complexity, and hassle of installation [8].

A significant obstacle within this portion of the housing market, however, is that it is typically the tenant who is responsible for the payment of energy usage [46,47]. This means that any savings resulting from the property owner's investment in retrofit technology are ultimately passed on to the tenant. To counter this split-incentive issue, tenants can also contribute to the retrofit investment, a practice seen in some countries, such as France, Germany, and the Netherlands [48]. A challenge in the UK, however, is that renting tenants are typically younger, with less savings and expendable income and typically stay in properties for shorter durations.

12.4.6 Social renting

In the socially rented sector, the main incentives for social landlords are fuel poverty and energy efficiency targets, with those living in such accommodation disproportionately on low incomes, disabled, or retired – high-risk groups for fuel poverty [49,50].

The sector is markedly different from private rented and owner-occupied housing. Housing providers will have asset management strategies and dedicated, qualified staff. Relatively speaking, homes will generally have been treated consistently, with a more limited range of fabrics and technologies. Compare this to a terrace of owner-occupied housing in Greater Manchester with a succession of different owners over 100 years, each with a different approach to maintenance and renovation. The providers will be able to access government funding, although this will often not have been tied to energy efficiency as such, as much as to kitchens, bathrooms, and health and safety.

This is not to say that the sector is without challenges. Many housing providers will have 10-to-15-year maintenance cycles, with windows being replaced one year, kitchens another, and doors another. This can make it difficult to conduct full coordinated whole-house retrofits [40]. In common with the rest of the retrofit sector, tradespeople working in, or with, social housing may not have been trained in the latest technologies, meaning that installations of heat pumps and insulation need to be outsourced to a supply chain with its own limitations. Multi-occupancy building projects are also impacted by right-to-buy, with some properties no longer centrally controlled [51].

The relationship with occupants is important. In many cases, housing providers will have excellent relationships with tenants and effective mechanisms in place to engage with them. On the other hand, the tenants are more likely to be vulnerable in a range of ways, whether physical or mental health conditions, meaning that particular care needs to be taken to ensure that work does not impact well-being. High rates of unemployment mean that people will spend more time in the home and be more vulnerable to cold and damp conditions. Low literacy and learning difficulties mean housing providers need to take extra care to ensure that occupants understand the options available, what is being done to their home, and how to get the best out of the new features and technologies.

Research into the experience of social rented tenants shows that the experience of retrofit can be less about choice and more about consent, although this does not mean there are no opportunities for effective consultation and co-design. Residents expressed a need for greater accredited information about the process, as well as opportunities to ask questions [52]. With inadequate information and support comes a level of uncertainty, with residents potentially feeling pushed into decisions that they are uncomfortable with. This is concerning given that self-determination is a fundamental element of well-being [15].

12.5 Living through retrofit

As discussed earlier, the complexity and stress involved with the installation of retrofit technology is one of the most commonly identified barriers to investment in such initiatives [8,52]. The installation of retrofit measures requires multiple parties to work in the home for an extended period of time, which may serve to tip the balance in terms of the apparent pros and cons of retrofit investment, particularly in the case of owner-occupied properties. Furthermore, regardless of tenure type, the installation of retrofit measures will have an impact on the lives of occupants.

12.5.1 Being retrofitted

In a study of residents' experiences of an urban regeneration in the North West of England, which involved retrofitting multi-occupancy buildings with a heat pump [53], the impact of the scale and duration of the renovation was highlighted: '*And they came along and for 18 months they went form the top floor and worked down. Working round the block. So every – all the time, you've got drilling. For 18 months.'* [*54*].

An exacerbating factor highlighted by the participants of this study was the lack of clarity on the programme of delivery, which added further to the disruptive nature of the implementation of measures: '*You never knew who was going to knock on your door, with what, when, how. You know, to do this job or another. There wasn't seen to be a set programme'* [54]. This is an example, then, of a process that had some intrinsic level of disruption and was made worse by a lack of clear communication with residents. Similar issues were highlighted in the regeneration project in the North West of England: '*But certainly the noise, that started. It was 18 months. Then they came and said "we're going to re-wire" and we were thrown out, as it were. We had to go and – there was no accommodation for us . . . '* [54].

Our research into retrofit programmes in Greater Manchester provides further evidence of the disruptive nature of the retrofit process. Whilst, as above, many of the issues reflect challenges intrinsic to retrofit, such as noise, dust, room clearances, and needing at times to vacate the house, it is hard not to conclude that issues could be more effectively mitigated through careful delivery, project management, and engagement with householders. To some extent, this implies a need to manage expectations. One householder told us, for example, *'Well, to be fair, I had this idea like on television where you have a makeover and you would have a lovely house afterwards! Nothing was further from the truth!' (43).* It also highlights the importance, if a retrofit is to become mainstream, of providing a service that people want to recommend to their friends and neighbours and maybe even show off on television.

Reported issues included scaffolding being left installed on a house for weeks, during which no work was being carried out. This left an elderly social housing tenant with young people climbing on the scaffolding and causing her to fear damage or intrusion. Another householder in a similar position noted the increased risk of burglary and the lack of natural light. In another case, householders were asked to clear their front room, the walls were stripped of plaster and then there were significant, but predictable, delays in the delivery of the internal wall insulation, leaving them with extreme cold during the winter months. Both of these indicate challenges with coordination and timing and highlight the impact on occupants.

In one example, a contractor did not, in the view of this householder, show respect for their property: '*The first thing they did was rip off a 180-year-old skirting-board and say, "We'll get you a new one, you don't want that old thing". I'm going, "That's my skirting-board!" They drilled holes in it and burnt it, and I went bananas' (43).*

During the conversations, householders described to us the implications of these stresses for them, including pressure on their mental health, accentuation of a diversity of conditions, such as ME and asthma, and difficulty sleeping and managing shift work. What this research again shows is the potential for the process of implementation to undermine the intended benefits of retrofit initiatives. Fear of intrusion or damage to one's property can undermine the psychological need for shelter and safety, two fundamental elements of human well-being [16].

The above examples relate to social housing, and in such schemes, occupants often feel that retrofit is being 'done to them'. Even when work is free or funded by Local Authority or Government schemes, there is a risk that contractors feel they can cut corners and this can be particularly problematic for homeowners: *'We're not renting this house, this is not a private association, I'm paying a mortgage on this house… if you don't do a proper job I'm going to have to pay to have that job done properly'*. Even when financial support is available, it may not cover all of the costs, and a household may be incompletely informed about the implications of this: '*We knew we had to do decoration, but no way the level of decoration that we ended up with, honestly. We had no idea of that much… It's costing us as much now to get the house right the way you want it as it has actually cost with the work concerned'* [42].

12.5.2 Mitigation: decanting

One approach to reducing the level of disruption is to relocate occupants to other buildings, referred to as decanting. Whilst in itself an upheaval, the exposure to noise, dust and other issues is minimised. In their Enerphit retrofit of two social housing blocks, One Manchester [55] decanted the tenants. Occupants appreciated the need for this, but our conversations with them show that decanting is not necessarily a simple, pain-free process. There was some frustration, for example, over communications, particularly in relation to the length of time that they would have to stay in their temporary homes and, therefore, the importance of the role of tenant liaison officer was recognised. These difficulties arose, in part, from unexpected issues with the works that meant the retrofit process lasted longer than expected.

The suitability of the temporary homes is a consideration. One occupant found themselves in an upper-floor property but had mobility issues that made climbing the stairs difficult. Others reflected on the difficulties of visiting, in particular, their regular pharmacies and doctors. Another was concerned about the plants in their garden and was assured that it would only be for a short amount of time and that the plants would be looked after, but they found on their return that the garden had not been treated with respect: *'But when we came back everything was gone, broken - just dumped in the truck and away it went!'* [55].

Similar findings were evident in the research on regeneration in the North West of England [53], with residents identifying the temporary accommodation as inadequate for human habitation: *'This water was coming into the living room. Which there was no carpet down or anything, but… so I quickly shot up, picked my – you know the plugs that you've got the sockets in… The floor was soaking. I thought: what do I do about this now?'*

We have outlined how the process of implementing retrofit measures may undermine the home's capacity to ensure well-being. Theoretically, decanting can ensure that alternate accommodation temporarily replaces the home as a source of shelter and safety but what these accounts highlight is that such accommodation measures must be adequate to elicit the desired effect. Unforeseen relocation with little prior notice removes people from their home – their primary source of shelter and safety – and limits the sense of control one has over one's circumstances.

12.6 Living with retrofit

12.6.1 A performance (and expectations) gap

At this point in the process, the installation of the retrofit measures is complete but the retrofit journey is far from finished. Although pre-performance testing may demonstrate the increased energy efficiency of the technology and fabric installed, the predicted improvements in performance are not guaranteed. In fact, a gap between predicted and actual performance is a common occurrence [56]. Understanding occupant behaviour is an essential part of getting the best out of the retrofit process in terms of both householder satisfaction and energy consumption.

Energy performance is a product of fabric, system, and operation, and patterns of energy use are mediated by human factors including social norms, habits, and values [57]. Whilst it is not uncommon to talk about occupant behaviour as a problem – 'wrong actions' and 'erroneous habits', for example [58] – the broader point is that retrofit processes need to anticipate the needs of occupants and take the opportunity to engage with them. Whilst commonly conceptualised as a performance risk, occupant behaviour is, conversely, an opportunity to maximise savings by taking the opportunity to engage with residents on energy. Energy-consuming activities tend to be the result of routines and habits, making it difficult for people to connect these activities with energy consumption [59]. Retrofit presents an opportunity to disrupt and challenge these habits.

An example of the performance gap and the role of occupant practices is provided by Wingfield [60]. Although the buildings in the commercial housing development they studied were predicted by building performance models to make energy savings possible in terms of structural performance, once occupied, the actual carbon emissions were approximately 20–25 per cent higher than anticipated. Exploring this issue further, the energy use of four households was monitored for over a year. The results revealed differing levels of energy performance to be related to the manner in which residents operated the heating system. For instance, higher levels of energy use were seen in households engaging in a more 'interventionist' approach to heating: turning the heating off for several days at a time, followed by intermittent days of use to boost home temperature. Conversely, households displaying greater levels of energy efficiency maintained fixed timings of heating use. Another observed difference was the level of moisture in homes, which was shown to be caused by different approaches to using ventilation through the additional opening of windows and external doors during winter months, a behaviour also contributing to greater levels of heat loss.

Retrofit may 'disrupt' energy use habits, and new technologies and levels of comfort may create new patterns of living in the home that have implications for energy use. The rebound effect, for example, describes the action of people using more energy because they believe that the building is more energy-efficient [61]. Occupants may choose to benefit from retrofit or reduced energy costs by increasing their levels of comfort, rather than necessarily reducing energy consumption – referred to as the 'take-back' effect [62] or 'taking comfort' [52]. People may choose to begin using parts of the house that were otherwise difficult to heat, described as 'spatial rebound'. Additionally, they may simply manage their retrofitted home in ways that appear wasteful in terms of energy consumption, such as opening windows instead of turning down the thermostat but in fact reflect personal preferences relating to, for example, ventilation and fresh air.

It is important to note that this is not about 'blaming' occupants for potential savings not being realised; it is about accepting that modelling potential savings is a technical exercise that cannot anticipate all the ways in which individuals will enact habitation in their new dwelling. Sherriff *et al.* [2] explore this in interviews with fuel poverty experts. They caution against a 'faulty consumers narrative', arguing that there is sometimes little that occupants can do to improve their energy efficiency. This is particularly true of people experiencing fuel poverty, those in rented homes, and those with medical conditions that require warm environments, specialist equipment, or frequent washing loads.

12.6.2 Information and handover

Another aspect of this challenge relates to the importance of providing information and guidance on living with new technology or other retrofit measures. This is particularly important in rented housing, since tenants are less likely to have been involved in researching and choosing the measures. Research, for example, has questioned the extent to which communication about new systems is effective [52], showing that occupants often have difficulty operating new systems efficiently. This may be problematic, with perceived behavioural control outlined as a significant factor in behavioural decision-making [63]. Those interviewees who had greater levels of engagement suggested that those overseeing the maintenance of the system should seek to prompt people to ask for advice, to make this process easier. Others reported that when dealing with teething issues, they were made to feel unintelligent by the people offering support. Furthermore, residents described how they often relied on relatives, friends, and neighbours for guidance on how to effectively use the system. Once more, psychological approaches to behavioural decision-making [64] would outline this as being problematic, with the attitudes of residents towards their ability to utilise technological changes crucial to the formation of their behavioural intentions and, in turn, their resulting behaviour.

12.6.3 Heat pumps: a particular challenge

Some of the projects we have studied involved the installation of heat pumps and the experiences of the residents in relation to these highlight the particular importance of information and ongoing handover. The technology is a radical change from the gas central heating that dominates the UK and requires a conceptual leap in terms of operation. If the rollout of heat pumps is to happen at the pace as planned, then there is a need to urgently understand how to effectively manage take-up and avoid a combination of ineffective heating and high energy bills.

When compared with the conventional central wet heating systems, to which people in the UK are accustomed, heat pumps require a different approach. The units require a conceptual shift in that they should, in general, be left on and controlled via the thermostat. If the units are switched off, then it can cost time and money for the units to return to operating level.

In a retrofit development in Salford's Pendleton, where air-source heat pumps had been installed across a large number of properties managing on tight budgets, there were challenges relating to the operation of the units and the financial implications of these [65]. Our research was focused on the idea of 'energy champions' but it quickly became clear that any energy programme had to first address the challenges people were experiencing with the newly installed air-source heat pumps. Whereas they would have been accustomed to turning off heating at night, when out of the house, or even at the end of the month if household budgets were running low, doing this could now result in extra expense to return the machines to operational temperature.

At the time on the social housing estate, word of mouth was causing residents to be concerned about receiving high energy bills and it was our impression that this was prompting some other residents to turn off the units, therefore accentuating the issue. It appeared that a combination of the imposition of a new technology, a relatively low level of information about how to operate it, and the spread of news about some cases of high bills created a perfect storm. In time, instructional videos were produced and further engagement was conducted with occupants, but staff involved might now reflect that they underestimated the amount of focused support that was needed for occupants to transition to heat pumps. This would apply in any housing situation, but the fact that occupants would have been dependent on benefits and/or in low-paid work meant that the margin of acceptable error was very small. Whilst a middle-class household would be able to withstand a moderate fluctuation in energy bills, for these tenants, ten pounds extra could have made a large dent in the household food budget.

In another housing project, the combination of unfamiliarity and concern about costs was particularly problematic in a development for recovering addicts. A relatively high turnover of people coming for short periods of support meant that air-source heat pumps provided a particular communications challenge:

> *'Then the next resident is coming in, turning it on, because it's been off for so long, and the way air source pumps work, it's working its socks off to try and get the property – the property is freezing, all the liquid in the*

system is freezing and it's working its socks off to get it up to that temperature again... That resident is... then saying 'It's costing me a fortune to get it up to temperature' but it's the way it is because it's not been maintained the way it should be...' [42]

A social landlord explained the challenges of installing heat pumps, referring to the dilemma of aiming to decarbonise, whilst at the same time not increasing tenants' energy expenditure:

... 'low-carbon doesn't mean low cost... our tenants were really unsure of [the heat pumps], and we spent a long time and some energy... People had been treating them like boilers and switching them on and off, which they shouldn't be doing, and they weren't changing the filters, which choked them up... It's no use just sticking in lots of nice blinging technology if we don't bring the tenants along for them to just have a voice and maybe just either leaving folk open to fuel poverty or opening them up to it.' [14,64]

12.6.4 Achieving comfort

In our conversations with occupants in recently retrofitted homes, we have received a lot of feedback of a positive nature and it is fair to say that the dominant thread relates to the benefits of the changes made to their home. In many cases, the disruption had been worth it, even if they would have preferred a smoother process. This is not to say that there are no problems, and this is something we come to below.

In cases where occupants are content with the retrofit, comments focus on the temperature and the associated increase in comfort. These relate not only to the temperature per se but also to a sense of control over the indoor climate:

'When the heating's on, the house heats up quicker and it stays for very much longer. Particularly overnight; like today, I came downstairs, so at seven o'clock in the morning... in here it was 16.' [42]

'The temperature of the house is decided by how much we want to spend on gas, rather than the limits of the house's fabric. So, if we wanted to, this place could be the Bahamas all day, every day.' [42]

A sense of control is positive for well-being [15] and such issues may inhibit occupants' capacity to determine their level of thermal comfort in the home, reducing its capacity to serve as a source of shelter.

These positive experiences were related, although not always, to a decrease in expenditure. Some people recognise the reduced environmental impact, but it is fair to say that this is not a dominant theme:

'Less damp, much more stable temperature, warmer generally and, yes, the saving money, those are the major things. Anything else? I think it's just comfort and money really, and obviously the consequence of less money, lower carbon footprint, which is nice.' [42]

In our study of a retrofit project in Zaragoza (43 pre-intervention survey responses and 33 post-intervention), we found that residents were much more likely to agree with the statements 'My home is warm enough in the winter' and 'My home is cool enough in the summer' following retrofit, and less likely to feel that their apartment is draughty [66]. Almost all residents said their flats were 'slightly warmer' or 'much warmer' and the process of controlling temperature was either 'slightly easier' or 'much easier'. Residents continued to be concerned about their electricity bills, but fewer said this was the case.

Amongst these generally positive responses, which tend to reinforce confidence in retrofit as an approach overall, some residents expressed concerns about specific aspects of their home following retrofit. Such concerns highlight the importance of understanding resident experiences, as they often contradict what the monitoring is showing. Occupants have told us about, for example, new draughts, an inability to achieve the desired temperature, unexpected cost increases and, in some cases, associated health impacts. These tend to be a minority of cases, but they highlight the importance of speaking to occupants to understand the ways in which retrofit is affecting them. Even if the temperature and humidity monitoring indicate that there are no issues, these may not provide the whole picture or may not take into account the specific needs of occupants.

12.6.5 Feeling the benefits

We mentioned spatial rebound above. This is something that some householders talked about. Where previously they used to gather in one room during cold periods, for example, it was now possible, through the use of heating and greater control over temperature, to use more of the house. In some cases, this reflected the new, lower cost of getting those spaces to a comfortable temperature, but in others it was simply a result of the ability to achieve those temperatures in rooms that had previously been too 'leaky' to be worth heating. This interviewee referred to being able to be more casual about how they used the house, no longer needing to plan rooms ahead and to warm up difficult-to-heat rooms in advance of use:

> *'We use this room more. This is our living room and we used to [use it] only use it with preparation. So we'd put the fire on and so on, in the winter anyway, before watching the telly or something. Now we use it much more casually, so we'll actually hang out in here just to hang out, rather than specifically to do something that requires being in here.'* [42]

Householders in the Outer Hebrides talked about how retrofit had affected how they use the house, illustrating the impact of this, especially when housing conditions were quite poor, on how people use their home. Being able to use the whole house more, as a result of the better heating and insultation, made this householder feel better about their ability to keep things organised better in the house:

> *'although the work has only just been completed, I can already see that I'll enjoy going upstairs much more... [this] should mean that I'll be able to keep things organised much better* [9].

We were astonished by the difference in temperature even on the first day of the works. It is marvellous to have the whole house warm instead of a cold upper floor and a warm ground floor. [9]

In a difficult climate, they also found that they were sheltered from the noise of storms: "I can now stay upstairs in my bed in a storm", whereas they were 'always having to flee downstairs to the sofa before…' [9]

As a result of help with household improvements and energy advice as part of this initiative, one occupant in the Outer Hebrides was able to save substantial funds on energy and, as a result, purchase a car. Given his remote location and the limited rural bus services, this made a significant difference to his ability to shop, see friends, and attend medical appointments, contributing to an improvement in overall well-being. Another had received help with managing debt and described to health workers how this, and knowing that telephone support is available, had helped with their stress and ability to sleep [9]. These accounts illustrate the potential that retrofitting has to contribute to broader elements of well-being. Not only can it enable a home to better support the need for shelter, safety, and health, but as a result of the associated financial savings, individuals experience greater freedom to engage in their desired level of social activity, satisfying the human need for psychological connectedness.

Outer Hebrides residents commented on well-being in a more general sense, particularly their anxiety about energy bills and the deterioration of the home, as this project stakeholder reflected:

It's a physical improvement, but it's also a mental improvement, because they know the house is warmer and they know they don't have to have the heating on as long, therefore they won't be hopefully spending as much on cost of fuel, whether it be oil or coal or whatever. So you're helping them twofold… [9]

In a particularly striking example that, whilst by no means representative of experiences across the board, evidences the extent of change that is possible, one couple in Greater Manchester told us that they had quit smoking as a result of the increased air tightness of the house and, due to cost savings from this and the greater energy efficiency, had improved their diet:

Interviewee 1: No, but even with the house being like it is, warmer. We feel the benefits of everything and, like I say, you are eating better because the simple reason is we're saving more money, we spend more money.
Interviewee 2: What we do now, we have a cooked, full cooked meal every day.
Interviewee 1: Every teatime, don't we?
Interviewee 2: Everything teatime. We'll go lamb chops, chicken.
Interviewee 1: Fish, whatever.
Interviewee 2: Every veg you can name is in there now… [42]

12.6.6 Beyond energy

It is worth noting that whilst energy researchers often talk about comfort in terms of temperature and humidity, occupants are likely to see it in a wider sense that includes the factors that help them feel 'at home'. In this sense, energy and its services are part of a process of making a home: Butler and Sherriff [10] found that young people placed importance on the capacity to make changes to a living space and 'make it your own'.

There are sometimes surprising factors that shape post-retrofit experiences. In an Enerphit retrofit of social housing in Manchester [55], residents were on the whole very happy with the temperature of their apartments, noting the impact of this on their comfort and their energy bills. They felt restricted in other ways, though, noting the restrictions in place on affixing shelves and pictures to walls, the lack of their own letterbox, and having to rely on window cleaning being provided by the housing association. One resident was very positive about the thermal performance of the flat but missed being able to install their own satellite dish to pick up channels from their country of origin. A communal satellite dish had been provided, but this was configured for UK channels.

For some, elements of retrofit had been an imposition. This is particularly the case in social housing, where the tenants are less likely than owner-occupiers to be involved in the design and specification of the schemes and also likely to have smaller living areas where the introduction of new technologies and heating and cooling units can have a more significant impact. Examples include large heat pumps and similar units that take up space in kitchens or storage cupboards. Even if they improve thermal comfort, they have detracted from other aspects of the home that relate to comfort in a broader sense.

12.7 Summary

Retrofit measures enhance thermal comfort and address fuel poverty through the improvement of homes. Following retrofit, occupants generally report enhanced comfort, greater control over indoor climate, reduced costs, and environmental benefits. Occupants are crucial stakeholders in retrofit processes, influencing decisions and financing. Meaningful engagement with occupants, considering their diverse financial situations, tenures, and housing pathways, is therefore essential in enabling the development of tailored strategies to increase the take-up of retrofit measures and the successful delivery of their intended benefits. Decision-making is shaped by personal, social, and economic factors.

Tenure matters; homeowners face obstacles such as capital costs, access to funding, and concerns about long-term savings and property value. Many landlords in the private rented sector are not motivated to engage with retrofit, especially when, as per the split-incentive problem, it is their tenants that stand to gain financially. With comparatively more consistent stock, maintenance staff, and asset management strategies, social housing providers are in a good position to retrofit at scale but may struggle to find funding, as well as contractors skilled in relevant trades.

The implementation of retrofit measures can be highly disruptive to the lives of occupants, particularly those who are vulnerable, thus requiring measures to alleviate stress. Concerns about disruption, including noise, air quality, and relocation needs, can also pose a barrier to take-up, further highlighting the importance of transparent communication and of building trust in contractors through high-quality work and reliable delivery.

Retrofit measures are essential for enhancing energy efficiency, and maximising their benefits hinges on understanding occupant behaviour. Following retrofit information and guidance on new technologies is vital: a reflection that energy performance is shaped not only by technical elements but also by personal preferences, habits, and social norms. Whilst effective engagement with occupant behaviour is often perceived as a challenge, it presents an opportunity to engage with residents and maximise energy savings. Nonetheless, retrofitting can disrupt habitual energy consumption patterns and, potentially, lead to rebound effects, as occupants seek to maximise their comfort. In rented housing in particular, having had little involvement in the details of retrofitting, occupants may struggle to optimise systems unless sufficient information and guidance are provided.

What the literature discussed in this chapter highlights is that retrofit solutions are not merely an engineering and technical challenge but a complex interaction of sociological, psychological, legislative, and infrastructural factors. To capitalise on the myriad potential benefits requires a firm commitment to interprofessional and interdisciplinary practice and, crucially, meaningful engagement with the most important of stakeholders: the occupants themselves.

References

[1] Boardman B. 'Fuel poverty synthesis: Lessons learnt, actions needed'. *Energy Policy*. 2012;49:143–8

[2] Sherriff G., Butler D., Brown P. '"The reduction of fuel poverty may be lost in the rush to decarbonise": Six research risks at the intersection of fuel poverty, climate change and decarbonisation'. *People, Place and Policy Online*. 2022;**16**(1):116–35

[3] Shove E. 'Beyond the ABC: Climate change policy and theories of social change'. *Environment and Planning A: Economy and Space*. 2010;**42**(6):1273–85

[4] James M., Ambrose M. 'Retrofit or behaviour change? Which has the greater impact on energy consumption in low income households?'. *Procedia Engineering*. 2017;180:1558–67

[5] Karvonen A. 'Towards systemic domestic retrofit: A social practices approach'. *Building Research & Information*. 2013;**41**(5):563–74

[6] Whitmarsh L., O'Neill S. 'Green identity, green living? The role of pro-environmental self-identity in determining consistency across diverse pro-environmental behaviours'. *Journal of Environmental Psychology*. 2010;**30**(3):305–14

[7] Achtnicht M., Madlener R. 'Factors influencing German house owners' preferences on energy retrofits'. *Energy Policy*. 2014;68:254–63

[8] Green Finance Institute. '*Financing energy efficient buildings: The path to retrofit at scale*' [online]. 2020. Available from https://www.greenfinanceinstitute.co.uk/wp-content/uploads/2020/06/Financing-energy-efficient-buildings-the-path-to-retrofit-at-scale.pdf [Accessed 15 December 2020]

[9] Sherriff G., Lawler C., Martin P.B., Butler D.E., Probin M., Brown P. *Reshaping Health Services and Fuel Poverty in the Outer Hebrides* [online]. Salford, UK: University of Salford; 2020. Available from https://salford-repository.worktribe.com/output/1358036 [Accessed 13 January 2021]

[10] Butler D., Sherriff G. '"It's normal to have damp": Using a qualitative psychological approach to analyse the lived experience of energy vulnerability among young adult households'. *Indoor and Built Environment*. 2017; **26**(7):964–79

[11] Brown P., Swan W., Chahal S. 'Retrofitting social housing: Reflections by tenants on adopting and living with retrofit technology'. *Energy Efficiency*. 2014;**7**(1):1–16

[12] Hulme J. 'England lessons from delivering Decent Homes and affordable warmth' in Nieboer N., Tsenkova S., Gruis V., van Hal A. (eds.). *Energy Efficiency in Housing Management: Policies and Practice in Eleven Countries*. London: Routledge; 2012, pp. 97–114

[13] Guy S., Brand R., Henshaw V., Karvonen A., Lewis A., Sherriff G. *Building Comfort for Older Age: Designing and Managing Thermal Comfort in Low Carbon Housing for Older People*. Manchester: University of Manchester; 2013.

[14] Simcock N., Frankowski J., Bouzarovski S. 'Rendered invisible: Institutional misrecognition and the reproduction of energy poverty'. *Geoforum*. 2021; 124:1–9

[15] Ryan R.M., Deci E.L. 'On happiness and human potentials: A review of research on hedonic and eudaimonic well-being'. *Annual Review of Psychology*. 2001;**52**(1):141–66

[16] Maslow A.H. *Motivation and Personality*. New York: Harper & Row; 1970.

[17] Howden-Chapman P., Matheson A., Crane J., *et al.* 'Effect of insulating existing houses on health inequality: Cluster randomised study in the community'. *British Medical Journal*. 2007;**334**(7591):460

[18] Liddell C., Morris C. 'Fuel poverty and human health: A review of recent evidence'. *Energy Policy*. 2010;**38**(6):2987–97

[19] University College London, Marmot Review Team, Friends of the Earth. *The Health Impacts of Cold Homes and Fuel Poverty* [online]. London: Friends of the Earth & the Marmot Review Team; 2011. Available from http://www.instituteofhealthequity.org/projects/the-health-impacts-of-cold-homes-and-fuel-poverty [Accessed 6 November 2018]

[20] Sherriff G. *Changes4Warmth: Lessons Learned & Directions Forward* [online]. Salford: University of Salford; 2016. Available from http://usir.salford.ac.uk/id/eprint/41565/ [Accessed 13 January 2021]

[21] Gilbertson J., Green G. *Warm Front Better Health: Health Impact Evaluation of the Warm Front Scheme*. Sheffield: Sheffield Hallam University; 2008.

[22] Sherriff G., Young D. *Better Housing Better Health: A Qualitative Study of Energy Advice and Support in Oxfordshire* [online]. Salford, UK: University of Salford; 2024. Available from https://salford-repository.worktribe.com/output/2674439 [Accessed 23 August 2026]

[23] Middlemiss L., Gillard R. 'Fuel poverty from the bottom-up: Characterising household energy vulnerability through the lived experience of the fuel poor'. *Energy Research & Social Science*. 2015;6:146–54

[24] Longhurst N., Hargreaves T. 'Emotions and fuel poverty: The lived experience of social housing tenants in the United Kingdom'. *Energy Research & Social Science*. 2019;56:101207

[25] Sherriff G. *I was Frightened to put the Heating on*. 'Evaluating the Changes4Warmth Approach to Cold Homes and Mental Health [online]. Salford: University of Salford; 2016, p. 18. Available from https://salford-repository.worktribe.com/output/1403718/ [Accessed 23 August 2026]

[26] Ambrose A.R. 'Improving energy efficiency in private rented housing: Why don't landlords act?'. *Indoor and Built Environment*. 2015;**24**(7):913–24

[27] Barnes M., Butt S., Tomaszewski W. *EU Energy Poverty Observatory. The Dynamics of Bad Housing: The Impact of Bad Housing on the Living Standards of Children* [online]. 2017. Available from https://www.fuelpovertylibrary.info/content/dynamics-bad-housing-living-standards-children-%E2%80%93-evidence-families-and-children-study-facs-1 [Accessed 12 January 2021]

[28] Howden-Chapman P., Pierse N., Nicholls S., *et al.* 'Effects of improved home heating on asthma in community dwelling children: Randomised controlled trial'. *British Medical Journal*. 2008;337:a1411

[29] Grey C.N.B., Schmieder-Gaite T., Jiang S., Nascimento C., Poortinga W. 'Cold homes, fuel poverty and energy efficiency improvements: A longitudinal focus group approach'. *Indoor and Built Environment*. 2017;**26**(7):902–13

[30] Bouzarovski S. *Energy Poverty: (Dis)Assembling Europe's Infrastructural Divide*. Cham, Switzerland: Palgrave Macmillan; 2018.

[31] Castaño-Rosa R., Sherriff G., Solís-Guzmán J., Marrero M. 'The validity of the index of vulnerable homes: Evidence from consumers vulnerable to energy poverty in the UK'. *Energy Sources, Part B: Economics, Planning, and Policy*. 2020:1–20

[32] Sherriff G., Moore T., Berry S., Ambrose A., Goodchild B., Maye-Banbury A. 'Coping with extremes, creating comfort: User experiences of 'low-energy' homes in Australia'. *Energy Research & Social Science*. 2019;51:44–54

[33] Nicholls D.L., McCann H., Strengers D.Y., Bosomworth D.K. '*Electricity pricing, heatwaves & household vulnerability in Australia*'. Melbourne: RMIT; 2017

[34] Sukanen H., Taylor J., Castaño-Rosa R., Pelsmakers S., Lehtinen T., Kaasalainen T. 'Passive mitigation of overheating in Finnish apartments under current and future climates'. *Indoor and Built Environment*. 2023;**32**(7):1372–92

[35] Anderson G.B., Oleson K.W., Jones B., Peng R.D. 'Classifying heatwaves: Developing health-based models to predict high-mortality versus moderate United States heatwaves'. *Climatic Change*. 2018;**146**(3):439–53

[36] Middlemiss L., Davis M., Brown D., *et al.* 'Developing a relational approach to energy demand: A methodological and conceptual guide'. *Energy Research & Social Science*. 2024;110:103441

[37] McMakin A.H., Malone E.L., Lundgren R.E. 'Motivating residents to conserve energy without financial incentives'. *Environment and Behavior*. 2002; **34**(6):848–63

[38] Sherriff G.A., Swan W. '*Greater Manchester green deal communities programme scheme exit paper*' [online]. Greater Manchester Combined Authority; 2016. Available from https://salford-repository.worktribe.com/output/1392673 [Accessed 13 January 2021]

[39] Sherriff G. '*Your home better: Qualitative customer research*' [online]. 2023. Available from https://salford-repository.worktribe.com/output/2273821/your-home-better-customer-research [Accessed 23 August 2026]

[40] Red. '*Scaling up better homes Yorkshire - Executive summary*' [online]. 2020. Available from https://westyorkshire.moderngov.co.uk/documents/s17777/Item%206%20-Appendix%202%20SUBHY_Full_Report.pdf [Accessed 9 June 2026]

[41] Pillai A., Tovar Reaños M., Curtis J. 'An examination of energy efficiency retrofit scheme applications by low-income households in Ireland'. *Heliyon*. 2021;**7**(10):e08205

[42] Sherriff G. 'Green deal go early: Occupant research'. Salford: University of Salford; 2013.

[43] Yau Y., Hou H., Yip K.C., Qian Q.K. 'Transaction cost and agency perspectives on eco-certification of existing buildings: A study of Hong Kong'. *Energies*. 2021;**14**(19):6375

[44] de Wilde M. 'The sustainable housing question: On the role of interpersonal, impersonal and professional trust in low-carbon retrofit decisions by homeowners'. *Energy Research & Social Science*. 2019;51:138–47

[45] Sherriff G. '*Your home better: Qualitative customer research (Additional Unpublished Research)*' [online]. 2023. Available from https://salford-repository.worktribe.com/output/2273821 [Accessed 9 June 2026]

[46] Bird S., Hernández D. 'Policy options for the split incentive: Increasing energy efficiency for low-income renters'. *Energy Policy*. 2012;48:506–14

[47] Lang M., Lane R., Zhao K., Tham S., Woolfe K., Raven R. 'Systematic review: Landlords' willingness to retrofit energy efficiency improvements'. *Journal of Cleaner Production*. 2021;303:127041

[48] European Commission. *Joint Research Centre. Institute for Energy and Transport. Overcoming the Split Incentive Barrier in the Building Sector:*

Workshop Summary [online]. Luxembourg: Publications Office; 2014. Available from https://data.europa.eu/doi/10.2790/30582 [Accessed 22 July 2021]

[49] Snell C., Bevan M., Thomson H. 'Justice, fuel poverty and disabled people in England'. *Energy Research & Social Science*. 2015;10:123–32

[50] Pitt J., Nolden C. 'Post-subsidy solar PV business models to tackle fuel poverty in multi-occupancy social housing'. *Energies*. 2020;**13**(18):4852

[51] Bright S., Weatherall D., Willis R. 'Exploring the complexities of energy retrofit in mixed tenure social housing: A case study from England, UK'. *Energy Efficiency*. 2019;**12**(1):157–74

[52] Chahal S., Swan W., Brown P. 'Tenant perceptions and experiences of retrofit'. *Proceedings of Retrofit 2012 Conference*, Salford, UK, Jan 2012. Salford: University of Salford; 2012, pp. 24–26

[53] Lomas M.J., Ayodeji E., Brown P. 'Experiences of place attachment and mental wellbeing in the context of urban regeneration'. *Health & Place*. 2021;70:102604

[54] Lomas M.J., Ayodeji E., Brown P. 'Experiences of place attachment and mental wellbeing in the context of urban regeneration (Additional unpublished research)'. *Health & Place*. 2021;70:102604

[55] Sherriff G., Martin P., Roberts B. *Erneley Close Passive House Retrofit: Resident Experiences and Building Performance in Retrofit to Passive House Standard* [online]. Salford: University of Salford; 2018. Available from http://usir.salford.ac.uk/id/eprint/46328/ [Accessed 13 January 2021]

[56] Salvia G., Morello E., Rotondo F., *et al.* 'Performance gap and occupant behavior in building retrofit: Focus on dynamics of change and continuity in the practice of indoor heating'. *Sustainability*. 2020;**12**(14):5820

[57] Summerfield A.J., Pathan A., Lowe R.J., Oreszczyn T. 'Changes in energy demand from low-energy homes'. *Building Research & Information*. 2010; **38**(1):42–9

[58] Ascione F., Bianco N., De Masi R.F., Mastellone M., Mauro G.M., Vanoli G.P. 'The role of the occupant behavior in affecting the feasibility of energy refurbishment of residential buildings: Typical effective retrofits compromised by typical wrong habits'. *Energy and Buildings*. 2020;223:110217

[59] Hargreaves T., Nye M., Burgess J. 'Making energy visible: A qualitative field study of how householders interact with feedback from smart energy monitors'. *Energy Policy*. 2010;**38**(10):6111–9

[60] Wingfield J., Bell M., Miles-Shenton D., South T., Lowe B. '*Evaluating the impact of an enhanced energy performance standard on load-bearing masonry domestic construction: Understanding the gap between designed and real performance: lessons from Stamford Brook*' [online] 2011. Available from https://eprints.leedsbeckett.ac.uk/id/eprint/889/ [Accessed 9 June 2026].

[61] Jenkins D.P. 'The value of retrofitting carbon-saving measures into fuel poor social housing'. *Energy Policy*. 2010;**38**(2):832–9

[62] Stafford A., Gorse C., Shao L. *The Retrofit Challenge: Delivering Low Carbon Buildings* [online]. Centre for Low Carbon Futures, York, 2011. Available from https://www.2020climategroup.org.uk/wp-content/uploads/2012/01/Retrofit_Challenge_2011.pdf [Accessed 9 June 2026]. 2011, pp. 1–32

[63] Ajzen I. 'The theory of planned behavior'. *Organizational Behavior and Human Decision Processes*. 1991;**50**(2):179–211

[64] Dillard J.P., Pfau M. (eds.). *The Persuasion Handbook: Developments in Theory and Practice.* Thousand Oaks, CA: SAGE Publications; 2002.

[65] Sherriff G., Martin P. *Championing Energy in Pendleton* [online]. Salford: University of Salford; 2016. Available from https://salford-repository.worktribe.com/output/1405753 [Accessed 14 January 2021]

[66] Sherriff A., Lomas M., Swan W. *Managing Human Factors in Retrofit Projects* [online]. University of Salford; 2020. Available from https://salford-repository.worktribe.com/output/1351045 [Accessed 13 January 2021]

Chapter 13

Designing and building a deep and circular retrofit of a cottage in Dublin

Joseph Little[1]

13.1 Introduction

Case studies that explore solutions to multi-faceted issues are always of interest, but especially so when practices need to change radically. Here, we set out a series of urgent challenges for building design professionals and their clients to meet in each building project, with clear impacts on national and global scales. Despite these challenges being well-publicised, there appears to be a dearth of published building-level case studies that explore the arc from challenge, through technical design and assessment, to practical implementation. There is also a lack of clarity on how and where they might best be presented to aid comparison, provide guidance and support dissemination.

This is an account of an ambitious, energetic retrofit of a particularly small, mid-terrace, urban cottage (Figure 13.1). There are thousands like it in Dublin. In this and the following chapter, the arc described above is set out, circling back to the question of compliance at the end of the following chapter. Did the project meet the challenges set and could the case study support future larger-scale projects and, indeed, how should those case studies be presented? The building's historic value, form, size and context, and the mixed quality of its building fabric greatly influenced and constrained specification, programme, and budget. Pragmatic solutions were sought to deliver a high-quality, comfortable, and energy-efficient home, while at the same time prioritising the use of low-carbon building materials, enhanced moisture management, and building systems that are as simple as possible.

As a one-off project on a highly constrained site, built during a worldwide supply chain crisis, it was relatively expensive. The context, constraints, ambitions, choices made, and data gathered provide abundant material for a 'warts and all' case study. As building performance was a central focus, numerous tests and

[1]School of Architecture, Building and Environment, Technical University Dublin, Ireland

Figure 13.1 Case study cottage

simulations were carried out. While this case study stands outside the context of Reduced Data Standard Assessment Procedure (RdSAP), PAS2035, and large-scale retrofit projects, it is hoped that the breadth and depth presented add to the conversation on appropriate retrofit of Britain's historic dwellings.

As shown graphically in Figure 13.2, this chapter sets out global challenges that the case study aims to address and presents the historical, spatial, and personal contexts. The specification and works completed are then explored qualitatively through the lenses of moisture, circularity, services, and the need for pragmatic decision-making.

13.1.1 Energetic retrofit: the multiple challenges of large-scale retrofit

Under the UN's Paris Agreement of 2015, 140 nations agreed to reach Net Zero by 2050, meaning that *'total greenhouse gas emissions would be equal to the emissions removed from the atmosphere, with the aim of limiting global warming and resultant climate change*' [1]. As a sector, construction is the world's largest emitter of greenhouse gases. In Europe, including the UK, this sector is responsible for 36 per cent of CO_2 emissions [2]. In the UK, the equivalent value is 40 per cent and in Ireland 37 per cent. In Ireland, two-thirds of the emissions relate to building operational energy use and the rest to the production of raw materials, transportation of materials, construction, retrofit, demolition, and infrastructure [3].

While it is true for any particular user or site that '*the greenest building is the one that already exists'* [4], the building fabric and systems of most buildings are inefficient [5], with a resultant high operational carbon impact. Ninety per cent of the buildings that will exist in 2050 are here now, so retrofitting to decarbonise is becoming a central focus of governments. There are ~198 million dwellings in the

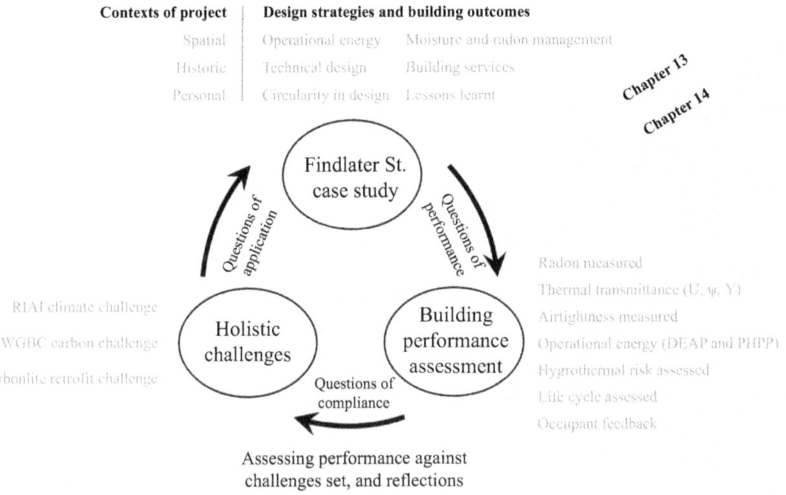

Figure 13.2 Case study structure

EU-27, 2.1 million of which are in Ireland. The current rate of retrofitting those dwellings is too slow. The European Commission reports that 0.4–1.2 per cent of the building stock, depending on the country, is renovated annually. In 2020, the Irish government set the ambitious target of retrofitting 550,000 dwellings to a B2 rating, or higher [6], by 2030. By the Spring of 2025, only 55,893 homes had been energetically retrofitted: at 11 per cent of the ambition, this is far from done. As an annual rate, it shows 0.5 per cent of Ireland's building stock was renovated in those years.

It is clear, therefore, that we must seek to reuse dwellings wherever possible, retrofitting them in a way that ensures a high quality of life *while also* meeting final operational and embodied carbon targets that enable us to reach Net Zero by 2050. For that goal to be credible, we must be well on the way by 2030. The task of retrofitting these dwellings to reach Net Zero is herculean, the greatest challenge to ever face the designers, builders, and users of the built environment. Research, including demonstrator projects and case studies, plays a key role in quickly identifying the best approaches to retrofitting the great diversity of our dwellings.

13.1.2 Personal context: a family plan and professional interest

The author is part-owner and architect of the case study project. Rather than buy a larger home, scaling up only to scale down a few decades later, the family chose to stay in a small, well-located dwelling in Dublin and purchase an even smaller dwelling that could be renovated and provided to their grown children in the future.

In the interim, the property would be rented. Dublin has a housing crisis, driving significant house price inflation. Given that a location within cycling distance of the family home was considered essential, purchasing a tiny dwelling, i.e., less than 50sqm, in poor condition was the only affordable option. The author's profession made a 'doer upper' more attractive than it might be for many families.

A mid-terrace cottage that fit the criteria, having a 42 sqm gross area, including an existing sub-standard extension, was purchased in 2017. It had been built in 1885–1886 as part of an innovative social housing initiative.

As a practising architect and researcher, the author has designed, built, measured, and simulated the retrofit of buildings, particularly dwellings, for decades. It was, therefore, tempting to use a family project as a vehicle to explore the retrofit problem. As developer, architect, and researcher, there were many moments where ambition and budget conflicted. Hopefully, reflection on that natural tension brings particular value to this case study.

Reflecting on the World Green Building Council (WGBC) graphic in Figure 13.3, case studies and demonstrator projects are a crucial first step in exploring and mainstreaming innovations. To succeed, collaboration is essential. If communicated well, they can raise awareness, build capacity, and support education in academia and industry, while showing compliant pathways to meet challenging national policies. The case study project engaged at each of these levels.

13.1.3 Outline of the retrofit project: property, project and team

After recent years of neglect, the terraced property was in a ruinous state. All elements needed repair, renewal, or replacement. Due to the need to replace a substandard extension, planning permission was sought. Given the presence of moisture damage and the need to understand what parts of the building fabric were in good order, layers upon layers of wall, floor and ceiling coverings, the evidence of a maintenance and refurbishment regime of many decades, were removed. Moisture-damaged surfaces were left to air dry. It was crucial that a high-quality, amenable builder was brought on board. COVID-19 and Lockdown then intervened to delay the commencement of building works, which started in May 2021, as outlined in Table 13.1.

The disruption to the worldwide supply chain and the steep increase in prices due to Lockdown had a significant impact on the project. Tradespeople were less available and material costs greatly increased. In some cases, the resolution adopted directly impacted on the project's embodied carbon. However, without doubt, the greatest impact on progress and cost was the extremely small size of the building and the lack of any useful curtilage for storage. No more than three tradespeople could work on site at a time and materials could only be supplied or removed in a piecemeal fashion. If a whole terrace of such dwellings could be deep retrofitted as a single building project, many issues would be more easily resolved due to the economies of scale, material supply, tradespeople availability, and quality assurance processes this would support.

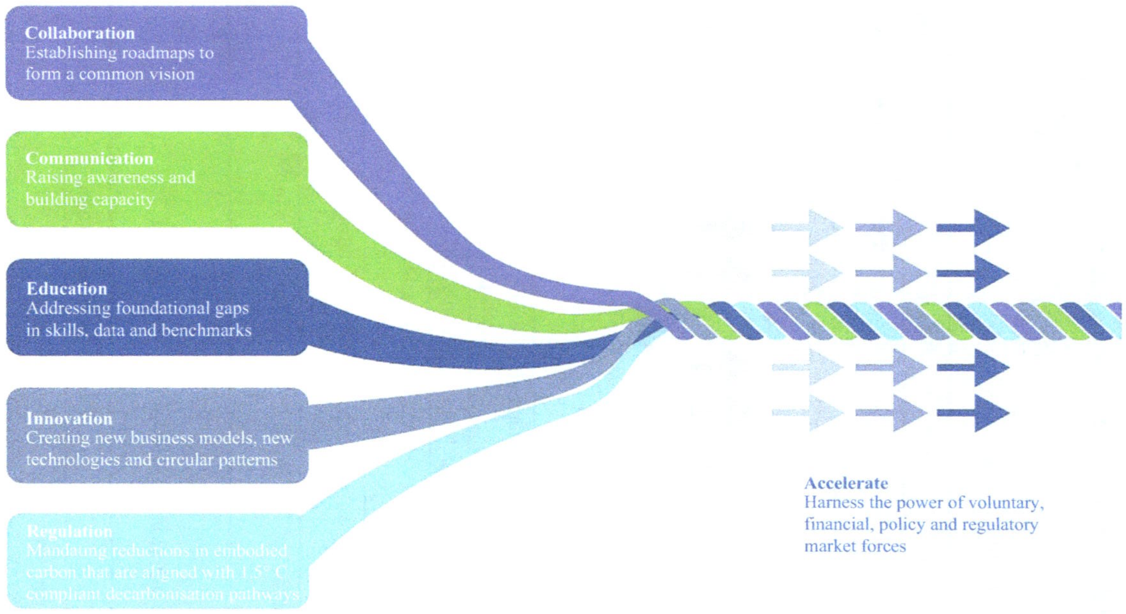

Figure 13.3 Making change happen in our sector (image provided courtesy of WGBC: Adams M et al. (2019). Bringing Embodied Carbon Upfront. WGBC

Table 13.1 shows a timeline of the project, while Table 13.2 shows the professionals and companies that supported the project during design and build stages. The author is very grateful for their support. Sirius Construction, in particular, showed a willingness to innovate and great patience, delivering a good project in a tightly constrained site. The team that supported assessment, calculation, and simulation is listed at the start of the following chapter.

Table 13.1 Brief timeline of the project

Phase	Date	Phase	Date
Planning permission granted	Oct 2017	New sub-frames and windows installed	May 2022
Acceptable tender returned	November 2020	Airtightness test	May
Demolitions and trial holes	May 2021	First fix electric	May
Drainage installed	June	Internal insulation and partitions	June
Steel and foundation engineering design	August	Plasterboard	July
Foundations for the extension installed	August	Skim coat, ext. cladding prep.	August
Steel portal frame	September	Fitout of shower room and kitchen and tiling	September
Timber frame erected	October	Paving and cladding	October
Floor installed, existing roof stripped and redone	December–January	Snagging: general builder leaves site	January 2023
Attic insulation and roof airtightness	March 2022	Client painting, decorating, furnishing and occupying	February–May
Existing windows and sub-frames removed	April	Tenant moves in	August 2023

Table 13.2 Key professionals involved in construction

Role	Person or organisation
Architect and developer	Joseph Little
Tender documentation	Brendan O'Connor
Engineer #1	Unnamed – substructure and portal frame only
Engineer #2	Roger Cagney (RCCE) – existing and new timber construction
General Builder	Sirius Construction Ltd
AVCL and cellulose installer	Clioma House
Airtightness tester	Greenbuild Services

13.1.4 Standards and targets

In May 2019, Ireland's parliament, Dáil Éireann, followed the UK Parliament in declaring a climate and biodiversity emergency. Under Ireland's Climate Action and Low Carbon Development (Amendment) Act 2021 [7], the government has committed *'Ireland to pursue and achieve no later than 2050, the transition to a climate resilient, biodiversity-rich, environmentally sustainable, and climate-neutral economy'.*

It is essential that early reductions in this transition are significant, given the extent to which building and construction contribute to greenhouse gas emissions; the length of time that greenhouse gases remain in the atmosphere once emitted; and the need to make the decarbonisation pathway to 2050 as smooth as possible. For those reasons, the Royal Institute of British Architects (RIBA), followed by the Royal Institute of the Architects of Ireland (RIAI), set its members a 'climate challenge' to meet by 2030.

A Domestic Client's Guide to the RIAI 2030 Climate Challenge [8] states that clients, architects, and other members of the design team are asked to work towards the key targets by 2030:

- Reduce operational energy by at least 33 per cent
- Reduce embodied carbon by at least 40 per cent
- Reduce potable water use by at least 24 per cent
- Improve site biodiversity by providing a net biodiversity gain
- Achieve core health and well-being targets

The architect's version of the same RIAI guide sets specific benchmark values that are tabulated in Chapter 14, alongside values achieved. Despite a limited budget and a disproportionately large thermal envelope, the author accepted the challenge to attempt to meet these targets. The lack of European or Irish benchmark values for embodied carbon of energetic retrofits made it harder to assess if a 40 per cent reduction was achieved.

The *Energy Performance of Building Directive* (EPBD) from 2002 to 2024 has done much to focus and energise thinking around energetic retrofit, or renovation, in Europe. The latest revision moves beyond a focus on operational carbon, requiring member states to publish a roadmap for the introduction of carbon limits, embodied and operational, by 2027 [9].

Short-term thinking, tight household budgets, and transient ownership can hamper the ability of the EU, or UK, to decarbonise the housing stock to the extent necessary. Increasingly, energetic retrofits that do not perform, lack ambition, or create 'lock-in' are considered problematic. In this context, lock-in refers to works completed that prevent, whether through budget or physical obstruction, later, more beneficial works from taking place. Policy responses to surmount these hurdles include provision of grant aid schemes for private and social housing retrofit works and, more recently, Building Renovation Passports (BRP) to assist and promote renovations staged over several years and, potentially, several ownerships. The

Irish government intends to introduce BRP, arising from the recast EPBD (2024), by May 2026.

Like many building projects, budget constraints impacted the ambition of the case study project. While some measures had to be omitted permanently, others were moved to a later phase but facilitated by works carried out in this phase.

13.1.4.1 CarbonLite retrofit standard

To satisfy the operational energy challenge above, the case study was assessed against the *CarbonLite Retrofit Standard* [10], which has been developed and managed by the Association for Environment Conscious Building (AECB) in the UK. This standard seemed a good fit in terms of ethos and suitability for assessment of an historic, solid-walled building. There is no Irish equivalent.

Table 13.3 sets out three voluntary retrofit standards for comparison. The first, *CarbonLite Retrofit Step-by-Step,* was created by AECB in 2023 to support those who seek to decarbonise but cannot yet take the step to a whole building deep retrofit. This approach decarbonises the dwelling's space heating and hot water systems, ensures good ventilation, and encourages owners to plan the next phase of works, thereby avoiding lock-in. The second, *CarbonLite Retrofit*, was developed in 2021. It supports whole-building retrofits and retrofit extensions. For both standards, an AECB certifier assesses information provided on a detailed building energy modelling tool and methodology known as the Passive House Planning Package (PHPP) [11] and supplementary photographs, drawings, and other information provided by the building designer. The third standard listed, *EnerPHit* [12], created by the Passive House Institute (PHI) in 2010, also draws from PHPP. Ironically, while the *CarbonLite Standard* was created in recognition that many well-designed and built energy retrofits in the UK were not meeting *EnerPHit*, the latter was created because of the number of deep retrofits not meeting the original *Passive House Standard.*

13.1.4.2 Embodied carbon and enhanced circularity

Bringing Embodied Carbon Upfront [13], published in 2019 by the WGBC, is a seminal text for the decarbonisation of the built environment. It sets out the challenges of the Net Zero Carbon Buildings Commitment, defines Net Zero Embodied Carbon, presents the tools that can be used, and challenges each stakeholder group to a series of actions by certain dates that lead to a compliant Net Zero world by 2050. The WGBC text states that *'net zero embodied carbon should be pursued as part of a whole life cycle approach to carbon reduction that includes net zero operational carbon'*.

The WGBC sets out 'foundational principles' to meet this challenge:

1. **Prevent** – avoid embodied carbon from the outset by considering alternative strategies to deliver the desired function.
2. **Reduce and optimise** – evaluate each design choice in terms of the upfront carbon reductions and as part of a whole life cycle approach.

Table 13.3 Comparison of three building energy retrofit standards

	CarbonLite retrofit step-by-step	**CarbonLite retrofit**	**EnerPHit Standard**
Building type	Existing dwellings	Existing dwellings	Existing buildings, including dwellings
Building energy modelling tool	PHPP	PHPP	PHPP
Delivered space heating (kWh/m^2.a)	report result	$\leq$ 50 kWh/m^2.a($\leq$ 100 kWh/m^2.a with certifier-approved exemption)	$\leq$ 25 kWh/m2.a (in cool-temperature climates)
Delivered space cooling (kWh/m^2.a)	report result	$\leq$ 15 kWh/m^2.a	$\leq$ 25 kWh/m^2.a
Either Primary Energy (PE, varies) or Renewable (PER) (kWh/m^2.a)	report result	report result	$\geq$ 60 kWh/m^2.a (variable)
Exterior insulation	Not stated	Not stated	0.15 W/m^2.K
Interior insulation	Not stated	Not stated	0.35 W/m^2.K
Windows	Not stated	Not stated	0.85 W/m^2.K (installed) Solar heat gain coeff. (Ug): g*1.6 $\leq$ 0
Ensure ventilation	Continuous MEV or MVHR must be installed: follow PAS 2035 Annexe C, or as req. by Part F of B. Reg's		MVHR (with min. 75 per cent heat recovery)
Air permeability (Q_{50}), airtightness (N_{50})	$\leq$ 5.0 m^3/m^2.hr (Q_{50})	$\leq$ 2.0 m^3/m^2.hr (Q_{50})	$\leq$ 1.0 air changes per hour @50 Pa (N_{50})
Thermal Bridges	If additional and significant fabric measures are being replaced or installed, certifiers will advise whether AECB Retrofit Standard level 2 reqs are applicable	Assumed to be less than 0.01 W/mK, or else accounted for in PHPP or for retrofits a default heat loss factor may be used	Thermal bridge effects must always be avoided or minimised
Surface condensation (f_{RSI}) assessed		f_{Rsi} to meet criteria in PHPP, or 0.75 (as Building Regulations/ PAS2035), or local standards, whichever is more onerous	
Heating system	Change existing fossil fuel (or direct electric) heating system to a heat pump	Existing heating systems may be retained, but a practical plan to allow for future low-carbon heating supply must be in place	
Thermal comfort	PHPP modelled overheating risk, <10 per cent Acceptable (Guidance: <5 per cent Good practice or <3 per cent Best practice)		Max 10 per cent > 25°C
Running cost comparison	Must be same/lower running costs than base case **	n/a	n/a

3. **Plan for the future** – take steps to avoid future embodied carbon during and at end of life.
4. **Offset** – as a last resort, offset residual embodied carbon emissions within the project or organisational boundary where possible or, if necessary, through verified offset schemes.

The types of decisions and trade-offs that were made to meet these principles can be seen later in the design phase.

Given the dearth of data on embodied carbon of energetic retrofits and the particularities of the case study project, a clear embodied carbon reduction target was not adopted beforehand. As both developer and designer, the author teamed up with skilled professionals to understand operational and embodied carbon impacts and committed to disclosing Life cycle assessment (LCA) data.

13.1.4.3 Life cycle assessment

Life cycle assessment (LCA) 'is a structured methodology for compiling and evaluating the environmental impacts of a product system throughout its life cycle. It can be applied to any product or process'.

The European normative standard EN 15978:2011 [14] sets out a calculation method for assessing the sustainability of construction works. It provides a modular approach to LCA based on stages. These stages, illustrated in Figure 13.4, cover material production (A1–3), also known as 'cradle-to-gate'; construction process (A4–5); building use (B1–7); and end of life (C1–4). The arc from Stages A1 to C4 is often known as 'cradle-to-grave'.

The standard also includes Module D, which reports on the environmental benefits and loads of materials once they have reached their end-of-waste status, through future reuse, recycling and energy recovery. Going forward, as the circularity of building materials and components increases, the value ascribed to Stages C3 and C4 will decline and that of Annex D will rise.

The WGBC has set the target, and the EU-27 are keen to engage. However, the lack of robust LCA data for buildings in member states hampers policy and implementation. In 2022, Rock *et al.* [15] reported that

> '... *The majority of EU Member States have low to no LCA data available for calculating bottom-up embodied carbon benchmarks, with only five Member States identified as having 50 or more LCA cases available ... within the countries included in the study, samples of sufficient size and quality could only be collected in Belgium, Denmark, Finland, France and the Netherlands.*'

For this reason, Ireland and two other countries received EU funding through the INDICATE project to '*increase the quality, quantity, and application of building-level LCA data, to manage and minimize the WLC footprint of buildings*' [16].

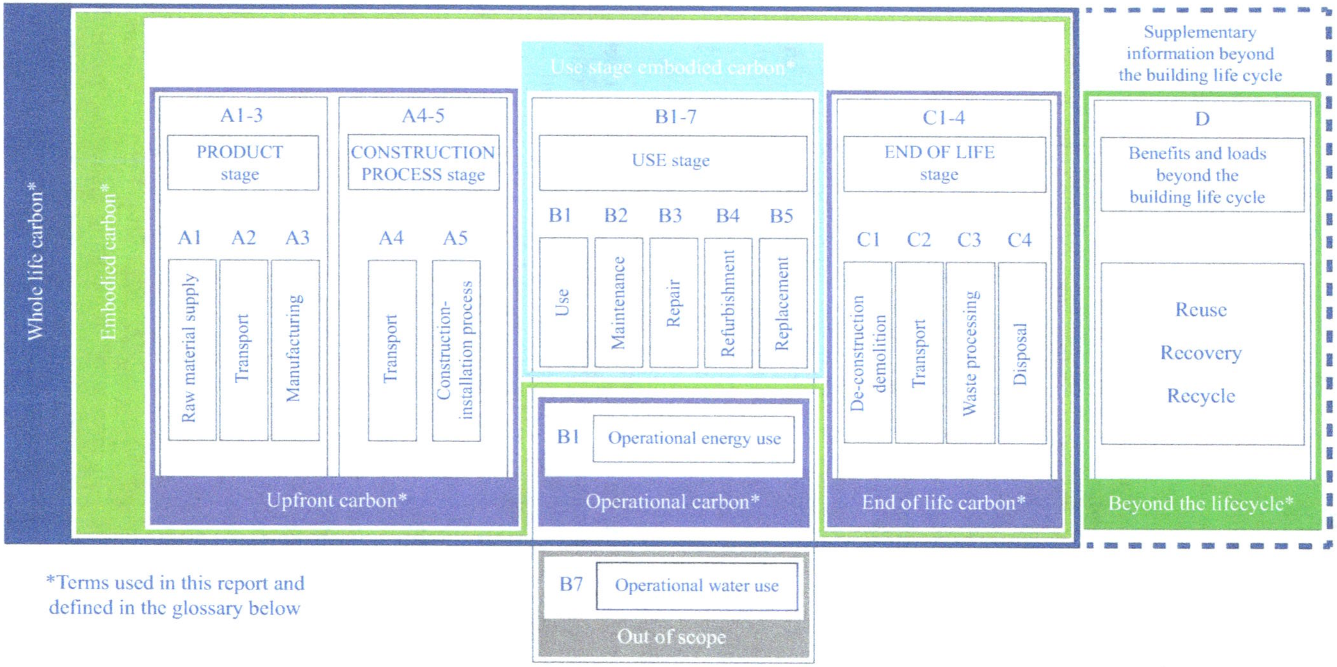

Figure 13.4 The stages of life cycle assessment image provided courtesy of WGBC: Adams M et al. (2019). Bringing Embodied Carbon Upfront. WGBC

Alongside this work, the Sustainable Energy Authority of Ireland (SEAI) has commenced work on a national methodology for calculating Whole Life Carbon (WLC) in support of national policy and the soon-to-be-transposed recast EPBD (2024).

Data on embodied carbon of energetic retrofits is a small but fast-growing subset. Bienert *et al.* [17] make clear the value of studies where a *'holistic and comprehensive approach'* is taken and a full LCA is created.

> *'The topic of embodied carbon in retrofits requires further research, as market participants currently lack a comprehensive and holistic view of this issue, and building materials change and decarbonise. In-depth research is needed for a better understanding of the scope and impact of embodied carbon in retrofits. A fundamental problem is the lack of data availability. Without sufficient and reliable data, it is difficult to make accurate statements about the actual embodied carbon in retrofit projects . . . By carefully examining the embodied carbon of retrofits, we can strive towards developing innovative and sustainable solutions that minimise the environmental impact associated with building retrofits.'*

The LCA and schedule of works from this case study project have been shared with the Irish Green Building Council to support future expansion of the Irish database of embodied carbon of energetic retrofits.

13.1.4.4 Circularity

Whereas life cycle assessment may be seen as an accounting activity, 'circularity' is a design and specification activity. It is a fascinating and fast-emerging field focused on the transformation of the current linear 'take - make – discard' business practice to one where the supply chain becomes 'circular' with increasingly less virgin material supplied. The methods used to bring materials together to form products and the way these products are fixed together to form building components greatly impact the ability to later repair, remanufacture, disassemble, or partially replace. Therefore, a fundamental rethink of processes, technology, and relationships is required. A new vocabulary, methodology, and even metrics are being created to judge and enhance building circularity.

A key moment in Europe was the passing of the Waste Framework Directive (WFD) [18], which introduced the concept of 'end-of-waste', in 2008. Ireland's National Waste Prevention Programme (NWPP), managed by the Environmental Protection Agency (EPA), commenced in 2004 and was reconfigured into the Circular Economy Programme in 2021, following the passing of Ireland's first *Whole of Government Circular Economy Strategy 2022–2023* [19]. For many years, the EPA has supported innovation in circularity of building materials, particularly cements and aggregates, due to their large mass, high carbon, and environmental impact. Despite this, the Irish government appears to have been less focused on the issue. In 2024, Ireland was reported to have one of the poorest circular material use rates in the EU-27 [20]. For ten years, it has hovered between 1.7 and 2.0 per cent.

The EU average was 11.2–11.6, per cent, while the Netherlands, best-in-class, varied between 25 and 29 per cent.

Two local authority areas in the Dublin region signed the Circular Cities Declaration in 2020, while the third, Dublin City Council, was selected to be one of the one hundred Climate Neutral and Smart Cities in the EU-27 in 2022 [21]. It will be supported in its efforts to become climate neutral by 2030 and tasked to serve as an innovation hub, testing cross-sectoral solutions and new governance models, to accelerate change and inspire all.

13.1.5 Historic and spatial context

To truly understand a building and its value, it is crucial to understand its spatial, historic, and technological context, as well as its specific design and build and the nature and quality of interventions during its lifetime. While many buildings are unique, most tend to be one of many, built on scale in a region and time, using similar materials and technology. In this way, the history of the case study cottage and the lessons learnt should be relevant to many dwellings that are yet to be retrofitted.

13.1.5.1 Historic context

By the late nineteenth century, Dublin was the most squalid city in the United Kingdom. The loss of political power to London in 1801, the economic collapse that followed the end of Napoleonic Wars, the Great Famine, and societal values and services that were strikingly different to modern times, contributed to a continuous growth in poverty and slums. Without the draw of employment in burgeoning factories and mines and the opportunities for self-advancement that stable employment can bring, both characteristic of Victorian Britain, Dublin's poorest remained unskilled, unhealthy, and badly housed. It is not surprising that slums grew until the late 1920s. In the process, many of Dublin's fine Georgian townhouses became decayed tenements. As an example, the 1911 Census recorded 850 people living in Henrietta Street, with more than 100 people living in one of its properties, which three generations before had been one of Dublin's grandest urban mansions.

This is the context in which the case study dwelling on Findlater Street was built. At the time, few local governments in Europe provided social housing. Influential citizens, who were already members of the Dublin Sanitary Association, formed the Dublin Artisans' Dwelling Company (DADCo) in 1876, following the passing of legislation that allowed local authorities to compulsorily purchase slum land. This semi-philanthropic private enterprise, run by Quaker and Protestant businessmen, set itself the task of replacing housing that was deemed *'highly injurious to the moral and physical welfare of inhabitants'*, while paying a modest dividend to shareholders. Amongst the founders were the Guinness family and wine merchant William Findlater. DADCo aimed to build housing for those in full-time employment in artisan occupations and other trades. Initially, it did this on compulsorily purchased slum land. It was understood that the evicted, unskilled lower working class would benefit somehow by finding new, better accommodation somewhere else, a vague process known as 'levelling-up'.

Between 1879 and 1908, the company built 3,300 homes for 16,000 tenants. The rents were always considered competitively high. In hindsight, the schemes it developed in the 1870s and 1880s were too small, aimed at the wrong cohort, and did little to address the housing and sanitary crisis. On the positive side, DADCo developments were, on the whole, well-designed and built and their provision did stoke national debate as to how Dublin's poorest should be housed. Within five years, the Dublin Corporation started building social housing. It was not until the early 1930s that the Irish Free State put in place adequate policies and resources to clear the slums and start building housing for Dublin's most vulnerable citizens [22]. Today, Dublin faces a new housing crisis and DADCo properties are much in demand.

The site DADCo developed, adjoining Infirmary Road, is one of the most densely settled parts of Dublin, with a housing density of ~86–91 dwellings per hectare. The site was apparently a field prior to start of construction in 1885 (Figure 13.5). Based on a piecemeal survey, it appears that its original surface, where undisturbed, still forms the gently sloping base of sub-floor voids.

When built, the site was, and still is, bounded to the north by mews houses and the back gardens of Victorian terraced houses facing onto the great arc of the North Circular Road; to the west by Infirmary Road and the 1750 acres of the Phoenix Park, and to the east and south by fields, beyond which lay Montpelier Hill and the Royal Barracks, one of Europe's largest. This barracks, with its associated coopers, stables, leather works, carriage builders, and munitions workshops, may have provided much of the employment for the tradesmen who moved their families to the development.

13.1.5.2 Design of the properties

The architect was Charles H. Ashworth and the builder was J. Beckett & Sons. Like other DADCo developments, the Infirmary Road site has a mix of two-storey and one-storey cottages arranged in terraces. Streets conform to bye-law widths, as laid out in the Public Health (Ireland) Act 1878. Aberdeen Street, the central avenue into the site, bisects two terraces at the front before forming a small crossroads with Sullivan Street, all two-storey, red brick cottages (Type-C). Surrounding this cruciform shape is a circle of streets, including Findlater Street, lined by single-storeyed rendered cottages (Type-E). What is striking, is the elevated architectural quality of the development.

The language of classical architecture was used to elevate the scheme and signal hierarchy. The four gables facing Infirmary Road are supported by shallow brick pilasters that corbel off the façade at the height of the first-floor windowsills (Figure 13.6). The bottom of the open pediment doubles as the capital of the pilasters. The result is a subtle evocation of a classical temple front. On the gable walls, at the mouth of Aberdeen Street, large chimney stacks protrude to the same extent as the pilasters, supported on corbels at the same height. Halfway up the stacks are

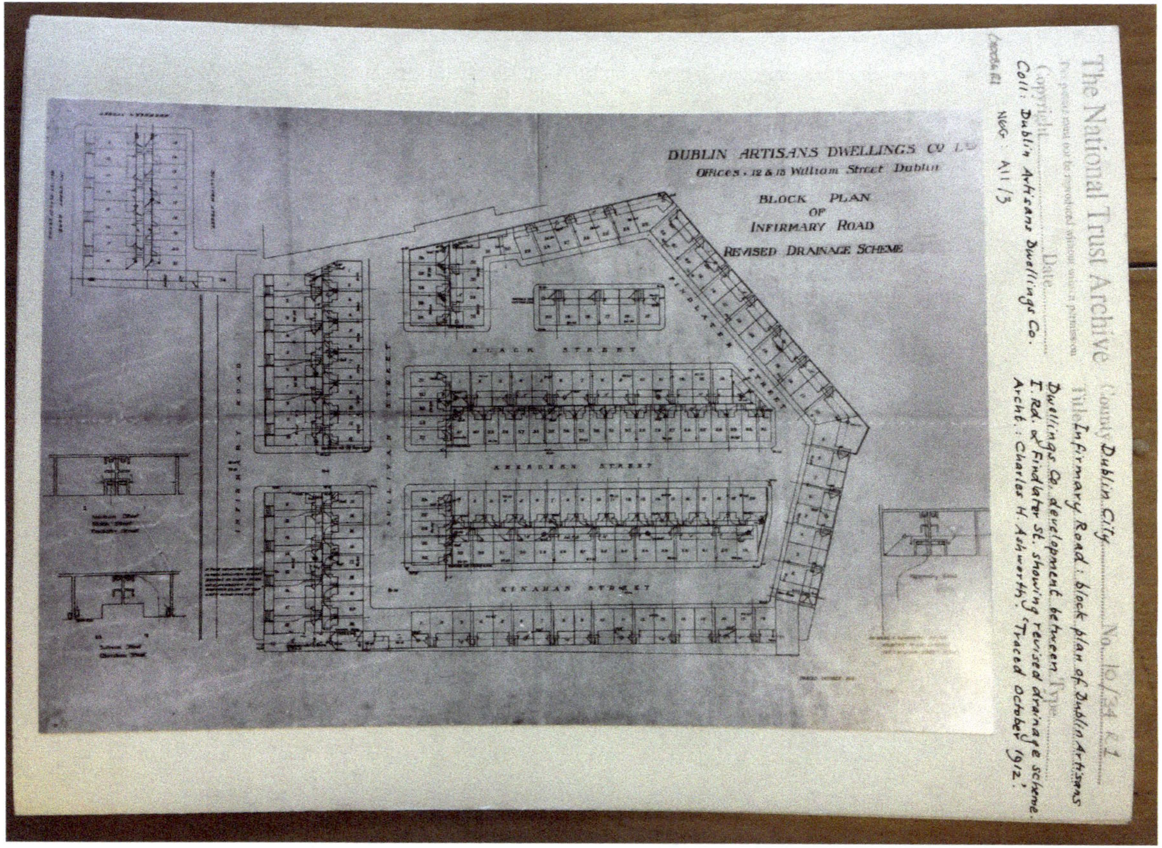

Figure 13.5 The site plan of the Infirmary Road development, 1885 (image provided courtesy of Irish Architectural Archive)

Figure 13.6 Architectural set piece of open pediment gables

recessed blind opes. The gable walls of the next line of cottages facing Sullivan Street feature another partially protruding chimney stack, but these are narrower and corbel out from the line of the base of the open pediment. The gable walls at the far ends of the front terraces are brick with a third type of protruding stack. Finally, the gable walls at the far end of the other terraces mark the start of the long terraces of rendered, single-storey cottages. Appropriately, these gable walls are also rendered, save for the brick pediments overhead. The architectural language of the development is an exposition of the values of Victorian Dublin; stratified, conservative, mercantile, and, at times, very elegant.

The Type-E cottages, including the case study dwelling, are simpler with few architectural features. The most obvious feature today is the brick detailing of the chimney stacks, complemented by terracotta ridge tiles. As can be seen in Figure 13.7, the original tongue-and-grooved doors are simpler than the six-panel doors of the Type-D cottages and have a smaller fanlight overhead. The cambered arched fanlight over the door lightens a room but also signals the importance of entrance. Without doubt, the key architectural feature of the cottages was intended to be the windows. These are metal with the upper portion, opening on a horizontal pivot. The complex pattern echoes a chinoiserie bamboo screen. It is likely these windows would have cost more than the sliding sashes of the grander Type-E houses. The appearance of these windows along a whole terrace of cottages must

Figure 13.7 (from left to right) A DADCo drawing of the façade of a Type-C cottage and a drawing of the façade of a Type-E cottage

have appeared quite ornate. The windows in the photograph of Figure 13.7 had been removed from a cottage in nearby Black Street. In a striking example of circularity, the new owners of a cottage in nearby Kirwan Street asked for them and had them reconditioned and fitted by a nearby joinery and foundry.

13.1.5.3 Original layout and construction of cottage

Figure 13.8, from the Irish Architectural Archive, shows the original layout of the property proposed for retrofit. The internal area of the cottage is only 30.5 sqm, excluding the coal store, toilet, and fire breast. This includes the scullery and internal partitions. Nonetheless, many large families grow up in these tiny cottages. The generous 3.1 m height of the main room would have given a sense of space, while also providing a 'well' of air to moderate indoor environmental quality.

The centrally located entrance door opens into an efficiently shaped, off-square lobby. It gives direct entry to the main room. Cooking was done over the only hearth. The left side of the hearth provided space for an armchair; the right side may have been the location of the table, near the scullery. Off the main room are a large and a smaller bedroom. To the rear is a door to a small scullery with a Belfast sink that also functioned as a lobby to the yard. Behind the scullery were a coal house and an outside toilet entered from the yard. The latter is so large that it may have been sized to facilitate craftwork or, perhaps, house a pig or chickens.

In DADCo's very first development in 1875, and again in this 1885 development, the client, architect, and builder chose to use cast concrete, a novel material at the time. An article in the Irish Builder journal of 1886 extolls its value. The claim that it was particularly suitable for housing the 'very poor' may explain its use for both external walls and internal partitions for the Type-E cottages.

> '... *We have sufficient data to know that if made from a non-porous aggregate* [Portland cement concrete] *is impervious to moisture, and yet at the same time ... will take up a sufficient quantity of moisture from the air to prevent condensation upon the surfaces of the walls. We also know that*

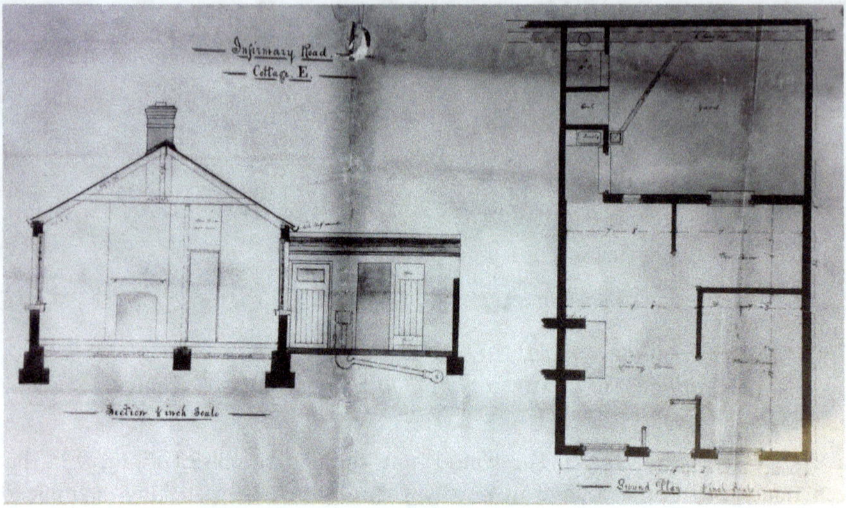

Figure 13.8 Original plan of the Type-E cottage (1885) (image provided courtesy of Irish Architectural Archive)

> *it not only resists the disintegrating influence of the atmosphere, but becomes even harder with the lapse of time... The use of concrete as a material for buildings will be found to meet all the defects set forth by practical people, as it may be made fire-proof, vermin-proof and nail-proof; and in dwellings for the very poor, will, therefore resist the usually destructive tendencies of the occupiers. Scarcely anything, therefore, could be better as a building material.'* [23]

As will be seen, the confidence of the writer quoted did not reflect the level of knowledge and skill of the tradesmen who cast the concrete walls of the Infirmary Road development. External walls were cast at 8" thick. Internal walls appear to have been only 4" thick, slimmer than earlier DADCo developments. The aggregate used was field stones with broken brick or clay pot inclusions. During opening-up works, it could be seen that an inadequate cement binder was used and consolidation was poor. As a result, surfaces were friable and there were many cracks, several on day joint lines and one continuous vertical crack between the main room and kitchen, large enough to fit one's hand.

The floor was traditional suspended timber, and the ceiling was traditional plaster and lathe. Baltic pine was selected for the A-profile cut-rafter roof structure. The natural slates were likely shipped from Wales. A large coat applied to the underside of the slates resisted wind suction. A stamp on the decorative terracotta ridge tiles identifies their source in County Somerset. A brick chimney stack, with feature corbelling, rises from the ridge above the left party wall. It houses two flue pipes, one for each property. It appears some attics were common, due to gable party walls to either side of chimney stacks being omitted.

Figure 13.9 Grating of gully in rear yard displaying the letters 'DAD'

Given the sanitary focus of DADCo, drainage was important. A shared private combined drain (Figure 13.9) was installed to drain surface and foul water.

Yards were surfaced in concrete so that there could be no percolation of waste materials into the soil. A dished surface drained the yard to a gully by the door, capped with a specially commissioned cast-iron grating displaying the letters of the developer. The specification makes clear the ambition and pride of the developers.

13.1.5.4 Post-construction

The Dublin Artisans' Dwelling Company continued to own its developments until the late twentieth century. It appears it has had a good maintenance programme, as witnessed by the layers of historic floor finishes and paints removed from the case study dwelling. Facing competition from local authority housing, DADCo started to sell off its houses in 1961. In some cases, multiple properties were purchased by new landlords. If the condition of the case study cottage in 2017 was general, some of the new landlords did not maintain the properties to a good standard. When this author purchased the cottage, maintenance had been abandoned, conditions were unsanitary, and the extension to rear was of very poor construction.

Today tenancy is mixed. It is unclear if owner-occupancy is higher in the two-storey than in the single-storey dwellings. Importantly, the Infirmary Road development was, and remains, a thriving, inner-city community. Benches, planters, flower baskets, and children's toys are common. Interestingly, some of the founders of DADCo [22] actively sought such signs of community and care, giving prizes for best kept house and window box, while associating pride of house with moral fortitude.

13.1.5.5 Current condition

The cottage is zoned 'R2 Existing Residential' in Dublin City Council's latest development plan. Its poor condition in 2017 would have turned many viewers away (Figure 13.10). There was significant damp on the rear wall due to wall cracks. In certain areas, much of the plaster had fallen or failed. Floorboards abutting the wall had been cut back due to rot, leaving a gap. Wire wool had been stuffed under skirting boards to resist vermin. The original tongued, grooved, and ledged internal door leaves, including strap hinges, were found to be serviceable once overclad wallboard was removed. The original lead water pipe that entered the cottage, which branched off to supply the neighbouring cottage, still provided the only water supply, a situation that needed to be quickly remedied.

Figure 13.10 *(clockwise from top left) Dropped ceiling in main room; original tongued and grooved internal doors revealed behind wallboard cladding; layers of paint; newspapers separating layers of linoleum and vinyl flooring; poor conditions and resistance to vermin at yard door; internal cast concrete partition skipped; condition of friable mass concrete walls (note: broken clay brick inclusions)*

A dropped ceiling had been installed, as an attempt to reduce the volume of air to heat, and clad with centimetre-thick expanded polystyrene tiles for thermal comfort. When this was taken down, it could be seen that the original plaster and lathe ceiling structure was badly cracked. After making a hole in the ceiling, it was possible to see that there was no insulation in the attic. The parging plaster had collapsed. Daylight was visible where some slates had slipped. In many attics, the internal gable party walls one would expect to see abutting chimney stacks, were missing, resulting in common attics and a clear fire risk that had to be addressed. Outside it could be seen that the roof had been partially repaired many times. Blueish stone slate and asbestos tiles had been used to supplement the original Welsh slates. Many slates were kept in place using strips of metal. The brick mortar, top haunching, and terracotta pots were in poor condition.

To the rear, an extension, possibly built in the 1960s after purchase from DADCo, was built on top of the yard surface with a floor finish 80 mm above the yard surface. Not surprisingly, mould and plant roots were found amongst the rotting floor battens. The original scullery, coal store, and toilet, including their lean-to slated roofs, were still present. The extension's roof membrane had been simply lapped over the slates, turning down on the far side of the party wall.

Displaying a striking lack of understanding or care, the extension's builder arranged that the rear roof of two cottages, the roof of the extension and the yard, all drained into the original dishing cast in the yard floor, which openly drained under the extension's floor to the original gully (Figure 13.11). In consequence, an inspection hatch had been created on the floor above the gully to clear blockages due to moss, leaves, and dirt. Clearly, vermin had easy access. It was a stark context in which to commence a design and specification that sought to retain as much of the original building as was possible, while creating an energy-efficient, low-carbon, and high-quality renovation-extension project. The first decision was to remove layers to reveal what of value was salvageable and to allow surfaces, which had been stressed by trapped moisture, some months to dry out.

13.2 Design and construction

It is crucial that a building be objectively appraised. What inherent qualities justify the purchase and what qualities should inform design decisions? The value of proximity to the city centre with amenities is significant. A strong local community is also of great value. Ironically, the building's poor condition also had an upside, as it resulted in a more affordable sale price and ensured there was no price uplift for already-completed retrofit works of dubious quality.

The existing building fabric and fittings, where salvageable, were seen to have triple benefits of reducing cost, preserving built heritage, and reducing the embodied carbon impact of the project. In support of operational carbon reduction, the roof of a new extension to the rear would have optimal orientation to south for generating solar energy and would allow a large panel size if the roof were built as a monopitch.

Figure 13.11 *(clockwise from top left) Collapsed parging; chimney stack in need of repair; slipped slates & multiple repairs; storm drainage of roof and yard combine to drain into dishing; plaster and lathe ceiling revealed above polystyrene ceiling tiles*

Thermally, the mid-terrace position reduces overall heat loss through the walls. Spatially, the uneven sub-floor, as much as 600 mm below the finished floor level, and deeper than shown in the historic section, provided an opportunity to super-insulate a new solid floor. The attic void provided another opportunity to super-insulate overhead (Figure 13.12). The author was keen to preserve the full floor-to-ceiling height, as it gave the main room a sense of generosity, but it can also contribute to indoor environmental quality (IEQ). This is because pollutants, whether humidity, carbon dioxide, or volatile organic compounds, take longer to reach steady-state concentration in larger room volumes (Figure 13.12).

13.2.1 Heat loss form factor and insulation choices

Calculating the heat loss form factor (HLFF) helps us understand the ratio of heat-losing surfaces, including floors, walls, roofs, windows and doors, to gross internal floor area. Despite being mid-terrace, the extended-retrofitted cottage has a surprisingly large HLFF. Two-storey detached houses generally have an HLFF of ~2.5, 2-storey mid-terrace houses have a HLFF of ~1.7, and detached cottages or bungalows have a value of ~3.0. This means that, proportionately, a detached

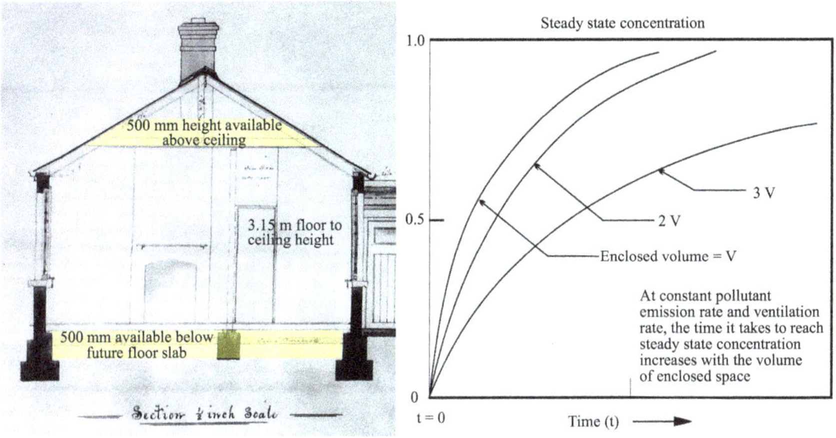

Figure 13.12 (left) Assessing spatial value in the section of the case study cottage; (right) the building as a 'fresh air' reservoir (from Figure 1.3 of A Guide to Energy Efficient Ventilation' by MW Liddament, courtesy of AECOM)

cottage has almost twice the amount of heat-losing surfaces as a typical mid-terrace house. While the actual area of heat-losing surface may be larger in the latter dwelling, the area of surfaces to be insulated, with associated cost, in the former dwelling *per square metre of usable floor surface enclosed* will be far higher.

Given that the heat-losing surfaces were calculated to be 156 m^2 and the gross internal floor area of the extended, insulated cottage is 37 m^2, the HLFF (156 / 37) is 4.2. Despite the thermal value of party walls that, notionally, do not need to be heated, this is significantly greater than the HLFF of a notional detached cottage. In Figure 13.13, the extended case study dwelling is an outlier amongst sixteen historic Dublin dwellings, which were explored in the Built to Last research project [30]. Interestingly, the two mid-terrace eighteenth-century townhouses – a large 3-bay, four-storeys-over-basement townhouse on North Great George's Street and a palatial urban mansion in Henrietta Street – have excellent HLFFs of 1.0 and 0.9 due to their efficient geometries, despite their great size and generous floor-to-ceiling heights. The HLFF makes clear that, no matter what material specifications were selected for the retrofit extension of the cottage, it would be comparatively expensive.

Buildings are often classified by their wall type. The insulating of walls also tends to receive most attention when solid-walled, so-called 'hard-to-heat', properties are energetically retrofitted. In Figure 13.14, the historic solid walls, forming part of the thermal envelope, make up only 11 per cent of the thermal envelope, excluding windows and doors, while the extended floor and roof make up 41 per cent of the whole, and the new timber frame wall is another 23 per cent. A purely thermal strategy can therefore be justified that proposes that floors and roofs be *super-insulated* wherever possible, the timber-frame walls be *well-insulated*, and

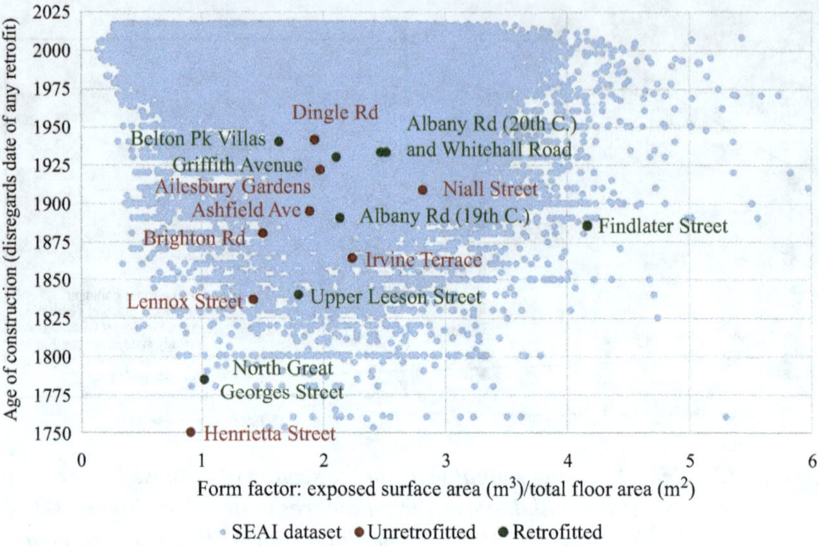

Figure 13.13 HLFF in context of age for sixteen historic dwellings overlaid on the same value for all EPC of dwellings in Dublin city

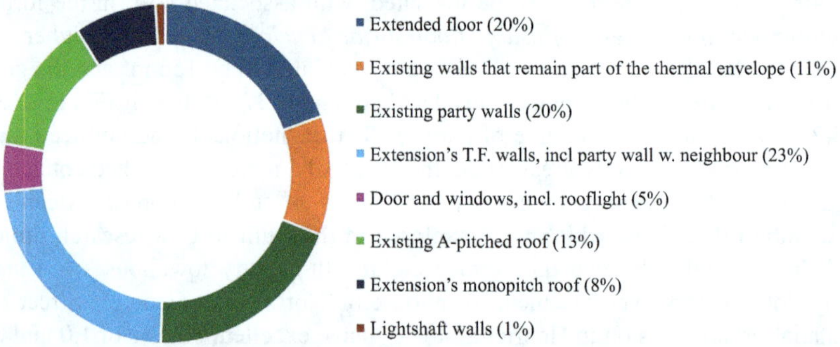

Figure 13.14 Percentage of elements in the thermal envelope of the case study

the existing solid walls, which remain part of the thermal envelope, be *sufficiently and appropriately* insulated. The floor and ceiling voids of the existing building and the nature of timber-frame walls lend themselves to this approach.

In designing the extension, the key challenge was to arrive at a construction system that could be built on the highly constrained site. It needed to insulate enough to meet the super-low energy brief and yet be narrow enough that it created an internal volume large enough to contain the brief requirements of a kitchen and shower room, with valuable storage overhead. It is the kind of challenge that designing tiny houses or extensions can create, which might not occur to the owner or architect of larger properties.

The window ope and granite sill of the cottage's second bedroom are functionally and aesthetically important. Therefore, the edge of the sill provided the ultimate constraint to the size of the extension. The extension would have three external walls. To the south, the wall facing the courtyard, to the east, the wall against the development's rear site wall, and to the north, the party line. During survey and opening-up works, it had become clear that the rear portion of the original northern party wall was structurally compromised. As the neighbour's extension was structurally separate, the old wall was taken down and the exposed face of the neighbour's wall insulated. After all demolitions, this left a zone of 3.8 x 3.6 m in size, 13.7 m^2, the size of a modest double bedroom, within which to extend. This zone included the future walls with cladding and the vertical ventilated cavity that would be required between the two extensions.

13.2.2 Build in timber or masonry?

The initial design intent for the extension had been a timber frame using Irish timber. A few months before site clearing was due to start, several timber frame companies advised delays of 12 months. As a result, a range of masonry strategies was explored. Masonry units have the advantage of being very portable, suiting constrained access, but are more thermally conductive than timber and have higher embodied carbon. Externally insulated aerated autoclave (AAC) blocks and insulated cellular terracotta blocks were considered. Walls that provided an acceptable U-value remained 'stubbornly' wide. See the rightmost column of Table 13.4.

The analysis showed that there was not sufficient value in moving away from timber frame in terms of U-value achieved or space saved. To move the project forward, an engineer-designed, 'stick-built' timber-frame structure using Irish timber was proposed in July. Finally a high-quality, timber frame company, Timbertech Homes, was identified that could deliver a quick turnaround. In addition, their preferred approach, using 'warm' studs and natural building materials, was also the author's preference.

Thinner timber studs (140 mm) were specified on two sides of the extension that would be sheltered, with wider studs (225 mm) used for the courtyard-facing wall and rafters for the roof. 140 mm wide timber frames are usual in the UK and Ireland. Locating them behind ventilated cavities and existing walls provides a thermal benefit, while maximising space available to the finished rooms within. Structurally, the roof required deeper (225 mm) rafters. It made sense to continue this to the wall facing the courtyard to compensate for the lower thermal performance of the patio door and window. The final gross internal floor area, including internal partitions, is 9.7 m. This means that four square metres of space, or 30 per cent of the overall 13.7 sqm available, is given over to wall build-ups of the extension, including the voids and rainscreens. While this seems punitive and *is* costly, it must be accepted that this is chiefly the workings of geometry. However,

Table 13.4 Comparison of different wall build-ups showing overall thermal performance and thickness

Wall facing courtyard	Description (from inside outwards)	U-value (W/m^2K)	Thickness (m)
Externally insulated AAC blocks	Plaster, Porotherm 300 T8 MW (300 mm), mineral wool batt external insl. (100 mm), render	0.15	430
Insulated cellular terracotta blocks (A)	Plaster, Mannox Aircrete Super (215 mm), mineral wool batt external insl. (160 mm), render	0.15	401
Insulated cellular terracotta blocks (B)	Plaster, Mannox Aircrete Super (150 mm), mineral wool batt external insl. (180 mm), render	0.15	355
Warm wide stud (225 mm) timber frame	Plasterboard, quilt insl. in service zone, racking board, 225 mm stud with Dämstatt paper cellulose insl., 50 mm Gutex Ultratherm wood fibre batt insl., battened void and rainscreen cladding*	0.14	378
Walls facing rear boundary and neighbour	**Description (from inside outwards)**	**U-value (W/m^2K)**	**Thickness (m)**
Warm stud (140 mm) timber frame	Plasterboard, quilt insl. in service zone, racking board, 140 mm stud with Dämstatt paper cellulose insl., 50 mm Gutex Ultratherm wood fibre batt insl., battened void and slates* *: excluded from Thermal Transmittance (U) calculation	0.18	348

the removal of a portion of the rear wall of the cottage, primarily due to the original scullery door being in the wrong place, gave 0.79 sqm of useful space, or half the width of a kitchen cabinet, back to the kitchen.

13.2.3 Spatial design

The author liked the original layout of the cottage, including the way Bedroom 2 pushes out, narrowing the living room, and the slight curve of that room's wall, which was cast in concrete. It may seem counter-intuitive, but cellular layouts suit small dwellings. This is because the partitions allow separation of functions that can only be attained in the open-plan layout of much larger buildings, while providing valuable wall surface for shelving and decoration. Isolation of sounds and smells from one room to another can also make small-house living more attractive. Retaining a second small room also makes sense, whether to be used as a bedroom, nursery, or home office. Given the unsatisfactory, friable nature of the cast concrete partition walls, the extent to which working on and around them would slow the builder, adding cost and the uncertainty as to whether the walls might crack again at any time in the future, a decision was made to remove these walls and replace them with non-structural stud partitions. While the new partitions resulted in additional cost and carbon impact, there were some savings by not repairing and plastering the old walls, nor installing edge insulation for the depth of the floor slab. A secondary benefit was the ability to adjust wall locations and ensure an acoustic specification.

The following is a summary of key spatial and functional measures (Figure 13.15)

- Internal 100 mm mass concrete walls were re-built as acoustically rated stud walls.
- Bedroom 1 was narrowed by ~100 mm while still accommodating a double bed (from one side), thereby widening the living room.
- Bedroom 2 was widened by ~100 mm, making it sufficient to take a single bed.
- A portion of the rear wall was removed to enable optimised locations for kitchen cabinets and the door. This had a secondary benefit of providing additional kitchen space.
- The location of radon sump and sump pipe ducted to and up the chimney flue.
- A rooflight brings eastern light directly into the living room. Light from different orientations can animate a space and reduce the need for artificial lighting. A simple design move was to centre the new rooflight with the pier between front door and window on one side of the room and the centre of the reused kitchen door on the other side.

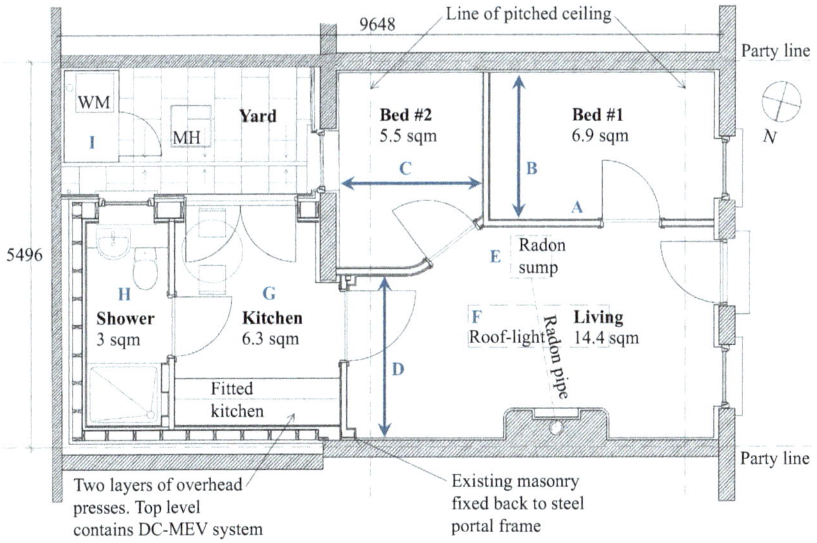

Figure 13.15 Plan layout highlighting key measures taken

- South-facing kitchen with a monopitch overhead creates a bright and airy room. The monopitch roof is ideally oriented for a future solar panel. The rear of the kitchen accommodates a double level of overhead cupboards. The top level of the cupboards accommodates the mechanical ventilation unit and ducting.
- The relatively generous shower room is naturally lit. Overhead is a storage space accessed using a ladder from kitchen.
- The re-location of freezer and washer/drying machines to a shed in the rear yard. This removes a source of noise and moisture from a very tight kitchen.

13.3 High-level specification decisions

The following two lists capture the high-level approach taken to specification, in the context of the design analysis and ambitions, to create an exemplary project in terms of circularity, moisture management, and the 2030 Challenge.

Principles to decarbonise and minimise operational energy use:

- Super-low U-values in floors, roofs and timber-frame extension.
- Appropriate U-values for solid walls.
- Minimising thermal bridging.
- Air permeability target: Q_{50} = <2.0 m3/m2.hr. (Q_{50} = 1.51 achieved.)
- Decarbonising space heating on a tight budget.
- Phasing to allow for future installation of heat pump or mechanical ventilation and heat recovery.

Principles to minimise embodied carbon and manage moisture and radon:

- Prioritising moisture management strategies.
- Maximising use of hygroscopic biomaterials.
- Use of rainscreen for rear solid wall, which faces dominant wind direction.
- Implementing circular principles, including the waste hierarchy.
- Reusing, repairing, and renewing wherever possible.
- Identifying where post-consumer waste technical materials can be used.
- Minimising petrochemical-based materials.
- Omitting radon barrier and damp-proof membrane (DPM). Install capillary breaks where needed.
- Installing radon sump ducted to reused chimney stack.

13.3.1 Managing moisture, radon and building fabric heat loss

13.3.1.1 Managing moisture

The World Health Organisation (WHO) believes that as many as 80 per cent of all building failures relate to moisture. This may be through surface or interstitial condensation and leaks, leading to rot, mould, corrosion, freeze-thaw damage, surface disrepair, and occupant ill-health [25,26]. This is especially true for the retrofit of historic, pre-1945, buildings, where modern materials and high occupant-generated moisture loads can be imposed with little understanding of their impact. One would think, therefore, that moisture management would be a primary focus of all governments, building designers, and builders, but it is not. Budgets are often tight, surveys may be piecemeal, and little formal design engagement may take place. This can be compounded by a lack of government guidance, trade association advice, and research. Only in the last fifteen years has increasing focus led to the commencement of dialogue to adequately support energetic retrofit [27].

It is tempting, but simplistic, to say that modern buildings *block* moisture, while historic buildings *regulate* or *manage* moisture. Modern buildings use capillary breaks, preventing the transfer of liquid water, seals, breather and vapour control membranes in a way that is well-understood. When designed and built at one time, this approach can be implemented in a coherent way. Historic buildings rely on measures that ensure the drying potential exceeds the wetting potential. There are likely to be fewer capillary breaks, and these will primarily be vapour-open air barriers. Roof overhangs, thick plaster and render coats, wall thickness, absorptivity of surface, the radiant heat of the sun, and fires and stoves all contribute to the drying potential. When present, liquid water and water vapour can move interchangeably through capillary-open, hygroscopic materials such as timber, concrete, and plaster. When surrounding relative humidity is sufficiently low, such surfaces facilitate distributed evaporation to the outside or inside [28]. When relative humidity is high, the same surfaces absorb moisture from the air, binding it to the material. In such a way, historic building materials regulate their own moisture levels but can also act as a moderating force, within limits, for the moisture present in a room.

13.3.1.2 Floor and radon

The wall's damp-proof course (DPC) and a solid floor's damp-proof membrane (DPM) act as capillary breaks, preventing liquid water movement. Air barriers are equally good capillary breaks but have the advantage of allowing water vapour to diffuse or move. A radon barrier is a substitute for a DPM, having the enhanced characteristic of blocking radon gas. It is best practice in new-build construction to have both membranes in place and continuously connected. However, a radon barrier can inhibit moisture management in historic buildings, as in the case study building, if such a barrier has been installed as part of a new solid floor build-up where the solid walls bounding the floor do not have capillary breaks.

In these cases, any moisture below the floor slab and within the rising walls has restricted drying by evaporation through the constrained surface areas of the walls, potentially leading to moisture stress. It may be argued that where the wall DPC is missing and measured radon levels are very low, it is beneficial for vapour to diffuse through the larger area of the floor.

For this reason, a floor build-up popularised by Mike Wye Ltd was specified. It is important to emphasise that radon gas escaping into a dwelling can kill. The approach taken was only possible because the levels of radon measured in the cottage were less than 1/7th of the threshold value for installation of a radon barrier, set out in the Irish Building Regulations, 200 Becquerels per cubic metre, Bq/m^3.

The floor assembly selected for the full floor of the extended cottage is a lightly compacted geocell, a foamed glass gravel, with geotextile membranes and a limecrete slab. The system, which is better known in Britain than in Ireland, has been approved by the Local Authority Building Control (LABC) of England and Wales for many years. The gravel is made from 100 per cent recycled glass. It can take a uniformly distributed load and is a capillary break, while also being highly vapour permeable. Given the depth of floor void available, it was possible to compact the subsoil, lay down 300 mm of the geocell on a geotextile, compact it down to 250 mm, and then repeat the process. A radon sump was installed as a precaution, flush to the top of the geocell. A pipe was connected to it and a geotextile was laid on top.

Gases move from high pressure to low pressure. The drainage pipe connected to the radon sump was, therefore, installed at a slight pitch so that gas could rise towards the fireplace, where it turned and rose up the chimney stack to just below the top of the flue (Figure 13.16). The intention is that any gas that might gather in

Figure 13.16 (from top left) Bags of foamed glass gravel delivered; agglomerated cork edge insulation & geotextile (above compacted gravel) prior to to pouring of limecrete slab; compacted foamed glass gravel and geotextile by door. Radon sump flush to top of consolidated foamed glass gravel; radon pipe rising up flue; bags of NHL lime; polymer fibres to add to limecrete mix. Limecrete mixer; limecrete slab and cork edge insulation beside portal frame; hessian spread on limecrete slab to ensure even cure

the sump would be drawn, through negative pressure, up the pipe and dispersed into the atmosphere beyond.

The builders had never used limecrete before and were justifiably cautious. The aftereffects of Lockdown made sourcing contractors and supplies more challenging. Quotes received from specialist sub-contractors were unreasonable, so the general builders made phone calls, read technical guidance, and installed the limecrete themselves. 50 × 50 mm lengths of agglomerated cork insulation were laid around the perimeter of the floor to lessen thermal bridges and to act as a form for the slab. Limecrete, with a binder of natural hydraulic lime (NHL 5), was specified with polymer fibres added to provide tensile reinforcing. Wetted hessian sheets were spread on the limecrete slab to ensure an even cure. The floor has worked well, without cracks or movement.

13.3.1.3 Repairing and replastering walls: revealing hidden secrets

Given the friable, poorly consolidated nature of the mass concrete walls of the cottage, it was decided they should be treated like rubble-built walls in terms of the moisture management and insulation strategy. As much of the original plaster had already failed and cracks needed fixing, it was judged appropriate to remove all plaster. It proved challenging to get back to a relatively dust-free substrate due to the friability of the concrete. Once removed, a far greater number of stress cracks could be seen in the concrete. As per good practice, helical bars were installed in a cementitious grout, literally strapping the wall together (Figure 13.17).

Figure 13.17 Without plaster the original construction of the fireplace and subsequent steps to reduce it can be seen. Note: cracks and repairs undertaken during retrofit

With the plaster removed, timber lintels and wall plates in reasonable condition could be seen. To the surprise of all, the timber sub-frames, to which the original mild steel windows would have been fixed, also became visible (Figure 13.17). As was common in Victorian times, materials could have several functions. In this case, the sub-frames acted as grounds for fixing the windows, but they also acted as permanent shuttering for the cast concrete window openings. A design change was needed before the replacement windows in storage could be fitted, with cost and carbon implications.

A good plaster scratch coat plays an important role in 'healing' a wall such as this, creating an airtightness line, and managing moisture. It is now well established that lime renders, not Portland cement renders, should be used internally and externally on historic masonry walls. Albeit, this wall was made of early Portland cement concrete, its poorly consolidated, friable nature suggested lime plaster would be beneficial here too. A natural hydraulic lime (NHL 3.5), commonly used in plastering historic walls, was selected and installed by a skilled plasterer.

13.3.1.4 Walls of existing dwelling

As the front wall abuts the public pavement, it could only be internally insulated. Having explored various options, it was decided to remove the remaining old plaster, allowing repair where necessary; reinstate and true-up new lime plaster and install a proprietary composite insulation system called Pavadry. This is made from wood fibre batts bonded to low-VOC oriented strand boards (OSB). This composite system is then fully adhered to the masonry wall with a notched trowel and mechanically fixed. As per specification, Baumit RK 70 Lime Plaster was the adhesive used. The Pavadry system has three advantages; it is fast to install with little drying time needed, it provides a smooth substrate for fitting the plasterboard room finish, and the whole assembly, including the adhesive, remains vapour permeable and capillary open, enhancing the drying capacity of the solid wall substrate. Figure 13.18 shows a relevant detail. Note how the Pavadry system passes in front of the window frame, reducing the thermal bridge at that location. Table 14.1 presents the U-values as calculated.

It was initially proposed that the rear wall of the original structure would be externally insulated and clad with a timber rainscreen. This would be beneficial from spatial and hygrothermal perspectives. External wall insulation of the right type is inherently lower risk than internal wall insulation. This is because the substrate remains warmer and drier and moisture within it remains in a vapour state, only cooling and condensing as it reaches the outer layers of the insulation beyond. Locating a rainscreen cladding in front of external insulation, where possible, has two additional benefits. It can facilitate fitting of insulation batts at a lower cost than that required for a full exterior insulated facade system (EIFS), and it removes the additional moisture stress of rain. However, the builders were so concerned about the friable nature of the concrete and its limited ability to take long fixings, effectively cantilevered through continuous slabs of external insulation to support cladding rails beyond, that a design change was agreed. The new approach

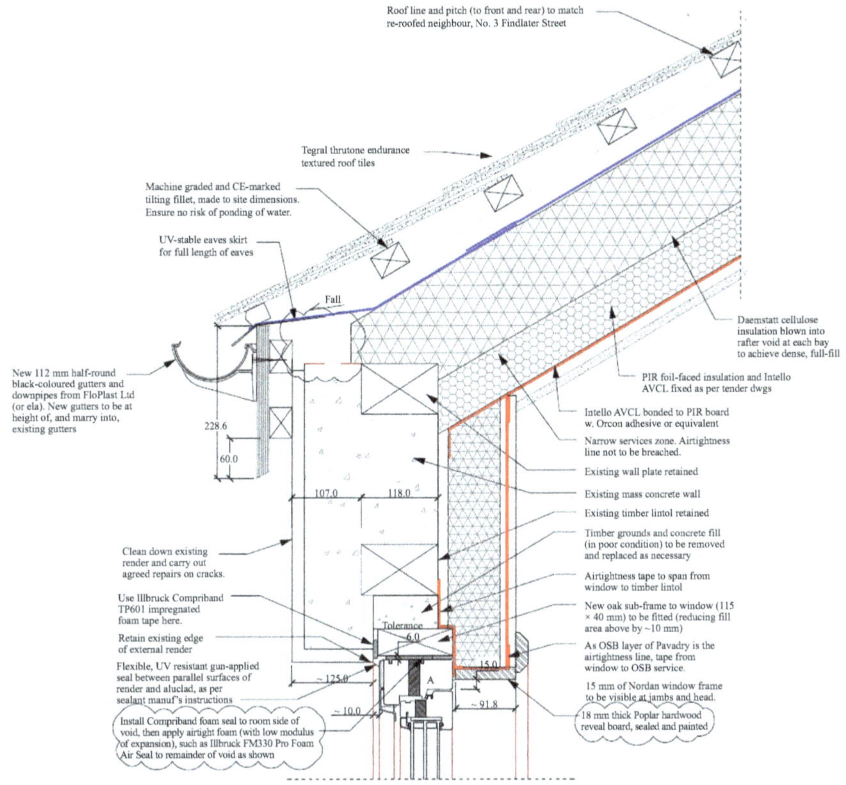

Figure 13.18 Eaves of front wall of cottage showing timber lintel and timber sub-frame to window, Pavadry composite internal insulation, blown cellulose between rafters and PIR board insulation slabbed below rafters

was to install a ventilated timber rainscreen on narrow, 25 mm thick, treated battens and internally insulate with the Pavadry system.

The supplier identified that the Pavadry system, stored in the UK, was available in only two thicknesses, 91 mm thick, made up of 80 mm wood fibre batts and 11 mm of OSB, and 51 mm, made up of 40 mm batts and 11 mm boards. Once the current supply was exhausted, prices were due to rise 30 per cent, so selecting one or both thicknesses was attractive.

The 91 mm composite board was selected for the front wall, resulting in a U-value of 0.41 W/m^{2}K. As 80 mm of internally applied thermal insulation will naturally cool a masonry substrate significantly, it was understood that the historic solid wall would experience an increase in moisture stress, regardless of climate. The 51 mm composite board was selected for the rear solid wall, resulting in a U-value of 0.65 W/m^{2}K. This narrower thickness of insulation would thermally isolate this wall less than the front wall. The use of a ventilated timber rainscreen

Figure 13.19 (from left to right) Two different thicknesses of Pavadry board stacked; boards being applied on a full bed of adhesive to the replastered walls with a toothed trowel; view of the main room showing the uninsulated fire breast; view of the insulated party wall and uninsulated fire breast adjacent to the front wall prior to installation of the composite insulation system

has the dual benefits of ensuring drier, less thermally conductive concrete and lower hygrothermal stresses due to the omission of liquid water from imbibed rain. It was difficult to simulate these risks at specification stage, as the only concrete datasets in the WUFI Pro material database at the time were for dense, consistent, modern concretes, quite unlike the one the cottage was built with.

In retrospect, the 71 mm thick version of the system, now available, may have been a better choice for both front and back walls. While the 51 mm composite board provided a space-saving value, the first occupants of the cottage claimed this was the only cool room in the cottage. The 91 mm composite board may also have cooled the substrate wall more than was desirable.

Finally, to counteract thermal bridging at various key junctions and maintain clean lines, the same thickness of composite boards was used on party walls. The fire breast was the only concrete wall left uninsulated for two reasons: its attractive curved corners could not be replicated with this insulation system and cladding it with insulation would have caused it to push even further into the room. From a thermal perspective, it was judged that the location of the fire breast deep in the plan, with large amounts of insulation above and below, minimised thermal bridges.

13.3.1.5 Walls and roof of the timber-frame extension

The author and the timber frame manufacturer that supplied the extension's wall and roof, share the same approach to moisture management. As can be seen in Figure 13.20, from inside to outside, the wall build-up features plasterboard, then a services zone with quilt insulation, then a racking board providing the airtightness

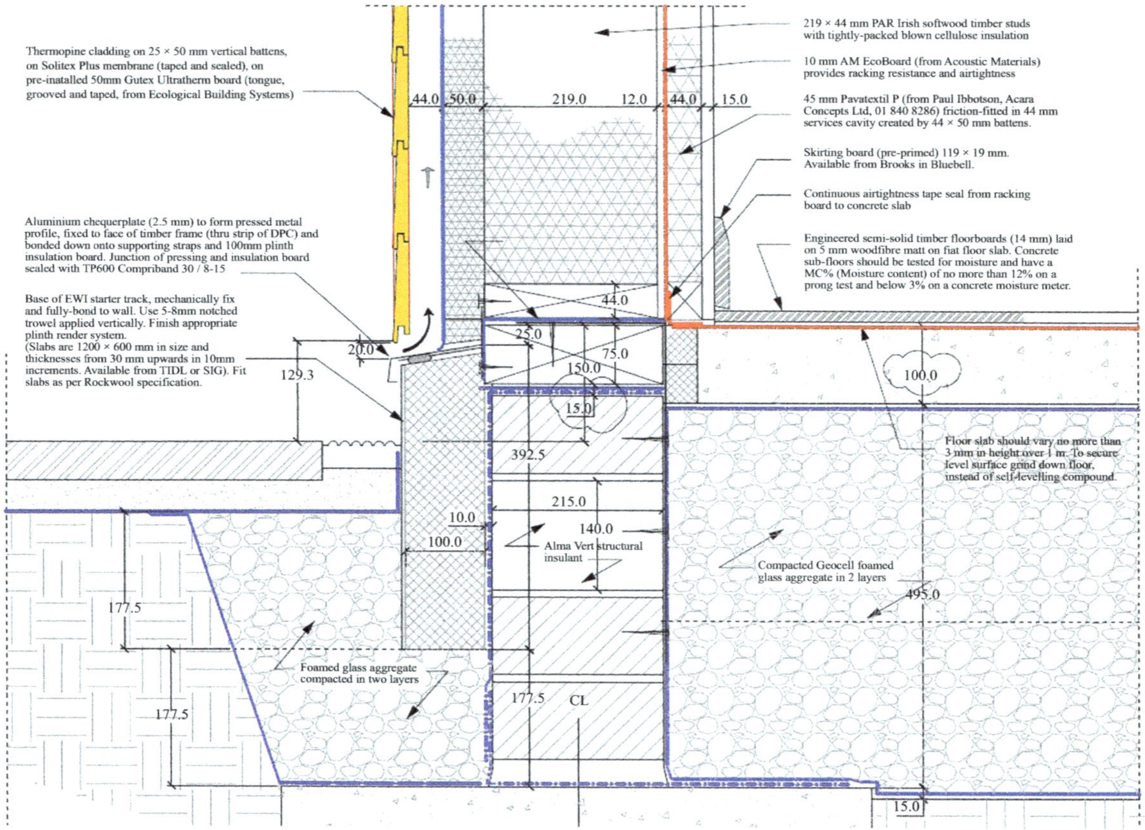

Figure 13.20 *Base of timber-frame extension structural insulation in rising wall and plinth insulation to reduce thermal bridge, foamed glass gravel floor internally and wood fibre insulation batts 'warming' the timber-frame*

line, shown in red, fixed to the inside face of the frame. Paper cellulose fibre insulation is tightly packed between the timber studs of the frame, and tongue-and-grooved wood fibre batts are fixed to the outside with taped breather membranes. Finally, a ventilated rainscreen void and shiplap timber cladding provide a weather shield.

Using these materials in this arrangement has benefits for both management of moisture and thermal performance. The airtight racking board minimises vapour migration into the wall from the room, while the materials outside of that, including studs, paper, cellulose, and wood fibre, allow any vapour present to diffuse easily to the outside through these vapour-permeable materials. If liquid water finds its way into the timber frame wall, the hygroscopic and capillary-active qualities of the cellulose and wood fibre, ensure that the water is wicked away from the primary structure. That water can continue to move as a liquid to the ventilated rainscreen cavity or evaporate at an earlier point and diffuse outwards. Finally, the rainscreen excludes the significant moisture load that driving rain would deliver. The extension's roof is similar except for the specification of a more durable breather membrane, battens, counter-battens, and roof tiles in lieu of timber cladding. The approaches taken for the wall and roof components may be considered robust and moisture-safe.

From a thermal perspective, the 'warm stud' approach of using a wood fibre layer to insulate the timber frame, keeps it uniformly warm and dry. From a structural perspective, it also ensures less movement. The quilt insulation in the services zone provides a useful top-up in thermal performance. However, the insulation on the inner and outer sides of the timber frame proved unexpectedly beneficial at lintel positions and at the narrow wall pier beside the patio door. In those locations, structural requirements resulted in the omission of all cellulose insulation in lieu of additional studs. While timber may be considered a weak insulator, it is very useful that those flanking layers are present to ensure adequate thermal performance in those locations.

Even where access to sophisticated simulation tools is limited, much can be learnt about moisture stress by listing and graphing the water vapour diffusion equivalent air layer thickness of layers that make up a wall assembly. The relevant metric is water vapour diffusion equivalent air layer thickness (Sd-value). Sd-value describes the resistance layer that a building component provides to vapour diffusion in terms of the thickness (in metres) by which a stagnant air layer would need to have in order to provide the same diffusion resistance. In Figure 13.21, these layers are listed and displayed for the internally insulated solid wall (above) and the south-facing wall of the timber-frame extension (below). The racking board of the timber frame is the most vapour-tight layer, yet it provides only enough vapour resistance to encourage moisture to migrate outwards. In contrast, the concrete substrate, whether it has characteristics more similar to concrete A or B, is many times more vapour-tight than the OSB layer of the Pavadry system, yet it is located where the ability of moisture to diffuse through it to the outside as a vapour is declining due to the substrate being impacted by cooler outdoor temperatures. When condensation occurs, liquid can remain longer or even migrate back towards the room.

	Cement render (20 mm)	Concrete - Option A or B (205 mm)	Plaster (15 mm)	Adhesive (5 mm)	Pavatherm-Combi batt (80 mm)	OSB (11 mm) (Airtightness layer)	Plasterboard (12.5 mm)
SD-value of materials within rendered solid wall (conc. A) with IWI	2.678	21.955	0.688	0.025	0.240	1.584	0.104
SD-value of materials within rendered solid wall (conc. B) with IWI	2.678	11.336	0.688	0.025	0.240	1.584	0.104

	Solitext Plus breather memberance (<1 mm)	Gutex Ultratherm woodfibre batt (50 mm)	Daemstatt paper cellulose within timber frame (219 mm)	AM EcoBoard (racking board, 10 mm)	Pavatextil services zone (44 mm)	Plasterboard (12.5 mm)
SD-value of materials within courtyard wall of TF extension	0.040	0.150	0.394	0.420	0.132	0.104

Figure 13.21 Representation of the Sd-value of the layers making up the internally insulated, west-facing, solid wall and the timber-frame wall facing the courtyard. (Colour heatmapping is used to visualise relative resistance to vapour diffusion. The thickness of layers shown is indicative.)

13.3.1.6 Airtightness

A misinformed advocate of historic buildings might declare that a focus on airtightness, or its converse, air permeability, is in opposition to breathability of the building fabric and occupant ventilation. However, this is fallacious. Airtightness is about the unintended flow of air through joints, cracks, and porous surfaces. These air flows increase the heating demand in winter, can transport unwanted moisture, and are conduits for noise. Breathability is about the extent to which moisture can move through building fabric, whether as a gas or liquid. Ventilation is about the intentional management of air within a building to ensure that high indoor air quality is provided to occupants. It is, therefore, quite reasonable and desirable to ensure that a new or retrofitted building featuring traditional building materials, or highly breathable building materials, also has a very low air leakage rate. For this to happen, a clear strategy must be in place.

An air permeability target of $Q_{50} \leq 2.0$ m^3/m^2.hr was adopted, measured using the method outlined in EN ISO 9972:2015. The airtightness line in the extension is the surface of the limecrete floor slab, the face of the AM Ecoboard, and the face of the window and door sets. The airtightness line in the existing cottage is the surface of the limecrete slab, the OSB face of the Pavadry composite boards, the window and doors, the air and vapour control layer (AVCL) glued to the foil face of the post-industrial recycled (PIR) rigid insulation on the pitched roof, and the AVCL membrane fixed under the original joists. All joints were connected using airtightness tape. The strategy was made clear in the specification, shown on all construction details, and discussed on site. The target airtightness value was written into the contract.

The top central image in Figure 13.22 shows taping along joints of the timber-frame panels, over all fixings, and over the large holes that were cut with a hole saw at the top of each panel in order to insert a hose through which paper cellulose insulation was blown. Initially, the hose was inserted down to the bottom of the void, then slowly raised as the insulation filled. This was done with impressive skill and consistency by Roman Szypura of Clioma House, an acknowledged expert.

Given that there was a new scratch coat on all the original concrete walls, it was hoped that the Pavadry composite boards that lined the party walls would not need to be taped (Figures 13.19 and 13.22), but use of a pencil-type anemometer during air leakage testing revealed that there was air movement between some of these boards, so eventually all joints were taped.

13.3.1.7 Roofs: form and layers

Once one moves away from traditional cold-pitched roofs, retrofitting roofs can be surprisingly complicated. A ‘cold’ pitched roof is one where the insulation is laid on the flat ceiling, with the cool or cold attic void overhead. A ‘warm’ pitched roof is where the insulation is laid on the pitch, following the line of the rafters, leaving a warmer void, or a dormer bedroom, below it. One can also talk about two hybrid conditions: hybrid ‘A’, where insulation locations transition from one to the other,

Figure 13.22 Airtightness work (from top left clockwise): tape to new floor slab before plaster application; tape used in timber frame at racking board joints, fixings and holes for blowing-in cellulose; membrane and tapes to rooflight and at portal frame to kitchen; at window sub-frame after plastering and taping of Pavadry boards to each other and taping over cables chased into the OSB layer of Pavadry boards

and hybrid ‘B’, where insulation has been installed above and below. The latter is generally an *ad hoc* approach that has only arisen in retrofit work in the last decade. It is generally a consequence of the original insulation being retained when insulation is installed on the pitch rather than a designed intention. The four types are shown in Figure 13.23. Given the small areas of pitched ceiling, also known as ‘coombes’, to either side of a flat ceiling in the original portion of the cottage and the monopitch roof extension, the case study cottage can be considered a hybrid ‘A’-type roof, with warm- and cold-pitched roof build-ups, as shown in Figure 13.24.

‘Cold’ roofs should be the easiest and lowest-cost major elements of a dwelling to insulate. It is disappointing to see how often this is done in a way that is thermally compromised, potentially resulting in heightened moisture stress. Homeowners can contribute to the underperformance by not clearing out their attics before work commences. Contractors looking to work quickly, in particular if they misapply spray foam technology, can also contribute to underperformance. Geometrically, insulating on the pitch requires more insulation than on the flat and it is also harder to get right. Finally, insulation on the pitch and on the flat, identified as Roof 4 in Figure 13.23, is very problematic given its ill-considered nature.

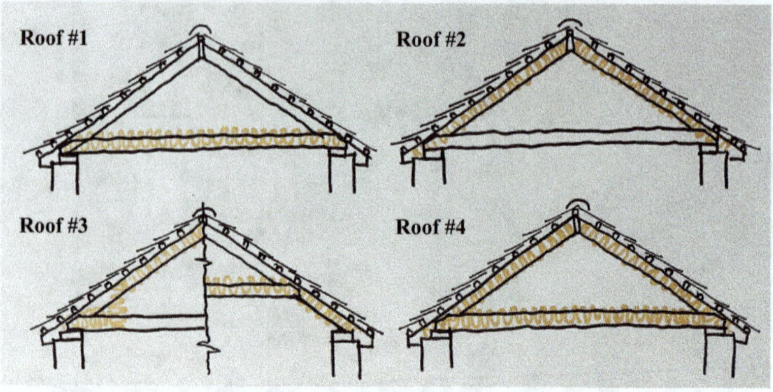

Figure 13.23 Roof 1: cold-pitched roof, Roof 2: warm-pitched roof, Roof 3: hybrid 'A' pitched roof, Roof 4: hybrid 'B' pitched roof

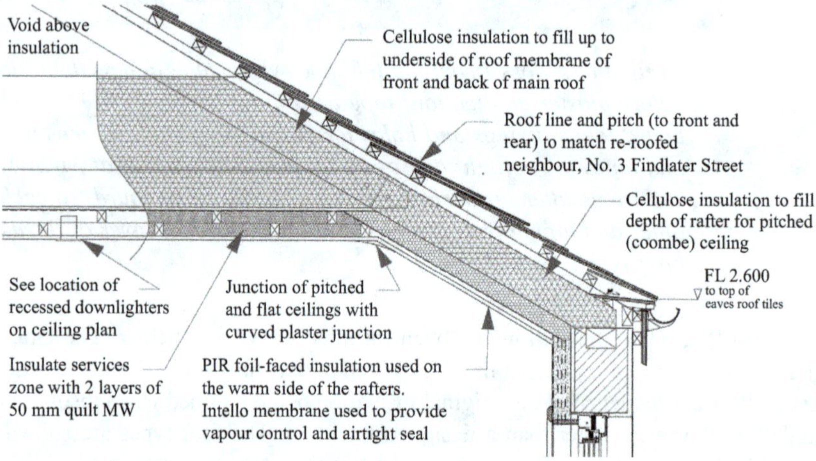

Figure 13.24 Cold-pitched roof and coombe of case study cottage

It lends itself to underperformance and is supported by neither technical guidance nor regulation.

To maximise thermal efficiency and minimise moisture risk, it matters whether the insulation is compatible with the primary structure. We need to consider the timber rafters and joists, whether a continuous vapour control layer has been installed on the warm side of the insulation, and what kind of underlay, either high or low resistance, has been installed on the cold side. We also need to consider what kind of roof ventilation is present, whether the roof tiles or slates are fixed to battens with counter-battens, or just to battens, and even whether the underlay sags [29].

13.4 Circular and carbon impacts of specification

The *take-make-discard* model of manufacturing and consumption that has characterised the last century is damaging our planet. Even though conventional supplies of some essential raw materials are expected to disappear within decades, we simply do not have the carbon budget to continue this approach. However, the impact is even greater, including devastation of local and global biodiversity, the weakening of indigenous communities, and built heritage. While alternative approaches have been discussed and researched for decades, and a significant shift in interest is occurring, it will take concerted effort and radical change to fully alter the current paradigm.

Circularity is an ideal to strive towards in all industries, sectors, and households. In its purest form, it prolongs the use of existing resources and restructures supply chains, radically reducing virgin resources and eliminating waste. Whereas efficiency can generally be delivered within a business-as-usual paradigm, a deep engagement with circularity and the circular economy will require fundamental change to business practice, the economy, and the passive role of the consumer. A focus on circularity naturally aligns with preservation of built heritage, a strengthened bioeconomy, adaptability, disassemblability, and the right to repair. In 2016, Geraedts captured this well at a building level when he wrote *'the longer a building can keep its functional life cycle instead of becoming vacant or being demolished, the more sustainable that building will be. The more a building is flexible and able to adapt to changing user demands, the longer it will keep its functional life cycle'* [30].

The mantra of 'reduce, reuse, recycle' is the simplified highlight of a sophisticated resource and waste hierarchy. Figure 13.25 shows a ten-step hierarchy [31].

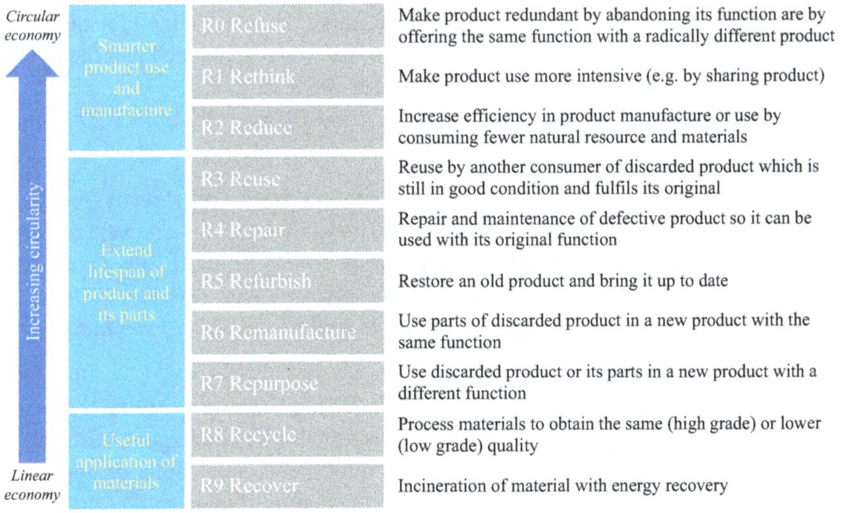

Figure 13.25 The 9R resource and waste hierarchy, adapted from Kirchherr et al. [31]

The pre-activity steps of *(R0) refuse* and *(R1) rethink* are systemically and strategically crucial. Once engaged in resource use, optimising *(R2) reduce, (R3) reuse*, and *(R4) repair* of products is crucial. *(R5) Refurbish, (R6) remanufacture*, and *(R7) repurpose* require increasingly great commitments of energy and material and, thus, carbon. *(R8) Recycle*, the strategy most celebrated, is one of the worst: it requires that products be broken down to their constituent materials, which are then further reduced to recyclate, such as grains, pellets, fibres, or dust, that can be fed into a new manufacturing process with a high energy and carbon penalty. Energy *recovery (R9)* from incineration is scarcely better than landfill; the ultimate, high-carbon failure of a system. Each step down the hierarchy requires additional processes and energy, adding to the carbon footprint of the product, so it should be avoided when possible.

When end-of-life *is* reached, elements of a building, components, products and, lastly, materials should be used in the order of preference listed [32]. To aid this, the Ellen MacArthur Foundation [33] recommends that materials be separated into 'biological' and 'technical nutrient' cycles. Biological nutrients are organic materials that can be easily returned to the biological cycle. Technical nutrients are manufactured products that should be reprocessed through industrial systems.

13.4.1 End-of-waste

There is something fascinating about the journey from a poorly regarded waste material to a consistent, highly regarded, premium building material. Where a material cannot be used at a higher level in the resource and waste hierarchy (i.e., R3–R5), the end-of-waste methodology structured by the European Union allows manufacturers to undertake a formal process whereby post-industrial recycled material (PIR) and post-consumer recycled (PCR) material can become the feedstock of new premium-quality building materials. By specifying PIR and PCR materials for new works, the demand for virgin materials is lessened and the overall embodied carbon impact of a building project is reduced.

The latest edition of the Irish EPA's *End-of-Waste – Guidance Document* states that

> *'Uncertainty over the point at which waste has been "fully recovered", and ceases to be classified as waste, has inhibited the development and marketing of waste-derived materials that could be used beneficially without damaging human health and the environment. This can in turn inhibit the recovery and recycling of waste and its diversion from landfill and other disposal routes.'* [34]

To counteract this, Article 6 of the EU's Waste Framework Directive states that something can be given 'non-waste' status when it has undergone recovery, is used for specific purposes, meets technical requirements and standards applicable to a product, and when a market exists for its new nature. It also should not lead to adverse health or environmental effects. Environmental protection agencies across Europe are working with innovators to turn waste streams, particularly those of

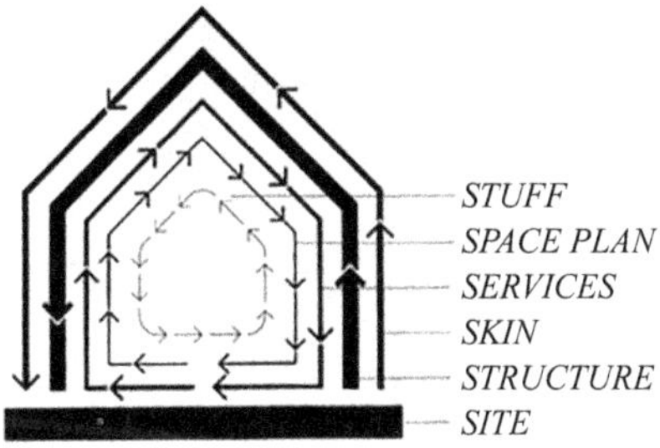

Figure 13.26 Frank Duffy's shearing layers as elaborated by Stewart Brand

greatest volume or environmental impact, into premium products using this framework. For this to succeed, those products must be valued and specified.

To support R5–R8, elements, building components, and products should be assembled in such a way that they can be disassembled with minimum loss or damage. In his 1994 book, *How Buildings Learn* [35], Brand writes about buildings 'always building and rebuilding' – a fluid state (Figure 13.26). Brand expanded on Frank Duffy's concept of a building's 'shearing layers': each layer, such as site, structure, skin, services, space plan, and 'stuff', changing at a different rate. One measure of the success of a building design is how each layer can shear and change without damage to adjacent, longer-lasting layers. To support this, there is emerging work, pioneered by Durmisevic, on how building products are best brought together [36], often with emphasis on fixings and compression seals over tapes and glues, to ensure they can be separated with the least loss to support multiple reuses [25].

13.4.2 Circular specification

A significant effort was made to ensure that building materials or products specified in the case study project were as circular as possible. The following three categories provide a quick overview: further detail is then provided for a selection of these. Figure 13.27 visually presents the materials and products listed below, excluding materials #1, #2, #3 and #5. Figures where the latter materials can be seen are shown and are listed against them.

Existing components were reused whenever possible. The extensive damage to the building in the decades prior to purchase resulted in much unavoidable loss. Materials reused include:

1. External mass concrete solid walls and foundations (Figure 13.8)
2. Granite door threshold and sills (Figure 13.17)
3. Internal doors and ironmongery (Figure 13.32)
4. Baltic pine roof structure (Figure 13.39)
5. Terracotta roof ridge tiles (Figure 13.39)

Proving that a wide range of biomaterials was already available in Ireland and could be appropriately combined in a single, high-quality building project was considered a key outcome for the building project. Too often, a perception

Figure 13.27 Circular materials used in the FS4 case study project

of non-availability prevents engagement. As building biomaterials are generally good at managing moisture, an elegant synergy of circularity and moisture management strategies is possible. The following products were specified:

6. Structural (CLS) timber in TF extension (from Glennon Brothers Timber Ltd)
7. Dämstatt wood cellulose fibre insulation (a biomaterial and post-consumer newspaper waste from Ecological Building Systems)
8. Agglomerated expanded cork insulation at edge of limecrete slab (from Roundtower Lime)
9. Pavadry wood fibre-OSB composite board (from Acara Concepts)
10. Ultratherm wood fibre sarking board (from Ecological Building Systems)
11. Thermopine external cladding without chemical preservatives (from Finsa Ireland)
12. New floorboards re-manufactured in Ireland from historic pine beams

The following new PIR and PCR materials were specified:

13. Foamed glass aggregate, ex-glass waste (from Geocell Ltd).
14. Alma Vert structural insulation, ex-plastic bottles (from Partel Ltd).
15. AM Ecoboard, ex-Tetrapak waste (from Acoustic Materials Ltd).
16. Pavatextil, ex-denim textile waste (from Acara Concepts).

To the author's knowledge, this modest project was the first use of AM Ecoboard as a racking board worldwide and the first use of Finsa's Thermopine external cladding in Ireland.

The new structural timber used was C16-grade Sitka spruce, grown in Ireland and processed by Glennon Brothers Timber Ltd. Strangely, the use of Irish timber in Irish timber-frame walls is uncommon and even more unusual in Irish timber trussed roof structures. This appears to be partly due to a prejudice against Irish timber and an outdated view of how it is processed in Ireland. However, it is also due to the less knotty appearance and higher strength grades of softwood timbers available at comparable prices from Nordic suppliers.

More innovative is the use of a construction board made from PCR material as the racking board. The waste is Tetrapak beverage cartons that are shredded and bonded under great pressure to form a board that is then lined in paper and traded under the name AM Ecoboard. The board has been used in Ireland for many years as a premium CE-marked acoustic product. Its use in the case study's timber-frame extension, following appropriate structural tests, is apparently its first use in Europe as a racking board.

Where carbon-intensive materials were used, efforts were made to reduce the quantities. The best example is the foundations. Structural engineer #1 designed concrete strip foundations 750 mm wide and 300 mm deep to support a modest single-storey extension, proposed at the time to be built of lightweight blocks, and to brace the rear boundary wall of the site (Figure 13.28). This had three negative impacts: financial, thermal (by reducing the depth of floor insulation), and carbon. Following discussion, the engineer omitted the extra concrete, replacing it with an extra layer of reinforcing mesh. This change reduced the concrete used in the project from 3.356 m^3 to 2.616 m^3, a reduction of 22 per cent and an embodied carbon saving of 32 tonnes of CO_2. This is interpolated from V3.0 of the ICE database, which assumes concrete of Class 30 MPa = ~0.132 $KgCO^2e/Kg$ and density of 325 kg per m^3.

Given structural engineer #1 signed off on the structural integrity of the final foundation design as installed, it is hard not to view the initial approach as significantly over-designed. Subsequently, after structural engineer #2 came on board, the limecrete floor slab was reduced from 150 mm to 100 mm, providing further financial and carbon savings. There is a lesson here for careful selection of design teams, as much as materials.

13.4.3 Missed opportunities

Any project will have missed opportunities where, with the benefit of hindsight, we might have made different choices. In the context of struggling to reduce the amount of concrete specified in the extension's foundations and keep the project on programme, the opportunity to reduce the embodied carbon footprint of the cement used was overlooked. A saving of as much as 48 per cent is estimated through use of 75 per cent ground granulated blast furnace slag (GGBS). This would have resulted in a further embodied carbon saving of 15.4 tonnes of CO_2.

The Baltic pine timber sub-frames, discovered when plaster was removed from the three windows and hall door, were in reasonable condition. However, Passive House Standard replacement windows, waiting in storage, had been sized based on a masonry reveal and some amount of fill. For the new windows to fit, either the old sub-frames would have to be removed, taken apart, milled to a thinner size and re-assembled without damage or weakening, which was judged unlikely, or new thinner sub-frames would need to be manufactured and installed. The latter approach was taken. Given the relatively harsh environment of being fixed into the concrete wall, a hardwood, American White Oak, was selected. The frames were

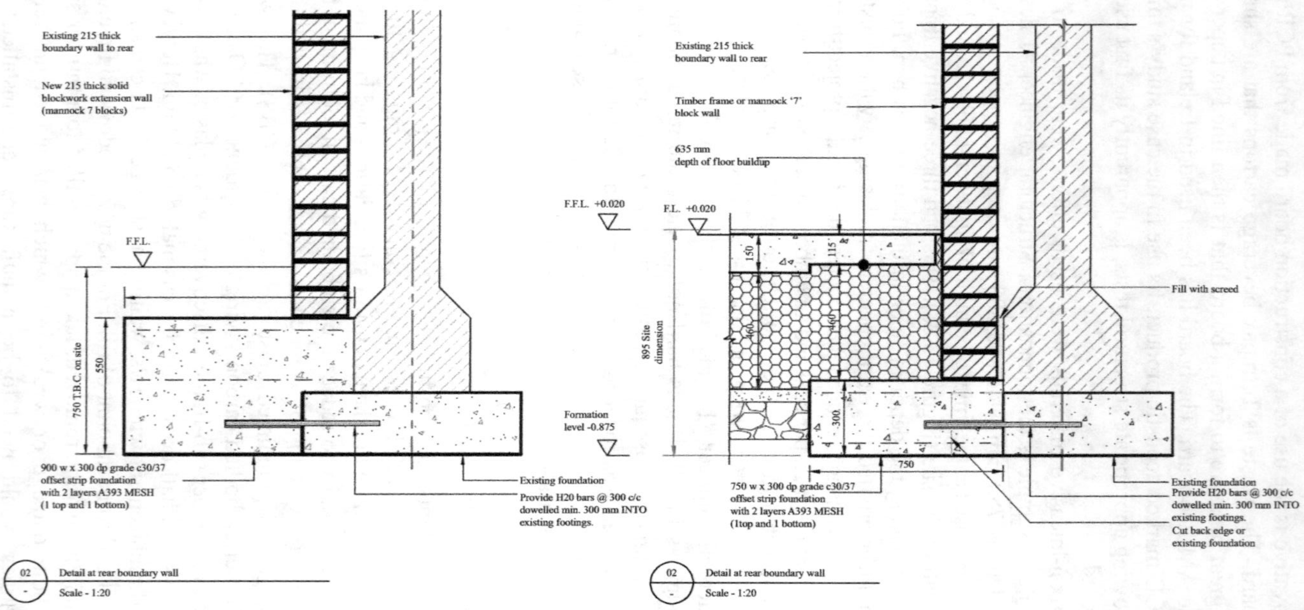

Figure 13.28 Revision A and D of structural engineer #1's sectional drawing of base of single-storey extension and underpinning of rear site wall

sized, manufactured and fitted. The materials cost was €1,000 extra. Of course, there was also an embodied carbon cost.

13.4.4 Implementing circularity

The Pavadry composite boards are made from Pavatex wood fibre boards bonded to oriented strand boards (OSB3) (Figure 13.19). The wood fibre boards are made entirely from waste wood chips that are heated and pressurised to break down into fibres, which are turned into rigid insulation boards using a dry process. As a natural product, Pavatex insulation can be recycled or composted at the end of its life. OSB have no added formaldehyde and are made from wood strands, layered and compressed with synthetic resin binders under heat and pressure. The composite system's environmental product declaration (EPD) demonstrates the composite system's minimal carbon footprint. Each board is fixed to the wall with two mechanical fixings and a full bed of adhesive for moisture management reasons.

With regard to disassemblability, due to the use of a full bed of adhesive to aid moisture management, removal of the composite boards would likely cause some damage to the layer of lime plaster behind them. While some boards might be reused, the wood fibre boards can be ground down as a recyclate in a re-manufacturing process, or locally composted. Due to the glues, OSB3 cannot be composted but could also be ground down for separate re-manufacturing as a small proportion of new OSB boards.

The design and installation of the timber frame may be considered a lesson in optimised circularity. Given the constrained access to site, TimberTech Homes divided the extension's walls into ten separate frames that could each be walked through the front door by two people, avoiding the need for a crane. Following site measurements, discussions, and sign-off of shop drawings, the panels were manufactured off-site. They were then delivered, installed, fixed, and weathered in a two-day period. Due to delays associated with Lockdown, the roofing took place more than two months later. The specification of a robust breather membrane, designed to manage direct rainfall, the specific characteristics of the wood fibre substrata, and the moisture-managing assembly ensured that this was not a concern.

The timber frames were made of 38 mm wide studs of C16-grade, Irish-grown, Sitka spruce. Most of the frames manufactured are full-height, 1120 mm wide, with a full-height central vertical rail braced at storey height, resulting in two large voids. 50 mm wood fibre batts were fixed to the outside of each frame and clad with the black breather membrane in the factory, as can be seen in Figure 13.29. The wood fibre gave the panels sufficient rigidity during delivery and installation. Once in place, the frames were bolted down to the timber wall plate fixed to the rising walls. The frames were then bolted together through the edge studs. Only then was the AM Ecoboard racking board fitted. To match the testing regime for the board, they were screwed to the timber studs instead of being nail-gun fixed.

After the wall frames were installed, the pre-cut rafters and junctions with the existing roof were installed and fixed to the frames. Racking boards were then fixed below the rafters, wood fibre insulation batts slabbed over, and a high-specification

Figure 13.29 (from top left clockwise) Timber-frame panels ready for transport, new timber frame meets old cut-rafter roof, wood fibre batts clad roof, interior showing AM Ecoboard as racking board

breather membrane installed, appropriate to the roof's likely higher exposure to rain and UV degradation. Once fitted, large holes were drilled through the racking board at the top of the voids in each frame to allow later blown installation of fibre paper cellulose insulation. It would be a few months before the insulation was blown in and airtightness tapes applied.

With regard to disassemblability, the plasterboard would likely be removed and skipped due to nail gun fixing and surface finishing. Equally, the services zone battens are only 25 mm thick and might not survive removal. However, the cellulose insulation can be fully removed by sucking through the original holes once reopened. It is reasonable to assume that the screw-fixed racking boards can be unfixed, the timber studs unbolted, and the tongued and grooved wood fibre batts removed. Membranes and airtightness tapes, by their nature, are single-use.

Thermopine from Finsa Ireland was fitted as the cladding of the rear wall of the cottage and extension: the first use in Ireland. Timber used as exterior cladding is treated to guard against degradation. This is usually done with chemicals, which create environmental problems while seeking to solve one. Instead, Thermopine is treated with a combination of pressure, temperature, and water vapour for up to one hundred hours. The process removes microscopic xylophages and fungi and residual moisture and resins that can cause decay. As the boards age, they take an attractive silver colour. Periodic coating with vapour-permeable oils slows this ageing process.

The use of a rainscreen cladding that is light and environmentally sound is very beneficial. The principal value, from a moisture management perspective, is that the cladding removes liquid water from the wall, enhancing its ability to manage

Figure 13.30 *(clockwise from top left) Battens and Thermopine cladding being installed (yard drainage and plinth insulation visible below), cladding adjusting to local geometries, cladding to side and back of monopitch roof, screw fixings and an Aereco wall vent*

vapour, remain structurally sound, dry, and insulating. To enhance disassemblability, screws were used instead of nails, as shown in Figure 13.30. Stainless steel Fischer screws were supplied by Masonry Fixings Services.

Perhaps, the hardest won carbon 'gain' was the preservation of the original timber roof structure. The first structural engineer and the builder's foreman tried to persuade the author to remove all the original timber, the former stating he would not certify the roof unless it had new structural timber, and the latter pressuring for a 'proper, long-lasting job'. Figure 13.31 shows (on the right) fast-growing, carbon-sequestering, C16-graded Irish Sitka spruce, which the extension was built with, and (on the left) slow-growth Baltic pine, whose 'second life' as a roof structure has outlasted its 'first life' as a plant. Roughly of equal size, one timber block contains eleven growth rings and the other about 120.

It was only when client and roofer were on the scaffold looking at the roof structure up close, that it was clear the original timbers were in excellent condition, including the wall plates. It would have been a grave mistake to replace the historic structural timber, even with the high-quality C16 timber shown. In contrast, the

Figure 13.31 (left) Slow-growth Baltic pine from the original roof and (right) C16-graded Irish Sitka spruce from the extension

fascia boards of the timber eaves were in sufficiently poor condition to merit replacement and the timber 'dooks' that had been cast into the inner face of concrete wall to take fixings near floor level were so decayed they powdered away when touched.

Knowing that there will always be some loss when old slates are removed, the author, like many architects, specified in the tender documents that the front roof should be re-slated with the original slates, while the rear and extension roofs should be slated with new fibre cement roof tiles. The slow emergence from Lockdown and the difficulty securing tradespeople and materials delayed the start of roofing works. The roofer, an excellent tradesman used to working on conservation projects, made clear in the first conversation that he could only take the job if the slates were already removed and stated 'don't even think of asking me to re-slate the front with the original blue bangers!'. As that work had been priced in the contract, a two-thousand-euro saving was agreed upon where the whole roof was slated with new roof tiles.

The author was anxious that the eleven terracotta ridge tiles, which visually link the red brick chimneys, would be retained. As three had been damaged, he sourced salvaged tiles of identical profile. The builder made clear that casting the ridge tiles in cement on a fibre-cement-clad roof would be a mistake. Cement drips would eat into the surface coat of the roof tiles, permanently defacing them. The builder recommended skipping them as a result. The author found a dry clip system that works with historic ridge tiles, which was then fitted. The absence of mortar joints between the ridge tiles visually emphasises their separation and colour differences, which is surprisingly pleasing (Figure 13.39). Inadvertently, the ridge tiles have become a subtle visual signal to the passerby that retention and reuse are valued there.

As seen in Figure 13.10, the two original internal door leaves were found still in place and operable. The v-joints of the boards had been filled with plaster to give a flush, mid-twentieth-century aesthetic and had been painted with multiple layers of oil paint. The door frames, architraves, and skirtings were in sufficiently poor condition that they could not be reused. The door leaves were removed for safety and restored. Impressively, they did not distort. The attractive profile of the architrave was measured, a replica made in American poplar, and a commercial skirting that matched the original profile was identified. While the original strap hinges were still in place and in good order, new surface-mounted sashlock and knob sets had to be purchased. When fitted in the door opes of the new partition walls, the refurbished door sets looked and performed well. As shown in Figure 13.32, panels over the original doors, which may originally have been fanlights but were solid when purchased, were re-instated and given a subtle decorative feel through use of a period, heavily textured wallpaper.

The historic 4-panel door that now gives access to the kitchen had the most quixotic journey of all. In 1971, this late Victorian hardwood door had been purchased by the author's father-in-law. He had replaced the solid panels with slim-profile double-glazed units and installed them as a porch door in the (c.1922) house he had recently purchased. Almost 50 years later, he removed it and replaced it

Figure 13.32 *(left) The original internal ledged doors with hinge straps doors were recovered under layers of filler, wallboard and paint, (centre) a late Victorian door repurposed as a glazed storm door in 1971, then (right) the same door on its third life as an internal door to the kitchen*

with a modern airtight door after an air pressure test indicated significant air leakage through the porch and hall doors. It was then identified as an ideal door for internal use in the cottage. After an attempt to remove the distortion in one corner of the door leaf failed, it was decided that it should be used as is, while a tight fit could be ensured by simply adjusting the bead of the new doorframe to fit it. While all walls and other doors were painted white, the three main doors of the house, hall door, kitchen door, and patio door, were painted blue, marking the main progression through the cottage.

There are two floor finishes in the cottage. The new fully vitrified floor tiles in the shower room are fixed with tile adhesive and finished with grout. The solid pine floorboards elsewhere in the cottage, were re-manufactured in Ireland from historic pine beams. Their width, flaming colour, and low carbon impact, due to the second life of a biomaterial, are very attractive. However, they are relatively soft, so they mark easily. The original ambition to recover and reuse at least some of the boards of the original suspended timber floor was quashed when the builders took them up. Many of them could not be saved; then one day, the rest disappeared. Architects and clients must understand that, at times, recovering materials can cost more in terms of effort and money than new materials.

The adhesive used to fix the pine boards to the concrete floor is TDS Bona 848, a single-coat, 'silane modified-prepolymer' applied evenly using a coarse notched trowel. The life cycle assessment flagged it as having a high carbon impact, comparable to the new aluminium windows, so we had to be careful in assessing how much of it was used.

Services have been accommodated throughout the case study building behind the plasterboard finish, ensuring they can be swapped out, or added to, without damage to structure, insulation, or airtightness. In the timber-frame extension, continuous service zones have been provided in the walls and ceiling, while in the main building, cabling has been run to sockets in narrow chases in the OSB layer of the Pavadry composite boards that were then sealed with airtightness tape. This prevented additional loss of space and airpaths along the line of the cables.

13.4.5 Building services

13.4.5.1 Foul drainage and water supply

In common with many terraced developments across the British Isles, the DADCo development features portions of private, combined drains running parallel to the street, laid under the rear of the properties. These are 'combined', in that storm and foul water are removed through the same pipe, and 'private', in that these pipes are owned and notionally managed by the property owners, unless vested in a management company. In this development, a manhole was in every fourth or fifth yard for maintenance purposes, including the cottage next to the case study property. As is common, these clay pipes can crack when residents unwittingly carry out building works, causing leakages or blockages. Once demolition allowed, the full extent of the pipe was exposed and its condition checked. Measures were taken to support and protect it, and a new manhole was created in the yard of the extended

Figure 13.33 (left to right) The original lead water pipes branching, the two new MDPE pipes supplying both cottages entering the sub-floor at the front and one branch away at start of party wall of extension

cottage to aid future inspection and maintenance. Given the tiny size of the site, it was not feasible to create a separate drainage system within the curtilage of the site, which local authorities often request.

Electrical power may first have been supplied to the cottage in the 1930s, but piped mains water was supplied *ab initio*. By design, water was supplied to each pair of cottages through a single lead pipe, with the second pipe branching away in the privy to rear to serve the adjoining cottage. While this has been regularised in many cottages on the estate, water was still supplied to the case study cottage and its neighbour through a single lead pipe at the time of purchase.

Clearly, the installation of separate, non-lead supply pipes was a priority. A range of supply routes and timings were considered, which would have the least impact on the neighbour. A formal agreement was drawn up, signed by the neighbour and notarised to formalise this new wayleave. The neighbour then applied to Irish Water for a separate supply. As agreed, after the new connections were made and all lead piping removed, two new medium-density polyethylene (MDPE) pipes were laid side-by-side at the base of the cottage's future floor build-up. At the junction with the extension, one was brought through to connect to the existing internal copper piping of the neighbour, while the other was made ready to connect to the future water pipes of the cottage (Figure 13.33).

13.4.5.2 An integrated energy and ventilation system

A central strategy of the project was to decarbonise energy systems on a budget, including phasing the works due to cost. The only energy supplied to the cottage is electricity, which will be augmented in the future by photovoltaic solar panels fitted to the extension's south-facing roof. A single mechanical system that could efficiently provide ventilation with space and water heating was considered ideal. Given the small size of the dwelling and the low space-heating demand due to the insulation strategy, it was judged that an air-source heat pump could provide space heating and domestic hot water.

The *Compact P Air 9*, provided by Nilan Ireland, was selected. This system features mechanical ventilation with heat recovery (MVHR), supplied and extracted through a ducted network, and a 2 kW heat pump that uses residual energy in the exhausted air to further reduce electricity demand for space heating and domestic hot water (DHW). The system can recover up to 95 per cent of thermal energy from the exhaust air, dumping it in the 180-litre cylinder, to reduce the energy demand for hot water. The specifications identify the efficiency of the main heating system at 558% and the efficiency of the main hot water system at 220%. A cooker hood, providing additional residual heat, can also be integrated into this system. Besides the thermal benefits, the fact that all items are integrated and commissioned at one time, with a number of local heat and plumbing tradespeople trained to maintain it, is very attractive. Figure 13.34 shows the ducting layout agreed at design stage. All ducting to 'dry' rooms would run in the dropped ceiling, while ducting to 'wet' rooms (shown in blue) would be located above the kitchen units.

The author did not proceed with the system for two reasons. Firstly, the 1sqm footprint of the cupboard that would house the unit was judged a significant loss in a dwelling as small as this one. Secondly, the supply cost of €12,750 excluding VAT quoted for the compact unit with ducting and cooker hood, less the €3,500 grant aid then available for exhaust air heat pumps, was judged too high to proceed. In lieu of this integrated approach, a simpler, low-tech strategy was adopted featuring space heating supplied using 'direct electric' heating, hot water supplied by an under-sink heater, and a separate electric shower. Ventilation was provided using demand-control mechanical extract ventilation.

This change in specification turned out to be momentous for the case study, as the lack of MVHR and the presence of 'direct electric' heating prevented compliance with the EnerPHit and AECB CarbonLite Retrofit standards. Given a little more time and hindsight, alternative approaches that enabled certification could have been taken:

a) A year later, the grant was increased to €6,500, a sum which might have enabled the original approach to proceed.
b) At about the same time, a comparable system with a more compact footprint, the *Compact S*, was introduced, which might have allayed concerns about the physical space the unit would occupy.
c) It was only later, when certification was sought, that the cost and space-saving value of a system that integrated ducted MVHR and HP space heating, without domestic hot water (DHW), became clear. Such systems are now available and can fit in a large kitchen cupboard. This approach would have facilitated certification of the building, while leaving hot water to be installed as was done.

Given that a dropped ceiling was installed, housing the recessed downlighters, it should be possible to install a ducted ventilation system in the future.

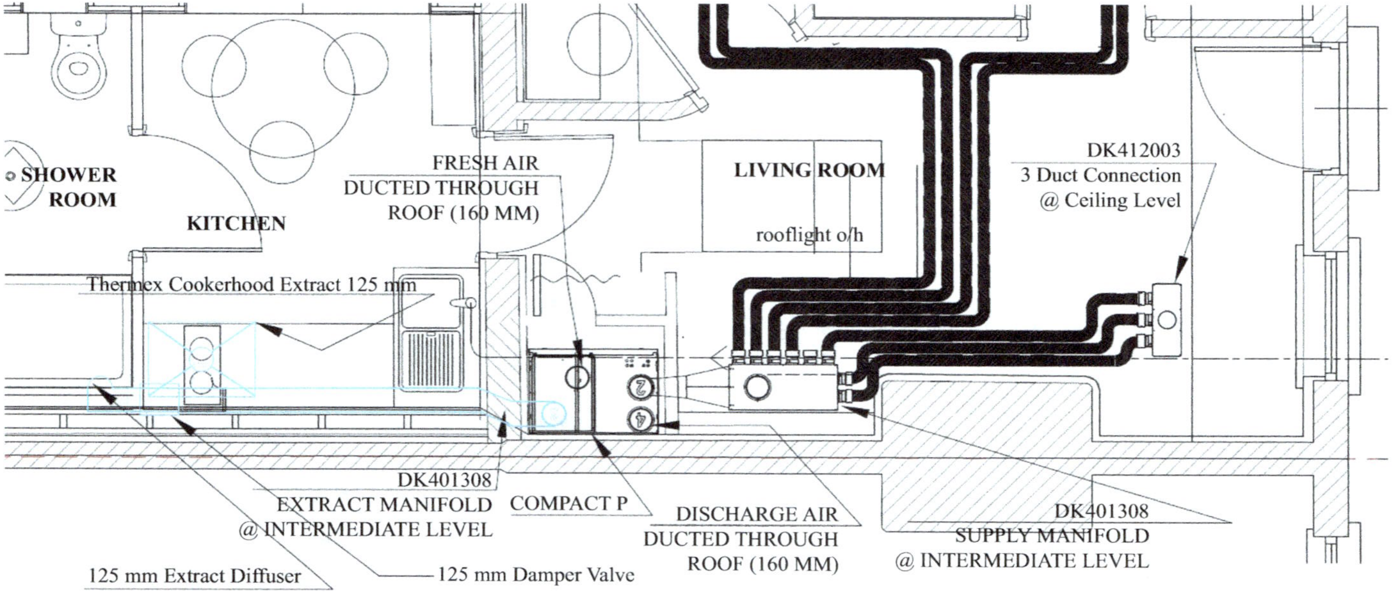

Figure 13.34 Initial proposed layout of compact P system

13.4.5.3 Power, data and renewables

The house was fully re-wired as part of the project. Modern services such as Ethernet, USB, TV points, and CO_2 and smoke sensors were installed for the first time. The future installation of photovoltaic (PV) panels was an important part of the project. It is intended that four PV panels, each with a capacity to generate 395 Watts peak and an overall footprint of ~8 sqm, will be installed on the extension's south-facing roof.

To facilitate this, a smart meter was fitted and an adequate power supply was provided to a small storage area just below the monopitch roof of the extension, accessed at a high level from the kitchen. From there, a 35 mm diameter pipe was installed through the roof build-up, filled with insulation and capped at both ends to facilitate easy installation of a power cable to future solar panels, without loss of insulation or airtightness in the future. Figure 13.35 shows the old meter and new smart meter; the consumer unit, and the thick white cable terminating at the DC isolator switch, which will draw power from the future inverter battery and PV panels. Up on the roof, the termination tile for the PV cable can be seen at 'A' and the rain sensor of the 'Z-wave controller' that operates the Fakro rooflight can be seen at 'B'.

Static battery storage can play a pivotal role in decarbonisation efforts when combined with direct electric heating and hot water heat pumps, particularly in low-energy homes where hot water is a larger load. Small hot water heat pumps, of 170 litres or less, are becoming available. When such strategies are configured and installed well, they can reward the consumer while helping to stabilise the national

Figure 13.35 (clockwise from top left) Original electricity meter, new ESB smart meter, consumer unit, cottage roof showing services and power control switch for future PV panel

grid through load shifting, thereby allowing better estimation of load and lower-carbon forms of electricity generation.

13.4.5.4 Water heating

A 2 kW under-sink water heater with an expansion vessel, that can heat and store 5 litres of water at a time, was fitted. Given that the electric shower is mains-fed and the occupancy of the cottage will be very low, there is little need to store more hot water. Figure 13.36 shows the under-sink heater, its time clock, and expansion vessel. To the rear, the mains water supply, a black hydrodare pipe with a blue handle, is visible. The image on the right shows a hot water pipe, which should be insulated but is not, running from right to left to serve the shower room's washhand basin, bypassing the cold pipe that has a T-branch to serve the shower below. This omission was recorded in PHPP, as it requires an input for the length of all hot water pipes and whether they are insulated. The Irish building energy modelling tool, Dwelling Assessment Procedure (DEAP), does not take account of this heat loss.

13.4.5.5 Space heating

Besides an electric towel heater in the Shower Room, direct-acting electric Q-Rad panel heaters from Dimplex were installed in living room, both bedrooms, and kitchen. These can be controlled locally using the intuitive heater controller or using a smartphone app via a hub, shown in Figure 13.37. All can be controlled

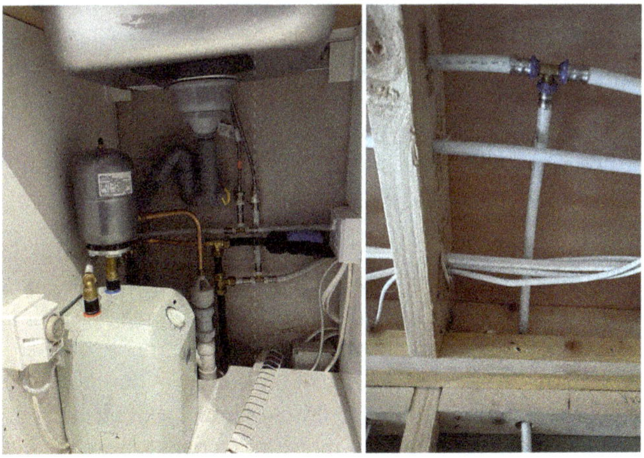

Figure 13.36 (left) Time clock-controlled under-counter water heater with expansion vessel. Mains water supply is visible at rear (black hydrodare pipe with blue handle), (right) uninsulated hot water pipe runs from right to left to serve washhand basin, bypassing the branched cold pipe to shower

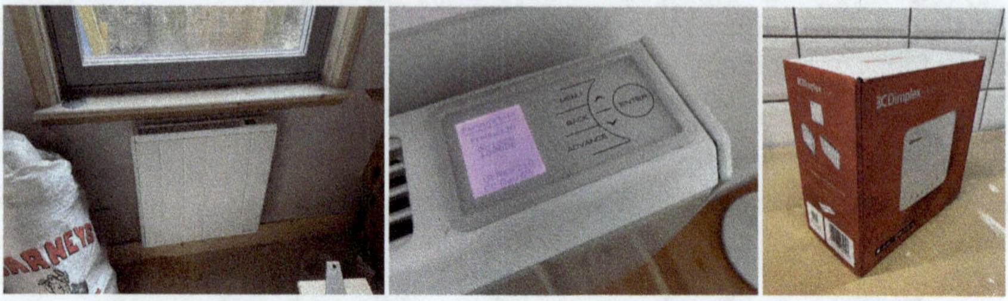

Figure 13.37 (from left) Q-Rad panel heaters from Dimplex, touch panel controller on panel heater and Dimplex Hub to enable smartphone control

seven days a week or boosted as needed. Smart controls can learn to turn heating on earlier or later than set if air temperatures justify it. Also, the heaters reduce heat quickly if a sudden loss of room temperature occurs, such as when a window is opened, and will remain so until that situation changes, with resultant energy savings.

There is an argument that direct-acting electric heating *is* an appropriate electrification heat strategy in super-small, low-energy dwellings such as this. They are a cost-effective option from a capital and maintenance point of view and 100 per cent efficient at the point of use. The simplicity of the separate space heaters and the reduction in materials used, compared to a fully piped wet heating system, are beneficial. There are also substantial embodied carbon savings by not investing in a wet heating system or mechanical ventilation with heat recovery and a heat pump. Advocates for the latter systems will argue that the operational carbon savings outweigh the embodied carbon cost, which makes sense for large houses (>100 sqm) but does this apply equally to tiny cottages or small apartments? Interestingly, direct electric heating is the only fuel that enables a BER to improve over time by doing nothing, as the government successfully implements its policy to decarbonise the grid. From an embodied carbon perspective, as the grid decarbonises further, the trend to install such heaters in super-small, low-energy dwellings, especially if combined with small heat pumps, is likely to grow.

13.4.5.6 Ventilation

In specifying and planning ventilation systems for dwellings, insufficient regard is often given to the fact that warm air rises and moisture, once released, can remain in the air, particularly at high levels in a room, for many hours. Therefore, a successful domestic ventilation system should (1) feature extract units at the highest point possible, (2) near to the source of moisture (separate from the kitchen cooker hood that extracts ventilation occasionally) that (3) extract moisture for as long as necessary to return conditions to background levels of humidity.

Due to the omission of an integrated energy and ventilation system, it was essential to install a lower-cost ventilation system that meets the three criteria above. The Aereco demand-control mechanical extract ventilation (DC-MEV) was specified. This system relies on isolated, acoustically baffled supply wall vents located in 'dry' rooms, which have a varying air intake depending on internal relative humidity levels, and extract vents in 'wet' rooms, such as bathrooms, which are ducted to a single fan with similarly varying airflow. These are usually located in an attic near an accessible attic hatch. These large fans are long-lasting, low-maintenance, and quiet.

Given that it was desirable to discourage access of any kind into the small attic of the case study cottage, the extract fan was located in a second level of overhead cupboards that the geometry of the monopitch ceiling of the kitchen accommodates. As can be seen in Figure 13.38, rigid (grey-coloured) ducts from the high-level room extract grille (a) and from the shower room (b) enter the fan (c) before the final exhaust duct rises (out of sight) to a roof tile ventilator. Several corners in the ducts were deliberately introduced to improve acoustic isolation. To the right, a vertical white cooker extract duct (d) can be seen rising straight to a roof tile ventilator. The top fascia panel also houses a smoke (e) and CO_2 (f) sensor.

Extract inlet units were located at the highest point in the Kitchen and Shower Room, near the hob and shower head, as shown in Figure 14.37. Importantly, these units are not linked to electricity supply, so there is no safety risk associated with nearby water. After installation, a testing company was retained to test that the installed air flows match the flows required by the system design. A certificate was provided validating air flow and compliance.

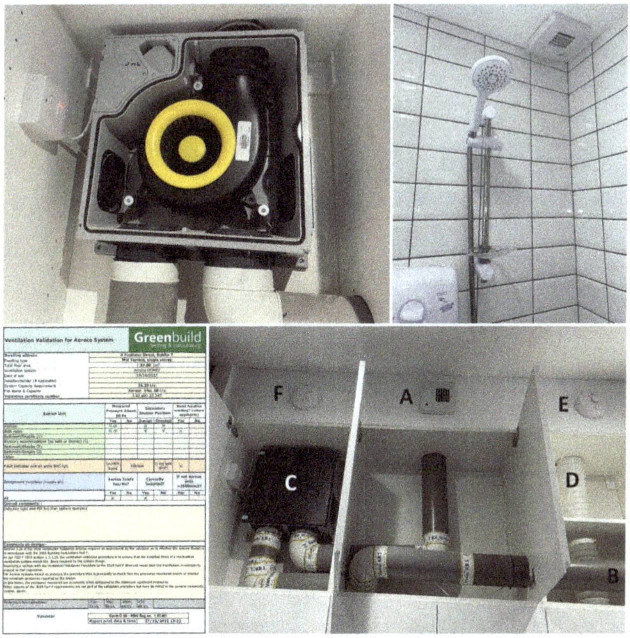

Figure 13.38 (from left clockwise) The Aereco V4A fan extract fan (cover removed), the extract unit in Shower Room directly above moisture source, upper cupboards in the Kitchen concealing the fan, cooker extract and ducting and the validation certificate

13.5 Completion and lessons learnt

13.5.1 Completion and occupancy

The final fitout and decoration took some time. The photographs selected in Figure 13.39 (from the top left, clockwise) show the refurbished brick chimney stack with new pots, haunching, and flush-and-brush lime mortar joints, the shower room, the main bedroom, the courtyard, the kitchen with double-height cupboards, the living room, and the front elevation. Direct light, from each side and above, and the reflected light of mirrors, animate and brighten the rooms. The colour scheme is simple: white walls and ceiling, and mid-grey external windows and door frame, while three mid-blue door leaves mark the route from the front door to the rear courtyard. The pine floor with its rich veins of colour and the soft furnishings give colour and warmth.

Figure 13.39 Views of the exterior and interior of the renovated cottage

The owners had the pleasure of being the first occupants in Spring 2023. Then the first tenants, a couple with a young child, moved in during the middle of Summer 2023. In Summer 2026, they are buying and moving to a new home nearby.

13.5.2 Retrofit – an act of negotiation

The completed cottage demonstrates that deep retrofit is not simply a technical exercise in reducing heat loss or improving an energy rating. It is an act of judgement, negotiation, and care. In this project, the ambition to deliver a low-energy, low-carbon, moisture-safe, and circular retrofit was continually tested by the realities of a very small site, an historic and imperfect building fabric, a constrained budget, a construction period shaped by Covid-19 supply-chain disruption, and inflation. The result is, therefore, not a flawless prototype, but something arguably more useful: a detailed account of what it means to retrofit responsibly when design aspirations encounter physical, financial, and regulatory limits.

The completed dwelling also confirms the value of starting retrofit design with the existing building rather than with a standardised package of measures. The cottage's modest area concealed a disproportionately large thermal envelope, a high heat-loss form factor, fragile mass-concrete walls, retained historic joinery, a valuable Baltic pine roof structure, limited storage space, and complex drainage and water-supply arrangements. These conditions required a bespoke response. Super-insulation was prioritised in the floors, roofs and timber-frame extension, while the historic walls were insulated more cautiously, using vapour-open materials and rainscreen protection where possible. This hierarchy of intervention is important: it recognises that energy efficiency must be pursued by understanding risk, particularly moisture risk, rather than by a simplistic assumption that more insulation is always better.

13.5.3 Key lessons – survey, circularity, heritage and collaboration

Several practical lessons arise. First, survey and opening-up works should be treated as a central stage of retrofit design, not as a preliminary formality. In this project, incomplete early knowledge led to repeated measurement exercises and late discoveries, including the original timber sub-frames at window openings and the true condition of the roof structure. A point-cloud or lidar survey at the outset would have provided a stable dimensional baseline against which layers could be removed, added, and modelled. For small dwellings, where every millimetre affects usable area, junction design, and cost, this level of accuracy is not a luxury but a form of risk management.

Second, the project shows that circularity and heritage conservation can be mutually reinforcing, but only where they are named as project priorities from the beginning. The retention of the Baltic pine roof structure, terracotta ridge tiles, granite threshold and sills, internal doors, and ironmongery reduced embodied carbon while preserving historic value. However, retention is not automatically cheaper or easier than replacement. Builders may reasonably prefer consistent new products that reduce uncertainty and programme risk. For this reason, tender documents should state explicitly where retention, repair, reuse, and low-carbon specification are core project values. Contractor selection should also consider prior experience in heritage work, airtightness, moisture-safe retrofit, and material reuse.

Third, this case study underlines the importance of collaboration. The project succeeded where the designer, builder, and specialist suppliers were willing to learn together: in the use of limecrete, foamed glass aggregate, cellulose, wood-fibre insulation, rainscreen cladding, airtightness tapes, recycled-content construction board, and modular timber-frame panels. The work was not the delivery of a fully standardised solution, but the careful assembly of a project-specific one. This matters for national retrofit policy because it suggests that scaling up should not mean flattening differences. Instead, the sector needs repeatable methods of investigation, modelling, procurement, and quality assurance that still allow building-specific judgement.

13.5.4 Key lessons – certification, phasing and a multi-faceted approach

A further lesson concerns certification and phasing. The project achieved the target air permeability value, but, as-built, it cannot be certified to EnerPHit, or the selected AECB CarbonLite Retrofit Standard, because the final services strategy omitted mechanical ventilation with heat recovery and a heat-pump-based heating system. This was a disappointing outcome, particularly after significant effort had been invested in fabric performance. Yet, it is also a useful lesson. Certification frameworks can provide discipline and comparability, but they must be considered early and protected through budget decisions. Where measures are deferred, the route to later compliance should be documented. The project was designed with this in mind: photovoltaic panels, MVHR, and a small heat pump remain feasible future interventions. This is precisely the kind of staged logic that Building Renovation Passports should support.

The completed cottage, therefore, has value beyond its immediate success as a bright, comfortable, very small home. It shows that an energetic retrofit must address operational energy, embodied carbon, heritage, circularity, moisture, radon, indoor air quality, cost, programme, and future adaptability together. These concerns are often separated in policy, guidance, and grant structures. In practice, they are inseparable.

13.6 Conclusions: value within the national retrofit debate

13.6.1 Historic dwellings are central to successful national retrofit

Ireland's national retrofit challenge is immense. The State has set a target of retrofitting 550,000 dwellings to B2 or better by 2030, yet progress remains far below the rate required. At the same time, the recast Energy Performance of Buildings Directive, emerging whole-life carbon methodologies, circular economy policy, and the RIAI 2030 Climate Challenge are expanding the definition of success. Retrofit can no longer be understood only as the improvement of operational energy performance. It must also reduce embodied carbon, avoid maladaptation, protect occupants, conserve existing resources, and support a just transition to a low-carbon built environment.

This case study contributes to that debate by making visible the complexity that sits behind, apparently, simple national targets. A small terraced cottage may appear marginal in scale, but Dublin contains many comparable historic dwellings, and the lessons are transferable. The project demonstrates that older, solid-walled, and architecturally modest buildings should not be treated as obstacles to decarbonisation. They are part of the national carbon asset. Their continued use, repair, and adaptation can avoid demolition, reduce demand for new construction, preserve social and urban continuity, and contribute to climate targets.

However, the project also warns against assuming that retrofit can be reduced to a universal package of measures. The cottage required a fabric-first approach, but not a fabric-only approach. It required selective insulation, with moisture awareness. It required airtightness, but alongside deliberate ventilation. It required low-carbon materials, as well as structural judgement and practical buildability. It required heritage retention, but not at the cost of future performance. These trade-offs are the real terrain of retrofit practice, and national policy must support practitioners and householders in navigating them.

13.6.2 The value of the particular in supporting the national challenge

The case study also highlights a gap in current retrofit discourse: the need to value evidence from completed projects. Ireland requires more building-level case studies that disclose design intentions, construction decisions, measured performance, embodied carbon data, costs, failures, compromises, and occupant experience. Without this evidence, policy risks being driven by modelled assumptions, product claims, or narrow metrics. With it, standards, grant schemes, and training programmes can become more intelligent, more context-sensitive and more credible.

Ultimately, the value of the Findlater Street project lies not in presenting a perfect answer, but in widening the question. It asks how a tiny historic dwelling can remain useful, healthy, beautiful, and low-energy while drawing as little as possible from new material resources. It asks how retrofit policy can reward care, repair, and staged improvement, as well as speed and scale. It asks how designers, builders, clients, and regulators can learn from particular buildings without losing sight of national urgency. In that sense, the project is both a home and an argument: that the next phase of retrofit must be technically rigorous, materially responsible, historically informed, and generous enough to accommodate the stubborn particularity of the existing built environment.

References

[1] United Nations. *Climate Action* [online]. Available from https://www.un.org/en/climatechange/net-zero-coalition [Accessed 11 November 2025]

[2] World Green Building Council. *BuildingLife* [online]. Available from https://worldgbc.org/buildinglife/ [Accessed 11 November 2025]

[3] Kinnane O., O'Hegarty R. 'Whole life carbon quantification of the built environment: Case study Ireland'. *Building and Environment*. 2022; 226:109730

[4] Elefante C. 'The greenest building is … one that is already built'. *Forum Journal*. 2007;**239**(4):26–38

[5] Coyne B., Denny E. 'Retrofit effectiveness: Evidence from a nationwide residential energy efficiency programme'. *Energy Policy*. 2022;159:112576

[6] Coyne B., Denny E. 'Mind the energy performance gap: Testing the accuracy of building energy performance certificates in Ireland'. *Energy Efficiency*. 2021;**14**(6):57

[7] Government of Ireland. *Climate Action Plan 2019* [online]. 2019. Available from https://www.gov.ie/en/publication/ccb2e0-the-climate-action-plan-2019 [Accessed 11 November 2025]

[8] Royal Institute of Architects of Ireland. *A Domestic Client's Guide to the RIAI 2030 Climate Challenge* [online]. Available from https://www.riai.ie/discover-architecture/climate-change [Accessed 11 November 2025]

[9] European Parliament and Council. *Directive (EU) 2024/1275 of the European Parliament and of the Council of 24 April 2024 on the energy performance of buildings (recast)*. Luxembourg: Official Journal of the European Union; 2024

[10] Association of Environmental Conscious Builders. *Carbonlite Retrofit Standard* [online]. Available from https://aecb.net/aecb-carbonlite-retrofit-standard/ [Accessed 11 November 2025]

[11] Passive House Institute. *Passive House Planning Package (PHPP)* [online]. Available from https://passivehouse.com/04_phpp/04_phpp.htm [Accessed 11 November 2025]

[12] Traynor J. *ENERPHIT: A Step-by-Step Guide to Low Energy Retrofit*. London: RIBA Publishing; 2020

[13] World Green Building Council. *Bringing Embodied Carbon Upfront*. London: WGBC; 2019

[14] European Committee for Standardization. *EN 15978: Sustainability of Construction Works—Assessment of Environmental Performance of Buildings—Calculation Method*. Brussels: ECS; 2011

[15] Röck M., Sørensen A., Steinmann J., *et al. Towards Embodied Carbon Benchmarks for Buildings in Europe: #1 Facing the Data Challenge*. Aalborg: Ramboll; 2022

[16] INDICATE *What is INDICATE?* [online]. Available from https://www.indicatedata.com/ [Accessed 3 December 2024]

[17] Bienert S., Kuhlwein H., Schmidt Y., Gloria B., Agbayir B. *Embodied Carbon of Retrofits. Ensuring the Ecological Payback of Energetic Retrofits.* Worl: IIÖ Institut für Immobilienökonomie GmbH; 2023

[18] Change to European Parliament and Council *Directive (EU) 2020/323 of the European Parliament and Council on Waste* Luxembourg: Official Journal of the European Union; 2020

[19] Government of Ireland. *Whole of Government Circular Economy Strategy 2022-2023* [online]. Available from https://assets.gov.ie/static/documents/whole-of-government-circular-economy-strategy-2022-2023.pdf [Accessed 11 November 2025]

[20] European Environment Agency. *Circular Economy Country Profile 2024 – Ireland: ETC-CE Report 2024 – Ireland* [online]. Available from https://www.eea.europa.eu/en/topics/in-depth/circular-economy/country-profiles-on-circular-economy/circular-economy-country-profiles-2024/ireland_2024-ce-country-profile_final.pdf/@@download/file [Accessed 11 November 2025]

[21] Dublin City Council. *Dublin City Council Receives Prestigious EU Climate Label* [online]. Available from https://www.dublincity.ie/node/15880 [Accessed 11 November 2025]

[22] Brady J., Simms A. *Dublin Through Space and Time*. Dublin: Four Courts Press; 2022

[23] 'Concrete'. *The Irish Builder*. 1886;XXVIII(626)

[24] Dublin City Council. *Build to Last Case Studies* [online]. Available from https://www.dublincity.ie/builttolast [Accessed 11 November 2025]

[25] Kumaraperumal, A. Sanders, C., Baker, P. (2006) 'Prediction of fabric moisture contents in a historic building using CFD and heat, air and moisture transfer modelling compared with full-scale measurements', Presented at 4th International Symposium of Computational Wind Engineering; Yokohama, Japan, 2006

[26] Bomberg M., Brown W. 'Building envelope: Heat, air and moisture interactions'. *Journal of Thermal Insulation and Building Envelopes*. 1993;**16** (4):306–11

[27] Sustainable Traditional Buildings Alliance. *Responsible Retrofit* [online]. Available from https://responsible-retrofit.org [Accessed 11 November 2025]

[28] Little J., Ferraro C., Arregi B. *Assessing Risks in Insulation Retrofits Using Hygrothermal Software Tools*. Historic Environment Scotland. Technical Paper 15; 2015

[29] Little, J. 'Counter Intuitive - Are counter battens the key to better pitched roof retrofits?' *Passive House Plus*. 2026;(52).

[30] Geraedts R.P. 'FLEX 4.0: A practical instrument to assess the adaptive capacity of buildings'. *Energy Procedia*. 2016;96:568–79

[31] Kirchherr J., Reike D., and Hekkert M. 'Conceptualizing the circular economy: An analysis of 114 definitions'. *Resources, Conservation and Recycling*. 2017;127:221–32

[32] BAMB *Buildings as Material Banks* [online]. Available from https://www.bamb2020.eu [Accessed 11 November 2025]
[33] Ellen McArthur Foundation *Making the Circular Economy Real* [online]. Available from https://www.ellenmacarthurfoundation.org/ [Accessed 11 November 2025]
[34] Environmental Protection Agency. *End-of-Waste - Guidance Document* [online]. EPA; 2020. Available from https://www.epa.ie/publications/corporate/consultations [Accessed 21 November 2025]
[35] Brand S. *How Buildings Learn: What Happens After They're Built*. London: Penguin; 1995
[36] Durmisevic E. *Circular Economy in Construction Design Strategies for Reversible Buildings* [online]. 2006. Available from https://www.bamb2020.eu/wp-content/uploads/2019/05/Reversible-Building-Design-Strateges.pdf [Accessed 11 November 2025]

Chapter 14

Assessing performance of a deep and circular retrofit of a cottage in Dublin

Joseph Little[1]

14.1 Introduction

This chapter continues the focus on energetic retrofit of a small cottage described in Chapter 13 (Figure 14.1), this time through the lens of building performance. Together, the two chapters form a larger case study outlined in Figure 14.2.

The extent to which technical design aspirations and global challenges have been achieved is assessed through a series of evaluations using thermal, hygrothermal, and carbon metrics standards. This process provides an opportunity to reflect on those standards and the support available to designers considering building performance evaluations of energetic retrofits. The chapter concludes by briefly discussing the occupants' experience in the first 12 months of living there.

This project is part of a wider study, initiated by Dublin City Council, called Built to Last. The *Built to Last Case Studies* [1] resource presents seven case studies of completed and seven proposed energetic retrofits of historic dwellings. The assessment and write-up of the former group were completed in 2013, while the latter group was updated in 2019, to reflect the latest revision to Part L of the Irish Building Regulations. This is considered the inception of the 'major renovation' standard for energetic retrofits and the increasing focus on decarbonising energy systems. One of the proposed case studies is the cottage assessed here. It is interesting to compare the initial proposal of 2019 with the works completed. A focus on preservation of built heritage, energy efficiency, moisture management, and buildability is found in all the case studies. More recent concerns, such as life cycle assessment, embodied carbon, and circularity, were not addressed and the level of detail provided overall is quite different from the assessment presented here.

[1]School of Architecture, Building and Environment, Technical University Dublin, Ireland

Figure 14.1 Street view of completed cottage

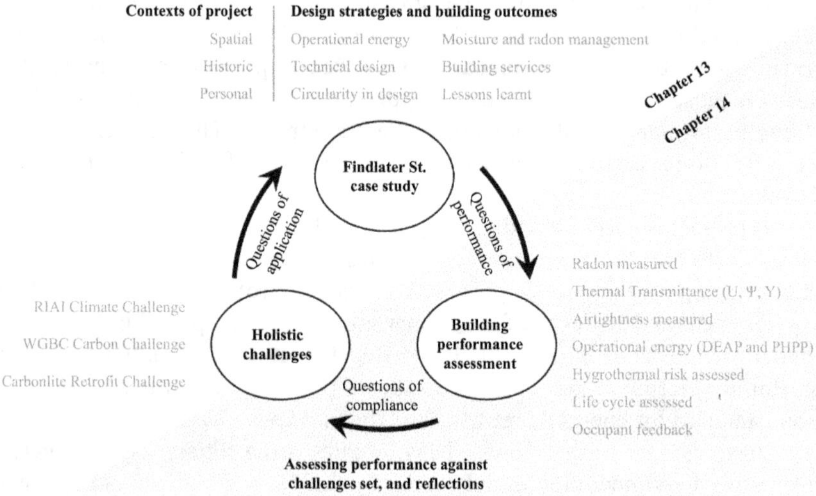

Figure 14.2 Case study structure

14.1.1 Knowledge, skills and building performance gaps

It is impossible to talk about building performance without talking about performance gaps. The phrase relates to the difference between the intended and realised performance of any technology or system. We apply it to the design and construction of a building or its control systems, but it can also be applied to their use post-occupancy or the technologies that measure performance and the assessors operating them [2]. One might identify and measure several different performance gaps in just one building.

The Irish Government and construction industry rarely address, or report on, questions of underperformance. It is fair to say that much of what Irish building designers know about performance gaps comes from research in the UK's construction industry. For example, the LowCarb4Real research project in 2008 [3] was innovative in implementing cross-sectoral engagement, structured knowledge exchange, a 'no-fault' disclosure process, measurement of underperformance, and a focus on cultural change within the sector. As a result, its detailed findings are unusually instructive.

> *'Underperformance, currently masked by cheap energy and oversized heating systems, will become more noticeable and less acceptable as energy prices rise and target emission rates are reduced. The carbon emission limits of Code Level 3 and beyond are likely to entail increasing complexity, less redundancy, faster rates of innovation and unfamiliar low and zero carbon technologies.*
>
> *Unless academia and industry are able to develop systems and a culture capable of managing these challenges, underperformance is likely to increase in relative and possibly even absolute terms ... The key to managing the process is likely to be the transformation of successful construction companies into learning organisations.'* [3]

How can constant learning and openness to assessment be integrated into the business practice and the culture of the construction industry? While continuous professional development courses have become normal amongst building design professionals, they are less common amongst the wider group of building professionals. Of those that are taken, how many explore innovation and unfamiliar technologies sufficiently, address building performance gaps, and culture change adequately?

Eight years after the conclusion of LowCarb4Real, the Bonfield Review [4] stated in 2016 that:

> *'The majority of ... failures are not thought to be due to intentional poor performance, but the result of gaps in standards or training provided.'*

In 2025, nine years after the Bonfield Review, the Retrofitting Homes for Net Zero [5] report focused on scaling high-quality retrofits to achieve national operational energy reduction. Its authors wrote about a lost decade, with complicated and ever-changing grant schemes, loss of confidence and skills, and knowledge gaps. Politicians and society in Ireland, and the UK, need progress and change *now*; it appears that many knowledge gaps and resolutions specific to the design and implementation of high-quality, historic retrofits are yet to be addressed.

Due to its multi-faceted assessment of building performance, this chapter shines a light on a lack of primary research, government guidance, late adoption of standards, targets focused on new builds but not retrofit, and questions of how to implement the great challenges we must meet.

14.1.2 *Achieving multi-faceted building performance criteria*

Too often, building performance is discussed in terms of energy performance coefficients, operational energy use, and renewable energy systems. Buildings are complex inter-related systems, with both building fabric and services, influenced by internal and external climates in a wider physical context, with the additional complexity that occupants and owners bring. They must be engaged with in their totality. For instance, the management of radon, moisture, and air quality is just as important as energy efficiency. Consider an energetic retrofit that results in either higher radon measurements, leading potentially to ill health, or heightened risk of timber decay, mould growth, or declining indoor environmental quality: surely it must be regarded as unacceptable. However, such projects with a narrow focus and significant unintended consequences have been identified many thousands of times, as proven by numerous UK and a few Irish studies.

Creating, curating, and challenging the information provided in this chapter has been a valuable learning experience. The key questions are always 'what are you trying to achieve, what could you have missed, were you successful, and to what extent?' and, of course, the challenge to 'prove it'. Presenting one's assessments and findings to a wider audience amplifies these questions, especially the final challenge. Presenting a range of concerns together shows the trade-offs and compromises that are required to address this complexity and deliver a retrofit that performs holistically. Here, we look at building performance metrics in detail.

Finally, as described in Chapter 13, since 2019, the World Green Building Council, the RIAI in Ireland, and the RIBA in the UK have requested that developers, building design professionals, and builders urgently engage with, and achieve, a series of increasingly ambitious and inter-related building performance challenges. However, published accounts of how this should best be done and where it has been achieved are harder to find. In the words of *LowCarb4Real*, it appears that *'the systems and a culture capable of managing these challenges'* have yet to be adequately considered and supported [3]. This is an attempt to contribute to that discussion.

14.2 Radon and thermodynamics

14.2.1 *Radon measurements*

Radon is a radioactive gas formed by the radioactive decay of uranium, which is present in all rocks and soils. Fifty-five per cent of the average annual radiation dose received by Irish residents is from radon. It is the second biggest cause of death from lung cancer after smoking. As a colourless, odourless gas, it can only be measured with special detectors. *'The National Reference Level (NRL) for long-term exposure to radon in dwellings is 200 Becquerels per cubic metre, or 200 Bq/m^3. Above this level the need for remedial action should be considered'* [6]. The Irish national average concentration for dwellings is 77 Bq/m^3.

The Environmental Protection Agency's (EPA) radon risk map shows that the case study cottage is in an area designated 'Areas other than High Radon Areas'. The EPA estimates that *'about 1 in 20 homes in this area is likely to have high radon levels'*. It is advisable to measure radon levels before occupancy of any dwelling commences. As the author wished to implement a moisture management strategy that would omit a radon barrier, measurements were taken before and after construction. In 2018, the test company estimated that the average radon concentration throughout the year would be 39 Bq/m^3, far below the threshold for concern. In 2023, six months after the works were completed and while the cottage was in regular use, the test company estimated that the new value was 27 Bq/m^3. This is 65 per cent below the national average concentration but, more importantly, 31 per cent below the initial value, a substantial reduction.

Given that the air permeability (Q_{50}) of the cottage reduced from 8.85 to 1.51 $m^3/m^2.hr$, an 83 per cent reduction, it would have been reasonable to expect an increase in radon levels, even if still below the threshold value. In that context, it is tempting to theorise that the reduction in measured radon is due to the greater resistance of the limecrete slab to radon migration, compared to an uninsulated suspended timber floor, and the installation of a radon sump ducted to an open stack. It is also possible that there are additional benefits from a demand-controlled mechanical extract ventilation (DC-MEV) system, despite the absence of balance. However, this is speculative and further studies should be undertaken.

While there are several studies of radon in energy-efficient dwellings fitted with mechanical ventilation with heat recovery (MVHR) systems, very few focus on retrofitted dwellings with mechanical extract ventilation (MEV), let alone DC-MEV. No research papers could be found with a focus on building fabric moisture management strategies in the context of low radon levels, identifying a possible research gap.

14.2.2 Thermodynamics

14.2.2.1 Thermal transmittance values achieved

Table 14.1 presents the calculated Thermal Transmittance (U) values of the main elements of the cottage's envelope. The values achieved fall into four distinct categories (i.e. A to D).

- The transparent plane elements – the triple-glazed Passivhaus windows and rooflight are an impressive 0.77 and 1.10 W/m^2K.
- Working on the different scale and performance possible for *opaque* building fabric elements, the internally insulated solid walls are 0.41 and 0.65 W/m^2K. Given that U-values lower than this increasingly cool down the masonry substrate, thereby raising its relative humidity – regardless of external weather conditions – U-values at this level make an important contribution to moisture management, while still raising surface temperatures sufficient to reduce the likelihood of surface condensation and conserving a significant amount of energy.

Table 14.1 Thermal transmittance of plane elements

	Element	Description	U-value (W/m^2K)
A	Windows and patio door	Nordan PH Scand. Pine, triple-glazed	0.77
	Front door	Nordan Scandinavian Pine	1.10
	Rooflight	Fakro Z-Wave FTU-V	1.33
B	Cottage front wall	Ext. render, mass concrete, 80 + 11 mm Pavadry composite board	0.41
	Cottage rear wall	Rainscreen, mass concrete, 40 + 11 mm Pavadry composite board	0.65
	Cottage rooflight shaft walls	Plasterboard, 200 mm paper cellulose	0.22
	Cottage and extension floor	510 mm foamed glass aggregate, 100 limecrete slab with 50 mm cork edge insulation	0.19
C	Cottage roof (pitched ceiling)	batten + counter batten, paper cellulose to depth of joists, PIR 70 mm	0.14
	Extension walls (270 mm)	Rainscreen, 50 mm wood fibre, 140 mm timber stud with paper cellulose, Pavatextil services zone	0.17
	Extension walls (350 mm)	Rainscreen, 50 mm wood fibre, 220 mm timber stud with paper cellulose, Pavatextil services zone	0.13
	Extension roof	Rainscreen, 50 mm wood fibre, timber stud with paper cellulose, Pavatextil services zone	0.13
D	Cottage roof (flat ceiling)	Attic void, services zone, 450 mm paper cellulose (collapsing to 500 mm)	0.07

- The highly insulated roof and walls of the timber frame extension come in between 0.17 and 0.13 W/m^2K. The floor of the whole cottage has a U-value of 0.19 W/m^2K.
- Finally, benefitting from the thickness available to it, the super-insulated 'cold' pitched roof comes in at 0.07 W/m^2K.

These U-values are calculated using the Passive House Planning Package (PHPP), rather than the Dwelling Energy Assessment Procedure (DEAP).

14.2.2.2 Thermal bridge heat loss and risk of surface condensation

A focus of the technical design was to minimise the non-uniform, or 'thermal bridge', heat loss that is associated with the linear junctions, where plane elements, such as walls, floors, roofs, doors and windows, meet. Formally, this heat loss is termed Linear Thermal Transmittance (ψ, psi, W/mK) and is calculated using standard EN ISO 10211 [7]. BR 497 [8] clarifies the calculation methodology used in the UK and Ireland, which includes the use of measurements taken between internal finished surfaces. Calculating non-uniform heat loss is invaluable, as it makes clear the extent to which non-uniform heat loss (i.e. ψ) has, or has not, been minimised by the architect's technical design and enables easy assessment of the likelihood of surface condensation forming at junctions through calculation of the Temperature Factor (f_{Rsi}). It is also a key input, when expressed as a Y-factor, in calculating a building energy rating and sizing a conventional heating system.

All junctions were designed with a focus on the minimisation of thermal bridge heat loss. A registered thermal modeller used the details and specifications provided to calculate the thermal bridge heat loss of nine key junctions of the original cottage and four of the extension. Figure 14.3 presents diagrams of the junctions, where the banded lines and colours represent isotherms – areas of the same temperature. Red represents temperatures increasingly close to the standard room air temperature of 20 °C, varying through yellows and greens to dark blue and the standard external air temperature of 0 °C.

Table 14.2 presents the junction names, the Linear Thermal Transmittance (ψ) at each junction, and the length of the junctions and the resulting heat flow rate. Any junction with a ψ-value less than 0.1 W/mK may be considered a low thermal bridging junction. The colours used in the table signal how well the junctions perform. Pink to red colours indicate thermally efficient junctions, which generally enable warmer surfaces. Blue-hued colours indicate greater heat loss and cooler surfaces. With some notable exceptions, thermal bridge heat loss has been successfully minimised. Some of the ψ values shown are negative. This indicates that either the overall heat loss at those junctions is *less* than the Thermal Transmittance (U) of the flanking plane elements, indicating a particularly well-designed junction, or geometrically, that it is an inner corner.

An individual junction may have high thermal bridge heat loss yet contribute relatively little to the overall heat loss due to the shortness of that junction, while another junction could contribute greatly due to its length. This can be seen for the party wall junctions (C09, C11, C12). Their combined length is 32.5 m. Taken together, they account for 71 per cent of all non-uniform heat flow through junctions associated with this cottage. Only 50 per cent of the thermal bridge heat loss that occurs at party wall junctions is directly associated with this cottage; the rest is associated with the neighbouring cottages. Junction C09 is the only party wall junction that has been calculated for this specific condition. Its linear thermal transmittance value is also far higher than the other two (0.368 W/mK). This is presumably because the disparity between super-insulated floors and attics on one side of the party wall and an assumed absence of insulation on the other is greater than normal. In that context, perhaps junctions C11 and C12 should also have been calculated in this way. To fully minimise thermal bridge heat loss, terraced properties should be energetically retrofitted in a coherent way, ideally at the same time.

14.2.2.3 Surface Temperature Factor

The Surface Temperature Factor (f_{Rsi}) is a dimensionless ratio that indicates how close the surface temperature of an internal building component is to the indoor air

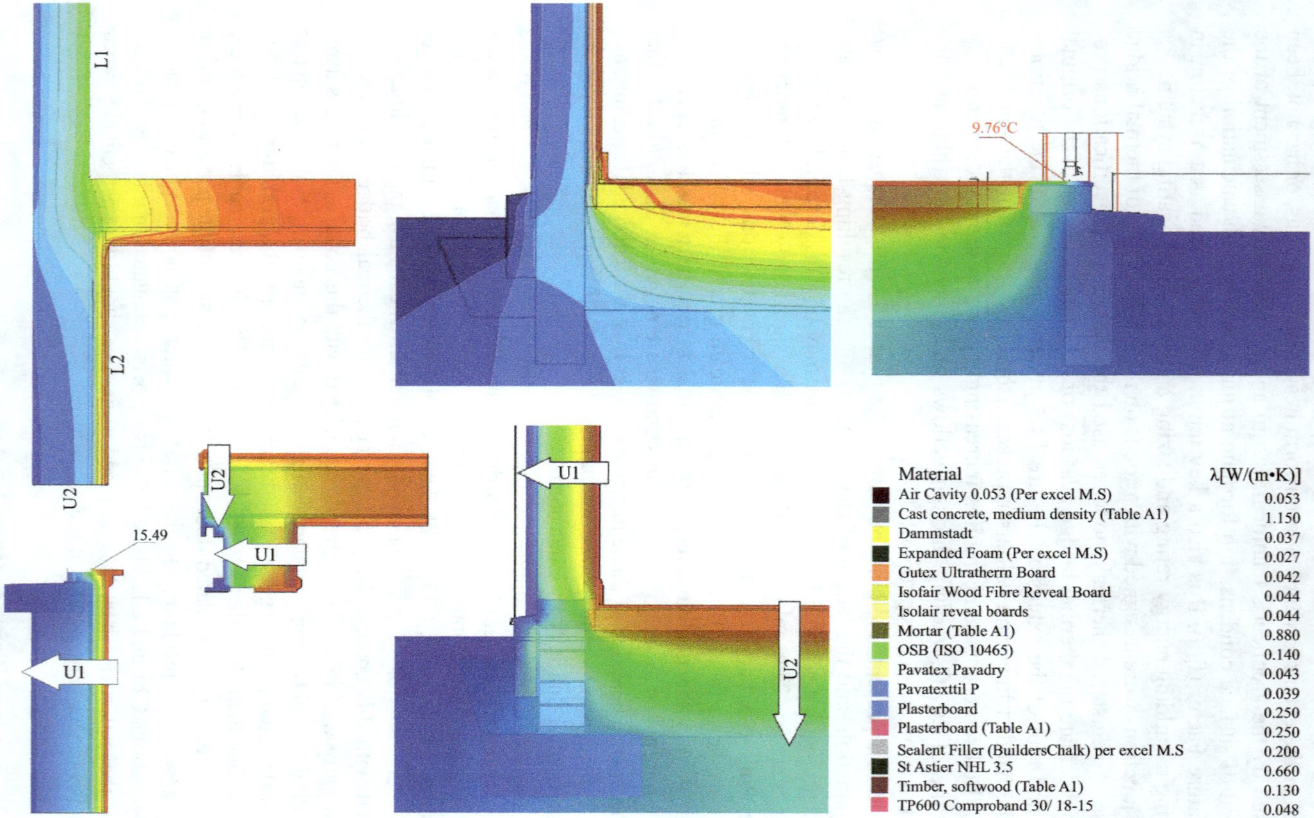

Figure 14.3 *(Clockwise from left) Isotherms of base of rear cottage wall (C04), Party Wall Rear (C08), Rear Windowsill – Wall (C05), Front Door Threshold (C01), Courtyard Wall Base – Floor (E01). Dewpoint line is shown in two images as boldened red lines.*

Table 14.2 Thermal bridge schedule used to calculate Y-value for DEAP of key junctions

Type	ψ value (W/mK)	Thermal bridge length (m)	Heat flow rate Psi x L (W/K)
C01: Front Door Threshold-Floor	0.389	0.96	0.373
C02: Front Window Sill-Wall	0.059	7.524	0.444
C03: Front Wall – Pitched Ceiling	0.017	10.237	0.174
C04: Pitched Ceiling – Flat Ceiling	0.043	10.237	0.44
C05: Rear Wall Base-Floor	0.085	6.244	0.531
C06: Rear Window Sill-Wall	0.046	0.945	0.043
C07: Rear Win. Head-Pitched Ceiling	0.002	2.889	0.006
C08: Rear Window Jamb-Wall	–0.001	9.48	–0.009
C09: Party Wall Rear	0.368	**10.542**	3.879
C11: Party Wall Floor	**0.12**	**10.797**	**1.296**
C12: Party Wall Head	**0.229**	**11.187**	**2.562**
C13: Rooflight walls corner	–0.015	9.951	–0.149
C14: Rooflight jamb	0.11	2.434	0.268
C to E: Internal corner	–0.015	2.624	–0.039
E01: Courtyard Wall Base-Floor	0.012	10.547	0.127
E02: Wall Head – Pitched Roof	0.027	3.757	0.101
E03: Window & Door Jamb-Wall	0.025	5.02	0.126
E04: Window Head	0.044	2.45	0.108
E05: Window Sill	0.034	0.8	0.027
E06: External Corner	0.03	10.113	0.303
E07: Verge	0.024	6.405	0.154
E08: Monopitch	0.04	3.757	0.15
Sum Psi x L: 68 m		138.901	10.914

temperature. It is calculated to give a broad sense of the risk of surface condensation and, thus, mould risk. The word 'broad' is used because the simple calculation is based on an assumed relative humidity. The chemistry of the materials is also not considered. The nearer the f_{Rsi} is to 1.00, the surface is increasingly likely to be close to room air temperature and surface condensation is unlikely. Table 14.3 indicates the results for the relevant components.

As can be seen in the right-hand column, most of the selected junctions have very low thermal bridges and very high temperature factors. However, four junctions of the existing cottage have temperature factors that are lower than the 0.75 f_{Rsi} advised for dwellings in Technical Guidance Document (TGD) L of the Irish Building Regulations. *C01 Front Door Threshold - Floor* has the greatest thermal bridge (0.389 W/mK) and risk of surface condensation (0.424), which is not surprising given that its historic granite threshold spans from inside to outside and lies just under the 15 mm thick, solid timber floor finish. It is a challenging junction to improve. The elevated risk of surface condensation at *C09 Party Wall Rear* is due to the party wall issues previously identified. *C02* and *C06*, the

Table 14.3 Surface Temperature Factors

Type	Heat flow rate Psi x L (W/K)	f_{Rsi} (-)
C01: Front Door Threshold-Floor	0.373	0.49
C02: Front Window Sill-Wall	0.444	0.61
C03: Front Wall – Pitched Ceiling	0.174	0.901
C04: Pitched Ceiling – Flat Ceiling	0.44	0.959
C05: Rear Wall Base-Floor	0.531	0.811
C06: Rear Window Sill-Wall	0.043	0.72
C07: Rear Win. Head-Pitched Ceiling	0.006	0.891
C08: Rear Window Jamb-Wall	–0.009	0.819
C09: Party Wall Rear	3.879	0.424
E01: Courtyard Wall Base-Floor	0.127	0.85
E02: Wall Head – Pitched Roof	0.101	0.93
E03: Window & Door Jamb-Wall	0.126	0.93

junctions for the windowsills with the front and rear internally insulated walls, have higher values, but still have unsatisfactory temperature factors because of the proximity of the top of the cold granite sills to the room surface of the window board.

14.2.2.4 Thermal Bridging Factor (y)

The Thermal Bridging Factor (y) is a supplementary value representing all heat flow associated with thermal bridges, divided by the full area of the building envelope. As it has the same units (W/m^{2}K) as plane element heat loss (U), the values for both can be added together to represent *all* conductive-radiative building fabric heat loss. This is an important input in creating a Building Energy Rating (BER) using the Irish DEAP or UK Standard Assessment Procedure (SAP) methodologies.

In the DEAP methodology, the default Y-factor of 0.15 W/m^{2}K should be used for all existing or new buildings that have not been thermally modelled and do not comply with the Acceptable Construction Details (ACDs), which support Technical Guidance Document L of the Irish Building Regulations [9]. Where adequate compliance with the ACDs can be demonstrated, a default Y-factor of 0.08 W/m^{2}K can be selected. Finally, calculated Y-factors, using the work of a registered thermal modeller, the values of a proprietary construction system, or the values set out in the ACDs, can also be accepted. This approach can allow values far lower, therefore higher-performing, than the 0.15 default value. Achieving a low Y-factor is harder in energetic retrofits [10]. Y-factors as high as 0.52 W/m^{2}K have been calculated for retrofit projects, where little attention was paid to thermal bridging. Such a figure could indicate higher levels of heat loss associated with junctions than might be associated with large plane elements.

In the absence of a calculated value, the default Y-factor of 0.15 W/m^{2}K was selected for use in the BER uploaded to the National Administration System (NAS) in December 2022. To further explore the value of the low thermal bridging approach to the design and to assess how it might improve the BER, the Y-factor was calculated post-work. The registered thermal modeller added seven details from the Acceptable Construction Details (ACDs) and two junctions from a catalogue to the thirteen calculated junctions shown in Table 14.3. All junctions were identified, as shown in Figure 14.4, and their lengths recorded. ψ-values were multiplied by their lengths and combined. They were then divided by the area of the full thermal envelope. This gave a Y-factor for the extended cottage of 0.089 W/m^{2}K, very close to the Thermal Bridging Factor (0.08) allowed for *new* dwellings built in accordance with the ACDs. This is almost half of the default Y-factor value.

14.2.2.5 Air permeability tests

The air permeability of the building envelope was tested before the works commenced and at three times during the project. Tests were carried out in accordance with EN ISO 9972:2015 [11]. A blower door unit, containing a fan, was fitted in one leaf of the rear patio door and the building was pressurised and depressurised in stages. In the three construction-stage tests, a vane anemometer and the palm of the hand were used to identify areas of leakage. Where possible, it is useful to have two teams available for this test. The first group is to identify air leakage and the second ready to carry out remedial work or apply airtightness tapes.

Despite being in poor condition, the initial air permeability rate (Q_{50}) of the cottage was 8.85 m^3/m^2.hr, better than expected, as shown in Figure 14.5. Due to the risk of damage to the badly cracked plaster-and-lathe ceiling, the test was undertaken in pressurisation mode only. Figure 14.6 shows some of the findings of the air pressure tester in this first test. It was interesting to see air leakage through a crack in the party wall and a surprising lack of air movement through a joint in the original suspended timber floor.

The first construction-stage air pressure test took place when the thermal envelope was complete. The cottage walls had been sealed with a lime plaster and windows and doors were fitted and taped. The studs of internal partitions were in place and the Pavadry insulation boards lined the party walls. The incomplete storage space, ventilation openings, and the fireplace opening were temporarily sealed in preparation. Importantly, the first fix of electrical and plumbing was programmed to commence after this test, as it is far easier to achieve a high level of airtightness before such complexity is introduced. The Q_{50} achieved was 1.56 m^3/m^2.hr. While there was a desire to achieve the EnerPHit target value of $Q_{50} \leq$ 0.69 m^3/m^2.hr (interpolated from $N_{50} \leq$ 1.0 given the building's geometry), comfort was taken that the result achieved was lower than the CarbonLite Retrofit Standard target of $Q_{50} \leq$ 2.0. In Figure 14.6, note the air leakage at the joint in the window system due to a slight curve in the frame and proof that the OSB face of the Pavadry composite boards is airtight.

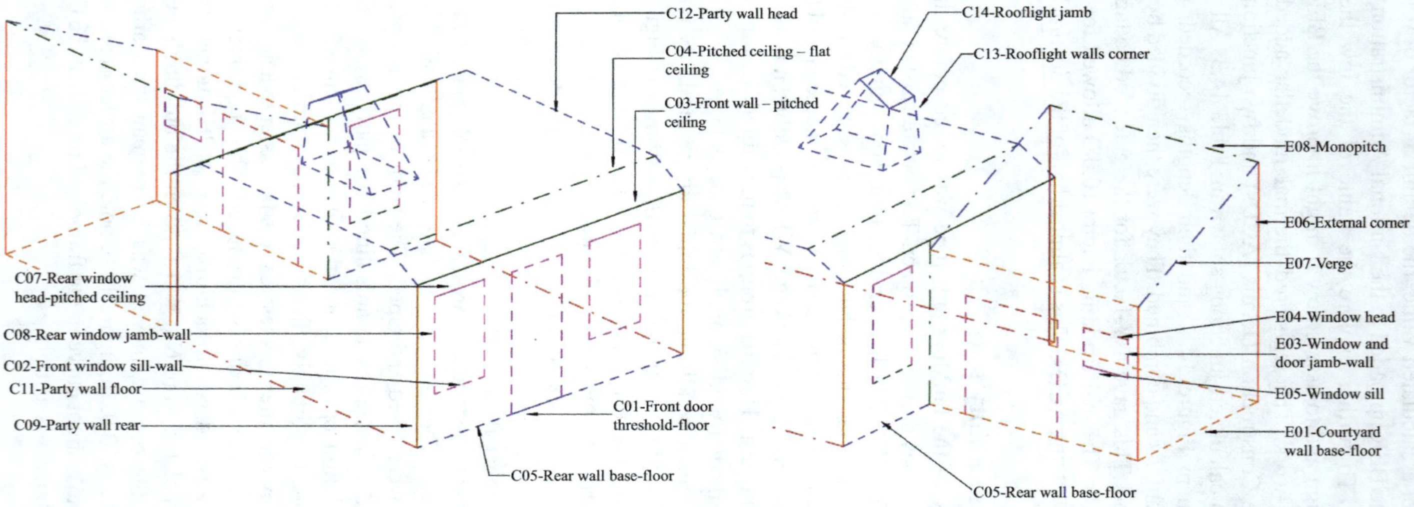

Figure 14.4 Identification of junctions of the extended cottage to aid calculation of a Y-factor (wireframe drawing courtesy of Shane Fenton)

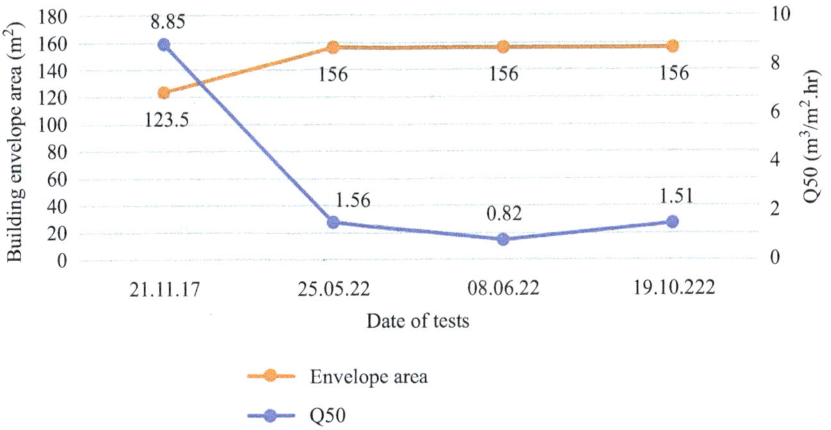

Figure 14.5 Building envelope areas and air permeability achieved

Prior to the second construction-stage air pressure test, internal partitions were completed. The Pavadry insulation boards were fixed to the front and rear walls, and the first fix of electrical cabling and plumbing was completed. A window ope also had to be adjusted. It is quite common that the first fix of services results in a slight increase in air leakage. However, an improved air permeability was recorded ($Q_{50} = 0.82$), exceeding the target value. On reflection, this may partly be because a small tetrahedron-shaped storage space was closed off for the test by agreement. A 'python' of cabling had been pulled through to this area and work was being done to seal the thermal envelope there. While the surface area of the building envelope being tested was temporarily reduced in this way, its size was left unchanged in the Q_{50} calculation, contributing to the improved calculated value.

The final air pressure test was in mid-October 2022. The Q_{50} achieved was 1.51 m^3/m^2.hr, complying with the CarbonLite Retrofit Standard target of ≤ 2.0. No doubt the comparatively large size and complexity of the thermal envelope frustrated efforts to reach a lower air permeability value. Innovative airtightness sprays, used while a property is pressurised, have since been identified on the market. The aerosol spray is naturally directed towards the areas of air leakage. Such a system would likely have enabled the more onerous EnerPHit target value to be reached.

14.3 Building energy modelling

14.3.1 Building energy model of case study using DEAP

The Sustainable Energy Authority of Ireland (SEAI) manages the Building Energy Rating (BER) scheme, the Irish version of the Energy Performance Certificate (EPC). The modelling tool used to create BERs for residential properties is the

Figure 14.6 (top row, left to right) Images from the last air pressure test (APT) underway, (middle rows) two insights from the first APT and the smoke test during depressurisation in the last APT, (bottom row) air leakage at junction of Pavadry boards and AVCL membrane, proof that Pavadry boards are airtight and air leakage due to slight curve in window frame

Domestic Energy Assessment Procedure (DEAP). Like the UK's Standard Assessment Procedure (SAP), it is used to calculate the theoretical annual primary energy usage and carbon dioxide emissions associated with the provision of space heating, water heating, ventilation, and lighting to a dwelling. Unregulated loads, such as consumer goods, are not considered.

The key values calculated are the BER, using a 15-point scale from A1 to G. In 2026 this 15-point scale was changed to a more consumer-friendly 8-point scale. The Building Energy Value (BEV) is calculated in terms of annual primary energy usage (kWh/m^2a) and $C0_2$ emissions ($kgCO_2/m^2a$). Additionally, the heat loss indicator (HLI) is reported, a measure of how quickly a home loses heat through conduction and convection, calculated as watts per square metre per degree kelvin (W/m^2K). The outputs generated are a consumer-facing BER Certificate and Home Energy Upgrade Advisory Report, as well as a technical Dwellings Details Report that lists all inputs and calculations.

The national building energy models were conceived in the first Energy Performance of Building Directive as asset rating tools used at point of sale or rent. The Irish Government, like most other governments in the EU, extended the use of DEAP to function as a regulatory compliance tool for new and retrofit building projects. For new build projects, this is done automatically by comparison of the outputs to a reference dwelling that has the same geometry and orientation and a fixed set of fabric and services specifications. Energy and carbon coefficients are then applied to assess compliance with the latest version of Part L of the Building Regulations. SEAI also uses DEAP to assess suitability for grant aid for energy-efficient retrofit works. SEAI has revised and improved DEAP many times. Their manual and supplementary guidance underpin DEAP and its inspectors audit BER assessors to ensure compliance.

Given the importance of BER/EPC as a support for EU and member state policy, it is of great importance that possible disparities between theoretical and calculated building energy use and carbon emissions are understood. There has been much research done in this area, especially since the introduction of Ireland's BER in 2005. There can be a *building performance gap* due to differences between the technical design and how the dwelling was built or retrofitted, a *user performance gap* due to actual users acting differently than the notional users in the model and, finally, a *model performance gap* due to simplifications and assumptions in the model itself [12,13].

Given its origin as a simplified asset rating tool, the model performance gap can be significant. For instance, simplifications include a single, fixed climate for Ireland, occupancy and water usage that relate largely to dwelling size, and a range of default values that can be selected where insufficient information exists. In accordance with the manual, it is possible for two BER assessors to create very different, yet compliant, BERs for the same dwelling, due to the first using a methodology that relies on visual inspection and default values, and the second using a methodology that accepts proofs that completed works followed the building designer's specification. Given that unregulated loads can account for between 18 and 50 per cent of actual household energy use, it is essential that these are quantified before an attempt is made to compare the calculated energy value with actual energy use and assess underperformance. Discrepancies between EPC assessors have also been observed in the UK [14].

In 2025, the Energy Security and Net Zero Committee of the UK's House of Commons issued the *Retrofitting Homes for Net Zero* report [5]. Interestingly, it strongly criticises the value and overuse of similar EPC assessments:

> *'Energy Performance Certificate assessments vary greatly in quality and consistency. It is therefore troubling that they are expected to perform such a range of functions that they were never initially designed for, such*

as determining eligibility for government support schemes. If the Government is to continue using [EPC] for such purposes, reform of the [EPC] metric must come quickly.

We are deeply concerned that [EPC], as currently calibrated, provide some consumers with poorly conceived advice and have a strong bias against low carbon, but initially more expensive, heating systems. As a cost-based metric, Energy Performance Certificates are not, as they stand, an effective tool for decarbonising UK homes. To the contrary, they are deterring consumers from making the switch to low carbon heating.' [5]

Some of the concerns raised in that report can be seen in the creation of a BER for the case study cottage.

14.3.1.1 Calculation of the building energy model

As expected, the pre-works BER of the semi-ruined cottage, identified as Version X in Figure 14.7, was 'G', at the very bottom of the BER scale, with a BEV = 725.44 kWh/m^2.yr. To explore the divergence that can occur using alternate BER methodologies, BER consultants IHER Energy Services carried out two post-work assessments in November 2022. The first of these, titled Version Y1, based on all defaults allowed in the DEAP methodology, resulted in a BER of 'F' and a BEV of 423.69 kWh/m^2.yr. The second, Version Y2, based on calculated U-values, specifications used, manufacturers' data sheets, and site photographs, resulted in a BER of 'C2' and a BEV of 181.91 kWh/m^2.yr. The latter BER was uploaded to the NAS as the official BER in November 2022.

The difference between these three values is notable. Version Y1 shows an uplift on Version X of only ~300 kWh/m^2.yr after significant effort and cost to transform a semi-ruin to a super-low energetic retrofit-extension. Version Y1 and the official Version Y2 show a similar difference of ~240 kWh/m^2.yr, despite being assessments for the *same works in the same dwelling*. To give a sense of how this divergence can occur, BER assessors are allowed to consider the very modern extension to be co-terminus with the late nineteenth-century cottage in terms of construction type and default U-values. While it is to be expected that the level of information available to a BER assessor about the construction and specification of buildings could vary greatly, are building owners informed of the methodology selected, what the divergence might be, and how that might affect their BER?

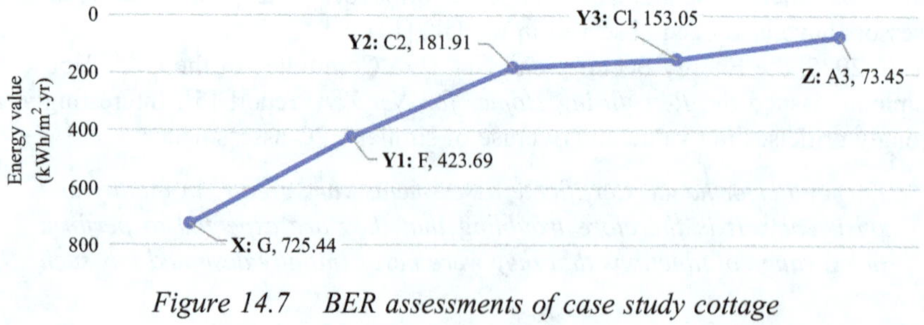

Figure 14.7 BER assessments of case study cottage

Energy performance of your home	NOW	POTENTIAL
Roofs	Good	No upgrade
Walls	Good	No upgrade
Windows	Very good	No upgrade
Floor	Very good	No upgrade
Space heating	Very poor	Very good
Water heating	Very poor	Very good
Renewables	Very poor	Very good

Figure 14.8 Excerpt from Home Energy Upgrade Advisory Report

Equally, does the Government quality assure such issues and have they considered what the impact of such divergences might be on national policy and forecasting?

It is clear that the measures taken to minimise building fabric heat loss significantly were generally successful, resulting in an excellent heat loss indicator (HLI) of 1.70 W/Km^2, which is recorded in the Dwellings Details Report for Version Y2. However, this high performance in building fabric is not reflected in that version's official BER of 'C2', because on-demand electricity, which the space heating and domestic hot water services use, is penalised in the DEAP methodology. The simple graphic in Figure 14.8, provided to homeowners as part of the Home Energy Upgrade Advisory Report, makes this clear.

14.3.1.2 DEAP: decarbonisation and future phases

As the carbon content in the national electricity mix continues to reduce, the primary energy conversion factor (PEF) used in DEAP for consumed electricity will also reduce, impacting the case study's BER. In June 2024, IHER reported that the PEF had reduced from 2.08 when the official BER was uploaded in late 2022 to 1.785 in January 2024, a 16 per cent reduction. If the BER were re-submitted, it would improve to 'C1', with a BEV of 153.05, without any change in the building fabric or systems. In Figure 14.6 this is shown as Version Y3.

The planned installation of a photovoltaic array in phase two of this energetic retrofit will provide an opportunity to re-submit an updated BER to NAS, incorporating the decarbonising value of site-generated renewable energy and a further-improved heat loss indicator (HLI) due to the calculated Y-factor. Using the

Table 14.4 Change in key BER metrics from official BER Y2 to proposed BER Y3

Performance type	Description	BER (Y2)	Proposed BER (Y3)
Thermal bridging factor (Y)	Default or calculated value representing all linear thermal transmittance expressed over the full thermal envelope	Default = 0.15 W/m^2K	Calculated = 0.089 W/m^2K
Air permeability achieved (Q_{50})	Measured airflow at a pressure of 50 pascals (pa)	Q_{50} = 1.51 $m^3/m^2.hr$	Q_{50} = 1.51 $m^3/m^2.hr$
Heat loss indicator (HLI)	Measure of how quickly a dwelling loses heat through its building fabric and ventilation	1.70 W/m^2K	1.51 W/m^2K
Solar PV generated	Predicted energy produced by 8 sqm PV array (4 × 395 W_p)	0.00 kWh/a	1357.5 kWh/a
Primary energy conversion factor (PEF)	PEF for consumed electricity is based on the carbon intensity of energies used to generate electricity used in Irish grid	2.08	1.785
Building energy value	The primary energy use per square metre per year (kWh/m^2a) calculated for space heating, water heating, ventilation and lighting under standard conditions	181.91	73.45
Building energy rating	An energy label, on a scale from A to G, that indicates a building's energy efficiency	C2	A3

existing published BER, Version Y2, as a baseline, the HLI will go from 1.70 to 1.51 W/m^2K, a 21 per cent reduction, due to the change in Y-factor, and the energy value will go from 181.91 to 73.45 kWh/a, a massive 60 per cent reduction, largely due to the offsetting value of on-site renewables.

Table 14.4 sets out the changes in key BER metrics that would result from verifying the Thermal Bridging Factor (Y) achieved and the installation of solar panels and a battery. The resulting BER would jump from C2 to A3. An impressive value for the energetic retrofit of a Victorian cottage with an unusually large heat loss form factor. If the PEF reduces further, due to another step in the decarbonisation of Ireland's national electricity mix, the improvements recorded by the BER will be greater still.

14.3.2 Building energy model of case study using PHPP

The leading alternate methodology for building energy modelling is the Passive House, or 'Passivhaus' Planning Package (PHPP). It is used throughout the world and has been validated many times. PHPP is designed to calculate, as accurately as possible, the energy demand based on the specific location, orientation, users, and equipment installed.

As the original building was hand-surveyed and drawn in two-dimensional CAD some years before, it was judged appropriate, post-works, to create a new three-dimensional *as-built* model that would better support accuracy in the creation of both the PHPP model and the life cycle assessment that would follow. The first step was the creation of a point cloud scan of the interior of the finished building, prior to the installation of the timber floor. A building information model (BIM) was built from the surface that the scan created using the author's technical drawings, specifications, and photographs. Figure 14.9 below shows outputs from the as-built PHPP building energy model. Criteria of the EnerPHit Standard are set out and whether these are fulfilled, taking the Component Method, is identified in the green column on the right.

As shown in Figure 14.10, energy is 'gained' from the consumption of electrical power for heating, shown in red, from various internal heat gains, solar heat gains, and from the reduction in building fabric heat loss at junctions of plane elements, when measured externally, due to negative psi-values. Balancing the gains are heat losses through the plane elements of building fabric and ventilation.

The only conventional energy supplied is electricity, primarily from renewable energy sources. It was calculated for this building that, per square metre per year, 50 per cent of the electricity contributes to heating, 20 per cent contributes to domestic hot water, 5 per cent to powering the DC-MEV ventilation system, 4 per cent to cooking, and the remaining 21 per cent to ancillary uses.

14.3.3 Contrasting the DEAP and PHPP assessments

The simplifications of DEAP and SAP have exorcised PHPP advocates for many years [15]. Here are some of the significant differences between the DEAP and PHPP assessments. The DEAP assessment estimates that the delivered energy for the cottage is 3,315 kWh, whereas the PHPP assessment estimates the delivered energy to be 6,285 kWh, which is significantly higher. This discrepancy is not unexpected, particularly in a retrofit project; however, there are several contributory factors.

- Heating periods: The DEAP and PHPP methodologies use different heating periods when determining the heating demand for a building and, as a result, the associated primary energy use. PHPP assumes a constant 20-degree ambient temperature, which contrasts sharply with DEAP's consideration of an 8-hour heating period, coupled with variable temperatures between the living areas and the rest of the dwelling.
- Climate: A site-specific climate was created to incorporate into the PHPP model, which better reflected the 'heat island' effect of the building's urban location. DEAP utilises the Dublin Airport climate dataset for all assessments nationwide. If the climate data of Dublin Airport were fed into the PHPP model, the primary energy demand would increase by $\sim$8 per cent. Localised climate datasets are crucial for the accurate assessment of a building's energy demand.

Specific building characteristics with reference to the treated floor area							
	Treated floor area	m²	36.0		Criteria	Alternative criteria	Fullfilled?[2]
Space heating	Heating demand	kWh/(m²a)	**89**	≤	-	-	**-**
	Heating load	W/m²	**28**	≤	-	-	
Space cooling	Cooling & dehum. demand	kWh/(m²a)	**-**	≤	-		**-**
	Frequency of overheating (> 25 °C)	%	**0**	≤	10		**Yes**
	Frequency of excessively high humidity (> 12 g/kg)	%	**0**	≤	20		**Yes**
Airtightness	Pressurisation test result n_{50}	1/h	**2.2**	≤	1.0		**No**
Non-renewable Primary Energy (PE)	PE demand	kWh/(m²a)	**378**	≤	-		**-**
Primary Energy Renewable (PER)	PER demand	kWh/(m²a)	**228**	≤	217	228	**Yes**
	Renew. energy generation (in rel. to projected building footprint area)	kWh/(m²a)	**0**	≥	-	-	

EnerPHit (retrofit): Component characteristics						
Building envelope to exterior air[1] (U-value)	W/(m²K)	**0.14**	≤	0.3		**Yes**
Building envelope towards ground (heat loss/load)	kWh/(m²a)	**8.0**	≤	18.3		**Yes**
Wall w/int. insulation in contact w/exterior air (U-value)	W/(m²K)	**0.41**	≤	0.5		**Yes**
Flat roof (SRI)	-	**72**	≥	-		**-**
Inclined and vertical external surface (SRI)	-	**72**	≥	-		**-**
Windows/Entrance doors ($U_{W/D,installed}$)	W/(m²K)	**0.98**	≤	1.09		**Yes**
Windows ($U_{W,installed}$)	W/(m²K)	**-**	≤	1.14		**-**
Windows ($U_{W,installed}$)	W/(m²K)	**1.21**	≤	1.24		**Yes**
Glazing (g-value)	-	**0.49**	≥	0.35		**Yes**
Glazing/sun protection (max. solar load)	kWh/(m²a)	**288**	≤	-		**-**
Ventilation (effective heat recovery efficiency)	%	**0**	≥	75		**No**
Ventilation (humidity recovery efficiency)	%		≥	-		**-**

[1] Without windows, doors and external walls with interior insulation
[2] Empty field: data missing; '-': No requirement

Figure 14.9 Key outputs in PHPP and component-based criteria for EnerPHit compliance

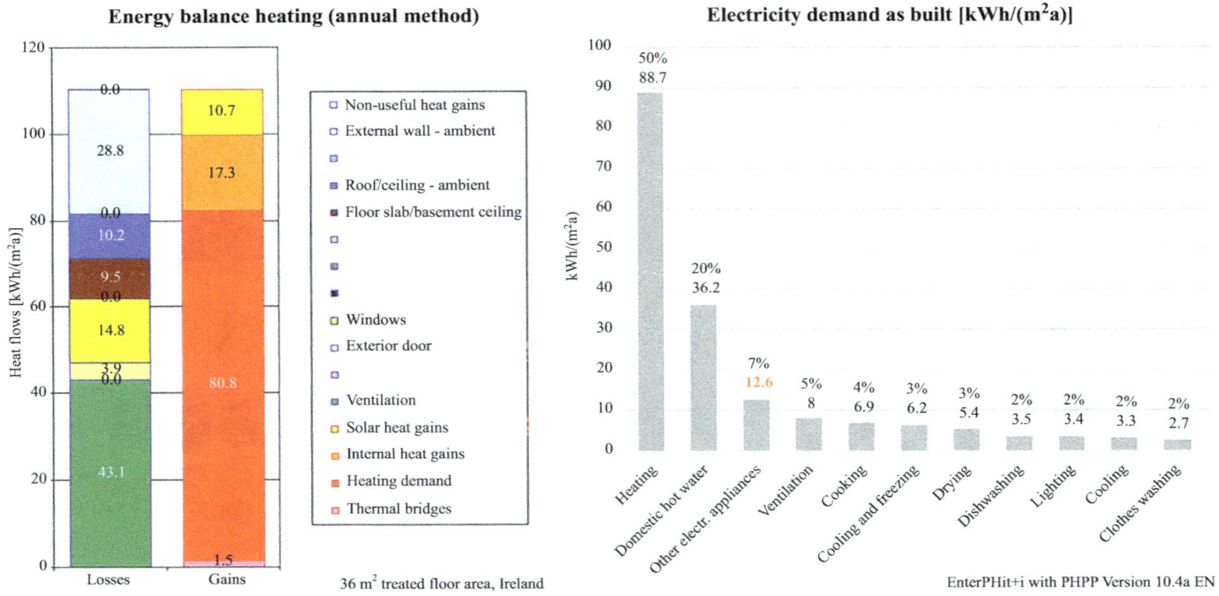

Figure 14.10 Energy balance and electricity demand per use type in PHPP

- Shading: The winter shading factor has a substantial impact on PHPP. In contrast, DEAP assumes no shading, as if the building were isolated on a flat plane.
- Overheating: PHPP enables relatively accurate modelling of overheating risk. Threshold values for compliance force designers to make changes that reduce risk and ensure comfort. Overheating is poorly modelled in DEAP and given marginal importance.
- Occupancy: DEAP's methodology has recently reduced the standard occupancy rate from 6.6 to 3.3, directly impacting the estimated domestic hot water usage in the dwelling, and it would appear conservative when compared to the PHPP method.
- Plug loads: General plug loads are not accounted for in DEAP, although there is an allowance for fixed lighting. PHPP considers general plug loads, including an allowance for cooking, washing up, televisions, and other small power loads, resulting in a more comprehensive representation of the primary energy associated with the dwelling.

Although the CarbonLite Retrofit Standard target of $Q_{50} \leq 2.00$ m^3/m^2.hr was achieved, it is interesting to assess what steps would need to be taken to attain compliance using the more onerous EnerPHit Component Method. Figure 14.11 shows that only two steps were necessary:

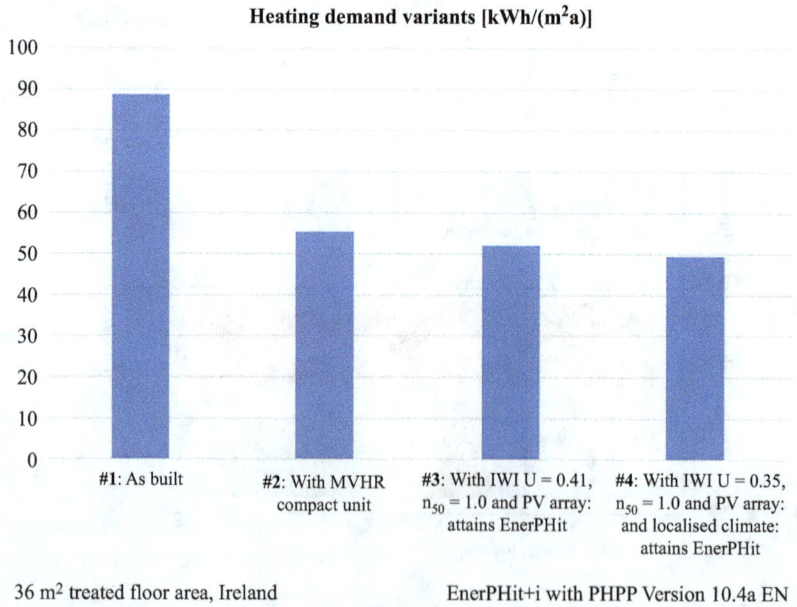

Figure 14.11 Possible interventions to achieve EnerPHit Standard

- Step #2: install a mechanical ventilation system with heat recovery (MVHR) in lieu of demand-controlled mechanical extract ventilation (DC-MEV), reducing heating demand from 88.7 to 55.3 kWh/m^2.a.
- Step #3: to this measure, add an increase in airtightness of N_{50} = 1.0 ACH (i.e. air permeability of $Q_{50} \leq 0.69$ m^3/m^2.hr) and install a photovoltaic array, reducing heating demand to 52 kWh/m^2.a.

A further suggested step is to increase insulation levels of the internally insulated walls to 0.35 W/m^{2}K and use a localised climate file that reduces the energy demand to 49.5 kWh/m^2.a, due to the urban heat island the cottage lies within.

A heat pump can be accommodated and may be fitted in a future phase. However, it is very unlikely that work will be done to further improve airtightness, given that the airtight layer is now hidden by room finishes. Equally, even if the Pavadry composite wall insulation were in sight, swapping out these boards to obtain a U-value as low as 0.35 W/m^{2}K would elevate moisture risk, which is inappropriate.

14.4 Hygrothermal risks: simulation and measurement

Hazards due to moisture, including rot, corrosion, mould growth, interstitial condensation, and decay of fixings or finishes, are the greatest causes of failure in buildings. Guidance and building design practice in the UK and Ireland have been slow to address this. This is partly due to the complexity of how heat and moisture are transported through building fabric. As intrinsically coupled physical phenomena, they cannot be looked at in isolation or quantified with simple calculations. The word 'hygrothermal', derived from the Greek words for moisture and heat, reflects this coupled nature.

> '. . . *Retrofitting building fabric to improve its thermal performance is likely to also impact on its moisture performance, particularly in traditional construction. Therefore, to avoid moisture-related deterioration of the building fabric and health risks, retrofit measures need to be assessed not only for their thermal benefits, but for their hygrothermal impacts too.'* [16]

14.4.1 Standards and regulations underpinning risk assessment

The steady-state Glaser method, an update of the 'dewpoint' method under standard EN ISO 13788: 2012 [17], is still the dominant hygrothermal risk assessment tool used by materials suppliers and building designers, despite significant limitations. The restrictive guidance of the standard's scope, such as the method only being applicable where *'capillary suction and liquid moisture transfer within materials' is* considered *'negligible'*, seems to be commonly ignored. This is unacceptable, given that it is now decades since advances in building physics and the availability of cheap microprocessors enabled the creation of computer programmes that can be used to transiently simulate a full range of hygrothermal conditions in building fabric. In this approach, equations are solved against each other for each cell of an assembly of materials and for each time step of the simulation's duration. WUFI Pro, developed by the Fraunhofer Institute for Building Physics (FIBP) in 1995, is the leading example of this new type of transient numerical simulation tool. It is used by building design professionals in over 100 countries and is extensively used for research.

The standard that underpins this more widely applicable form of hygrothermal risk assessment, EN 15026 [18], was first published in 2007 but was only adequately referenced in the UK's *Code of Practice for Control of Moisture in Buildings* (BS 5250) [19] in 2021. This revision was implemented after many years of lobbying by bodies working under the banner of 'responsible retrofit', led by Sustainable Traditional Buildings Alliance (STBA), which culminated in a white paper in 2017 [20]. Key foci of the revision are the integration of systems thinking and assessment of buildings 'as-built and in-service', as shown in Figure 14.12.

Ireland's Technical Guidance Document (TGD) L [21], supporting the building regulations for conservation of fuel and energy, which has referred to EN 15026 since 2011, might be viewed as ambiguous. Building designers are guided to carry out *'full checks … on the likelihood of … interstitial condensation'* in accordance with EN ISO 13788 and further guided that *'EN 15026 can also be used'*. BS 5250 is only referenced in Irish technical guidance in relation to condensation risk in roofs. The technical design of energetic retrofits in Ireland would benefit if the guidance of the updated BS 5250 were integrated into TGD-L and SR 54 *Code of Practice for the Energy Efficient Retrofit of Existing Domestic Dwellings* [22].

A trained user of transient numerical simulation tools can explore the impact of different types, thicknesses, and positions of material layers, different climates or boundary conditions and stressors, such as driving rain, failed seals, exposure, construction moisture, or leaky air barriers. While accepting that this, and all assessment tools, have limitations, parametric assessments of this kind can enable an assessor to identify and mitigate moisture risks of an existing or future building, as-built and in-service, in a way that is not possible with simpler assessment tools or through physical trialling alone. The knowledge and rigour used by the assessor

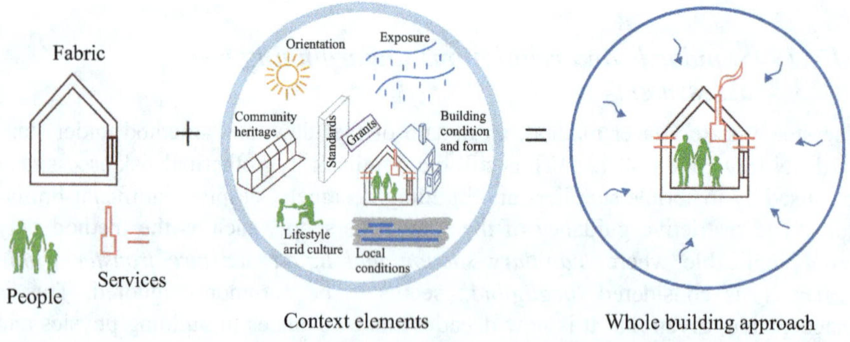

Figure 14.12 STBA's whole-building approach [20]

to select materials, climates, and parameters and how to interpret the results are crucial. As the material database increases and research into appropriate use of this type of assessment grows, the ability to make accurate assessments should improve.

14.4.2 Hygrothermal risk assessments

Parametric hygrothermal risk assessments were carried out to guide the specification of the case study cottage's walls and coombed roofs prior to commencement of works. The hygric properties of many materials, in particular the existing materials, were unknown. This need not hinder a well-prepared parametric study if the datasets of a sufficient range of similar materials are available from which to carefully select [16]. However, at the time, there were only two concretes in the database that represented anything other than modern, well-consolidated concrete in good condition. These had the enigmatic titles '*Concrete, C12/15 cracked, aged*' and '*C35/45 cracked, aged*'. While their use is guided for *'worst case considerations with cracked or weathered surfaces'*, it was not clear if these were historic concrete or poorly consolidated. As a result, the hygrothermal risk assessments carried out were judged to be of uncertain value to steer an appropriate specification. Given concerns about interstitial conditions, but also to support future research, eleven sensors were installed in various parts of the roof, wall, and floor to record hourly temperature and humidity levels at interfaces. They are still recording and should contribute to future research.

14.4.2.1 Improving understanding of materials in the model

The assessment of hygrothermal risks was re-engaged post-construction with the release of new data. University College Dublin's groundbreaking *FabTrads* research project, funded by the Sustainable Energy Authority of Ireland, has resulted in hygrothermal data for a significant range of historic Irish bricks, stones, and plasters being made available on the WUFI material database in Spring 2025. At the time of simulating and writing in Spring 2026, that team was close to publishing additional data on five historic Irish concretes, sections of which are shown in Figure 14.13. The three concretes from County Kerry are oldest, dating to (A) 1900–10, (B) 1910–15, and (C) ~1930. The two from Dublin County are from (D) 1950s and (E) 1930s. Like the walls of the case study cottage, the three Kerry samples are poorly consolidated with pebble-based aggregate and inadequate amounts of cement. The Trad Fabs team kindly provided pre-publication access to these datasets. In correspondence, Dr Rosanne Walker of that team highlighted important differences between the Irish historic concrete and the concrete already in the WUFI database,

> *'... The WUFI concretes have much higher sorption capacity (moisture content) between 50–95 % RH, although they have similar free saturation values. It is hypothesized that the modern concretes likely have a higher percentage of smaller pore diameters due to their greater hydraulicity, contributing to their higher adsorption.'*

Figure 14.13 (Clockwise to top left) Screengrab of horizon mask created in Meteonorm using a 360-degree image, Karsten tubes measuring liquid water absorptivity of painted finish and representative images of the five Trad Fabs concretes A–E

In the absence of measurement and given the importance of the hygric characteristics of concrete for this internally insulated solid wall, a 'bracketing' approach, involving the use of two concretes, was judged necessary for the revised parametric risk assessment. This approach acknowledges the presence of significant unknowns and should encourage a cautious approach in reaching conclusions. Concrete A has the highest Reference Water Content (W_{80}) and Concrete B the lowest of the five concretes, while B has a Water Absorption Coefficient (A_w) (or capillarity) that is 60% more than that of Concrete A. As these are the key values that determine how much vapour and liquid water are transported in hygroscopic, open-pore materials, it was judged reasonable to select those concretes, assuming that the real performance of the cottage's concrete is likely to lie somewhere in between.

In assessing the surface of the wall, Karsten tube measurements on-site [23], shown in Figure 14.13, confirmed that the paint was capillary-closed and established and that the render beneath the paint had a Water Absorption Coefficient, or capillarity, of 0.016 $kg/m^2\sqrt{s}$. The R3 historic cement-based render from the FabTrads database, which has an A-value of 0.030, was judged sufficiently close to use. Site measurement and a timely Irish research project increased the accuracy of key materials selections.

14.4.2.2 Developing accurate climate files

A major gap in the risk assessment toolbox is the non-availability of national hygrothermal reference year (HRY) climate files. This has been problematic for modellers in the UK and Ireland for many years. Tens of climates are available for selection in WUFI Pro for many other European countries, while only one is available for Ireland. In their absence, assessors generally use synthetic HRY files created using the proprietary Meteonorm app [24]. It has been established that files created with this tool correlate well with temperature and humidity but less well with driving rain in Maritime Europe. One compensating feature of the Meteonorm app is that it incorporates topographical obstructions, such as distant hills, and allows the input of a horizon 'mask' for local obstructions, such as buildings or trees. The app can then create synthetic climate files with associated reductions in the solar radiation reaching the plane element being assessed. Such shading can lead to reduced surface evaporation and reverse diffusion effects at certain times of the day and year. When localising a climate file in this way, the wind speed or driving rain must be manually reduced.

Architect John Morehead (WMA), a leading advocate of the use of site-specific climate files, supported this research project by creating a site-specific climate file. A 360-degree photograph was taken at the door of the cottage, using the Sunseeker app to ensure the image was centred due south and level. The top-left image in Figure 14.13 shows how the eastern and lower portions of the sky are obstructed by the west-facing façade and nearby buildings. A horizon mask was created, indicated by the red outline in image, in Meteonorm and this outputted a synthetic HRY climate with reduced solar radiation, better reflecting the real conditions on site. Finally, wind speed was lowered by 20% to reflect Findlater Street's urban conditions without changing the overall rainfall. With less wind, there will be less wind-driven rain. This approach acknowledges that the site's urban context is quite different from the exposed conditions commonly found around weather stations and takes account of current research on wind-driven rain modelling [25]. The modified climate file has a Normal Rain Sum of 809.3 mm/a, Counter Radiation Sum of 2,883.5 kWh/m^{2}a and a Mean Wind Speed of 3.5 m/s.

For naturally ventilated buildings, WUFI Pro derives the variation of indoor air temperature and relative humidity from the outdoor climate file using specified transfer functions. Several variants are available in relation to relative humidity. In this case, a medium moisture load of +5 per cent was selected for the internal climate. The increase of five percentage points on the relative humidity for normal residential cases, raising the transfer function to 35–65 per cent, is intended to provide an additional safety margin for hygric design purposes.

14.4.3 Risk assessment of embedded timber sub-frames

Given the novelty of the timber sub-frames of the original windows that were discovered embedded in concrete, long after the original steel windows they had supported were replaced, and the potential vulnerability of the replacement sub-frames relatively near the outside of the wall, it was judged of value to focus the risk assessment on these elements.

This was done by assessing the moisture content in a 20 mm zone of the concrete wall that corresponds to the location of the outer portion of the oak

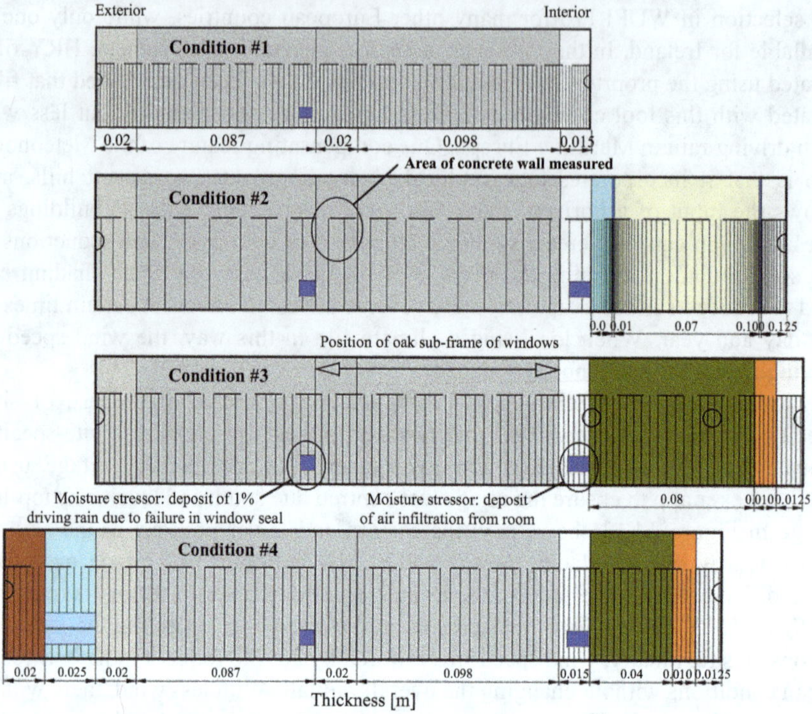

Figure 14.14 Representation of the calculation cells of the wall layers assessed in WUFI Pro

sub-frame. Four different wall conditions were considered. Variations within those groups resulted in twenty cases being simulated. The conditions are shown graphically in Figure 14.14, and the cases are tabulated in Table 14.5. Many other cases were also simulated, using different concretes, renders, and climates. They were judged to be insufficiently representative of the completed works or made little difference to the results.

The first condition assessed is the west-facing uninsulated wall with two different concrete types, A and B. The next group shows the west-facing wall as if it were internally insulated with a conventional insulated plasterboard system. Again, two concretes resulted in two cases. The next group shows the west-facing wall with the wood fibre OSB composite board internal wall insulation (IWI) system that was specified and installed. While 80 mm of wood fibre was installed here, it is interesting to assess risk at the location of the sub-frame for the three thicknesses of wood fibre that are available in this insulation system (i.e. 40, 60, 80 mm) with two concretes. This results in six cases. Finally, the east-facing wall is shown. 40 mm of wood fibre was installed there. Given that the residents

Table 14.5 Parametric cases for internal insulation of solid walls

Context	Case	Concrete type	Orientation	Insulation	Thickness (mm)	Comment
West-facing, uninsulated	Case #01	A	West	n/a	n/a	Pre-works
	Case #02	B				Pre-works
West-facing, conventional IWI	Case #03	A	West	Phenolic board, foil-faced	70	–
	Case #04	B			70	–
West-facing wood fibre IWI	Case #05	A	West	Wood fibre OSB composite board. (Thickness of WF listed)	40	–
	Case #06				60	–
	Case #07				**80**	**As built**
	Case #08	B			40	–
	Case #09				60	–
	Case #10				**80**	**As built**
East-facing wood fibre IWI w. rainscreen	Case #11	A	East	Pavadry	**40**	**As built**
	Case #12				60	–
	Case #13	B			**40**	**As built**
	Case #14				60	–
West-facing wood fibre IWI	Case #15	A	West	Rain turned off	80	–
	Case #16	B				–
	Case #17	A		No paint, no rain leak		–
	Case #18	B				–
	Case #19	A		No paint		–
	Case #20	B				–

mentioned that it is the only room that they find cool, it was assessed with both 40 and 60 mm, resulting in four cases. The remaining six cases will be discussed in relation to stressors.

14.4.3.1 Selection of stressors

Risk assessment is all about 'what ifs'. If the building were stressed by a likely condition, what would happen? Some building fabric assemblies will continue to dry out, despite the application of a particular stressor, while others, which had functioned acceptably beforehand, may move to failure. It is crucial to assess such likelihoods, reflect on acceptability, and explore what specification changes, if any, might be prudent. If we load an unrealistic number or severity of stressors, many assemblies will start to fail, despite such a situation being highly unlikely. A balance must be struck.

The following stressors were used in all cases:

- *Climate*. Modifications to internal and external climates are addressed in the text above.
- *Moisture ingress due to driving rain penetration*. In accordance with the ASHRAE 160 [26] requirement for moisture-safe design, one per cent of the rain hitting the surface is assumed to pass through the outer surface. In this case, it is hypothesised that this moisture is deposited at the vertical junction of the window-to-wall reveal. The usual gun-applied seals here are narrow and can fail for many reasons, not least due to lack of maintenance. The location of that junction is shown as the left blue square in each of the four conditions in Figure 14.14.
- *Moisture infiltration from the room*. The Fraunhofer Institute for Building Physics has determined that the *'mean flow density of infiltrated water through a standard cross section of a component is about a factor of 15 times smaller than the mean flow density through the whole building envelope'* [27]. Thus, a value can be determined for this wall based on the dwelling's measured air permeability, Q50 = 1.51 m^3/m^2.hr. By convention, moisture infiltration is deposited on the first continuous surface behind the retrofitted measures, in this case, the room side of the wet plaster finish.

The top two graphs and the bottom left graph in Figure 14.16 are based on the materials, boundary conditions, and stressors set out above. The final six cases listed in Table 14.5 return to the third wall condition, as was installed in the case study project with 80 mm wood fibre (Cases #07 and #10). They explore the following questions related to stressors.

- Cases #15 and #16: How well would the west-facing wall perform if the moisture ingress due to the 1 per cent driving rain penetration effect were omitted?
- Cases #17 and #18: How would the removal of the capillary-closed paint finish affect the west-facing wall if the 1 per cent driving rain penetration function were re-instated *but* there was no driving rain?
- Cases #19 and #20: How would the removal of the capillary-closed paint finish affect the west-facing wall if a 1 per cent driving rain penetration function were re-instated *and* driving rain occurred as before?

14.4.3.2 Selection of materials

Table 14.6 shows a selection of the hygric values of the materials selected from the database. Note that some are single values, such as a material's bulk density, while others are key values extracted from a hygrothermal function, such as the Reference Water Content (W_{80}), which is a characteristic value of hygroscopic, open-pored materials and is found on the Moisture Storage Function curve. The latter records the fixed relationship between relative humidity and moisture content for such materials. These functions are shown in Figure 14.15.

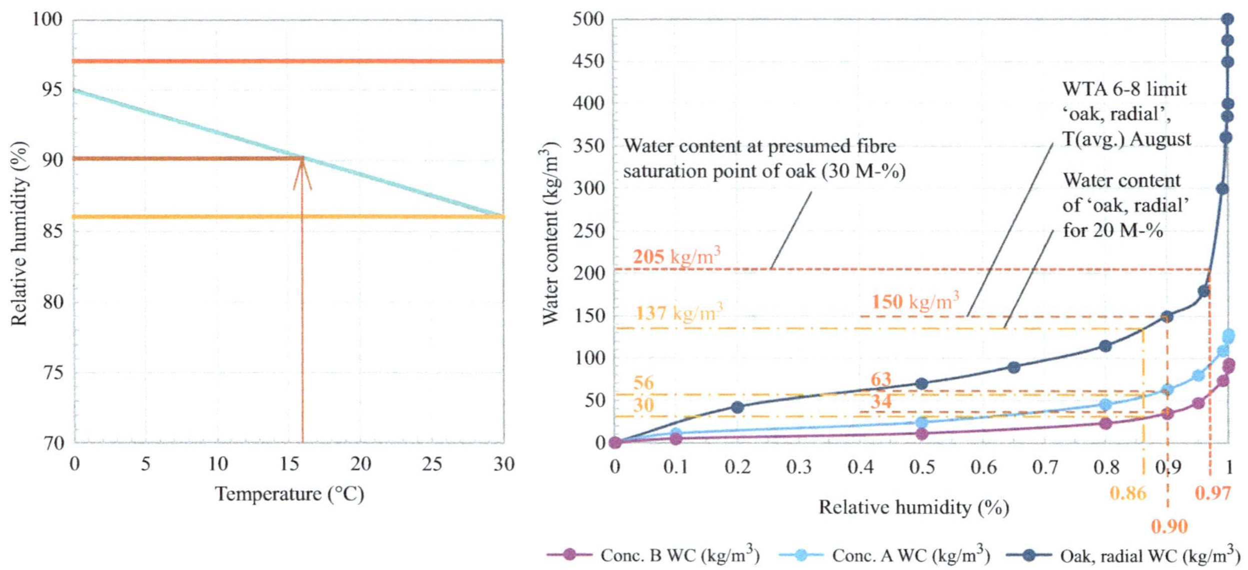

Figure 14.15 *On the left, the fixed risk limit for 'Oak, radial' based on 20 M-% (orange line); the risk limit for 'Oak, radial' in August (brown) based on the WTA 6–8 temperature-dependent risk limit curve for softwood (cyan) and the presumed FSP for 'oak, radial' 30 M-% (red). On the right, interpolated values using Moisture Storage Function curves of materials 'Concrete A', 'B' and 'Oak, radial'. Note that relative humidity values obtained in the right graph (e.g. 0.97 = 97%) are used in the left graph.*

It is useful to explore some of the material values tabulated. The wide concrete layer has a great influence on the performance of the whole wall and the hygro-thermal stress on both the embedded timber sub-frame and the internal insulation system. Concrete A and B were discussed as attempts to make a more accurate representation of the material in the model. In Table 14.6, Concrete A has a Water Vapour Diffusion Resistance Factor (μ) and Reference Water Content (W_{80}) that is almost twice that of Concrete B. Therefore, A is more vapour permeable and can store more moisture in service conditions. However, Concrete A has a Water Absorption Coefficient (A_w) 60 per cent lower than B, which means it will not only imbibe liquid water less vigorously but also lose this liquid water more slowly. These three values go a long way to explain why cases that feature Concrete A are consistently more stressed.

American White Oak was used to make the replacement timber sub-frames. 'Oak, radial' was chosen as the best fit from the WUFI Pro material database. The 'radial' refers to moisture transport being assessed perpendicular to fibres, which is appropriate given the fabrication of the frames. The guidance in the material database states that the material is *'to be used for wooden parts in direct contact with rainwater (e.g. framework constructions) and for analysing drying processes at high moisture contents. Should, however, be used with caution and only for approximative calculations, because of large uncertainties'*. In the absence of an alternative, it was selected.

The Sd-values of the masonry walls were calculated by multiplying the thicknesses of layers by the Water Vapour Diffusion Resistance Factor (μ) listed in Table 14.6.

Noteworthy are the high resistance to diffusion of water through the vapour retarding membrane (μ = 100,000) tabulated, compared to other materials listed, and the difference in vapour diffusion resistance (μ) and water absorption (A_w) between the lime cement mortar and the adjacent wood fibre insulation board.

The only material modified by the author is the exterior paint. A vapour-permeable, non-hygroscopic, non-capillary membrane was used as the base material to create the paint that was then adjusted using manufacturer data for Bulk Density and Water Vapour Diffusion Resistance Factor (μ). By convention, very thin materials such as membranes are modelled at 1 mm thick. The μ-value of the paint in the table above is ~3.7 times greater than the render behind it. Creating this layer was judged a more accurate approach than the alternative of assigning a modified Sd value to the exterior surface of the next layer, in this case, a coat of render.

14.4.3.3 Moisture in timber: design values and fibre saturation

Once cases have been 'built' within the WUFI Pro interface, the thresholds against which they will be judged must be clarified. For timber, fibre saturation is the greatest concern. Wood, the living organic, fibrous cell structure of trees and

Table 14.6 Hygric value of materials used in WUFI Pro simulations (Data measured by Trad Fabs Research Project)*

	Thickness	**Bulk Density (ρ)**	**Porosity (ϕ)**	**Reference Water Content (w_{80})**	**Free Water Saturation (w_e)**	**Water Absorption Coefficient (A_w)**	**Water Vapour Diffusion Resistance Factor (μ)**	**Liquid Transport Coeff. Redistribution (Dww_{80})**	**Liquid Transport Coeff. Redistribution (Dww_{100})**	**Thermal Conductivity (λ_{dry})**	**Thermal Conductivity (λ_{sat})**
Units	**mm**	**kg/m^3**	**%**	**Kg/m^3**	**Kg/m^3**	**Kg/m$^2\sqrt{^s}$**	**–**	**m^2/s**	**m^2/s**	**W/mK**	**W/mK**
Exterior paint	1	1280	60	–	–	–	500	–	–	2.3	–
Render R3 *	20	1950	24.2	38.28	160.18	0.032	133.89	–	–	1.29	–
Concrete A *	205	2133	16.0	45.03	128.1	0.033	107.1	2E-10	2E-8	1.42	2.33
Concrete B *	205	2200	16.3	22.83	93.6	0.055	55.25	4E-9	8E-7	1.51	2.26
Oak, radial	–	685	72	115	500	0.007	140	–	–	0.13	–
Lime Cement Mortar, Course	15	1910	25	24.65	200.01	0.085	45.89	–	–	0.8	–
Wood fibre Insl. Board	40, 60 and 80	155	98.1	19	980	0.007	3	–	–	0.042	–
Plywood (Agepan OSB/3 PUR)	12	616	52	85.6	395	0.0015	144	–	–	0.13	–
Air layer	10	1.3	99	0	0	0	0.73	–	–	0.071	–
Vapour Retarder	1	130	0.1	–	–	–	100000	–	–	2.3	–
Phenolic Foam	70	31.8	98	1.2	483	0.002	221	–	–	0.021	–
Gypsum Board	12.5	850	65	6.3	400	0.287	8.3	–	–	0.2	–

woody plants, contains a high proportion of water [28]. Softwood trees can vary between 40 and over 200 per cent moisture content by mass, and hardwoods 40 and over 100 per cent. After being cut down and processed as timber, a 'dead' material, moisture content must be greatly reduced and maintained within certain bounds to ensure that the final product's strength characteristics, dimensions, and durability remain within acceptable bounds. BS EN 942 [29] states that external joinery should have a moisture content of 12–19 per cent by mass and internal joinery should have a moisture content of 9–13 per cent. EN 336 [30] assumes the moisture content of timber is at 20 per cent for measurement purposes and lists dimensional increases for each 1 per cent increase up to the fibre saturation point (FSP). If conditions in the mass concrete walls of the case study were acceptable, it would be reasonable to assume that the moisture content of the embedded oak sub-frames would be in the range of values for external joinery, given the timber is located in the cooler portion of the solid wall, isolated from direct wetting of rain, as well as the drying effect of wind and sun.

Berry and Roderick [28] identified that the FSP was first defined by Tiemann as the moisture content of a piece of timber at which the cell cavities contain no water or sap, but the cell walls are fully saturated with liquid moisture. They note that FSP may also be considered the maximum amount of water that can be 'bound' to the cell wall matrix. As a hygroscopic material, timber can get wet and dry many times without degradation. However, once the FSP of a particular species of timber is reached for an extended period, free water in the cells will enable fungal decay to commence. Trada states the FSP for softwood species is 28–30 per cent moisture by mass [31]. A greater proportion of hemicellulose and a lower proportion of lignin in oak, as compared to, for instance, spruce, raises the FSP. Oak is commonly listed as having an FSP of ~30 per cent, a value that oak joinery needs to remain below.

14.4.3.4 Use of one-dimensional tools and risk thresholds to avoid fibre saturation

Figure 14.14 is a graphical representation from WUFI Pro of the one-dimensional calculation cells that make up each layer of the wall variants. Due to the timber sub-frames being discontinuous elements, they cannot be shown. Despite being one-dimensional, WUFI Pro may be used to assess conditions in discontinuous elements, as long as both bridged and bridging elements, such as concrete and timber, are open-pored and hygroscopic. This can be done because all such materials have unique moisture storage functions and will have a very similar relative humidity if they are located at the same place in the model. The right image in Figure 14.15 shows the three curves of 'Concrete A', 'B' and 'Oak, radial' with relative humidity on the X-axis and moisture content on the Y-axis. It is immediately clear that for the same relative humidity, each curve results in a different moisture content.

Risk thresholds, or limits, are measurable conditions used to manage the risk of a hazard, such as timber decay, manifesting. Risk exists above *and* below the line: when below the line, they are judged to be acceptable and when above, they should be mitigated or avoided. It has become common practice in hygrothermal risk assessment to use 20 per cent moisture by mass (M-%) as a fixed risk limit for solid timbers and 18 per cent for plywood. While this limit is stated in guidance [32], it was first stated in an early edition of (German standard) DIN 68800-2 [33]. In the UK, there is no mention in BS 5250 but Trada's advice [31] that *'timber below approximately 20% moisture content is not considered to be at risk from fungal decay'* is similar. This fixed value is now considered quite conservative, given that it reflects neither the species of wood, as previously highlighted, nor temperature dependency, as the FSP for timber in a cool location will be higher than when it is warm.

If a risk limit is set too low, it may indicate that a risk is elevated when it is not. For this reason, the Wissenschaftlich-Technische Arbeitsgemeinschaft für Bauwerkserhaltung und Denkmalpflege (WTA) [34], the European building science association that leads the development of hygrothermal numerical simulation internationally, introduced a temperature-dependent risk limit curve for softwoods [35]. The left graph of Figure 14.15 shows the WTA's temperature-dependent risk limit curve as a cyan-coloured line. This has since been included in DIN 68800-2 [33].

There has been a slowness to adopt this approach, partly due to the difficulty of graphing daily average temperatures of an element being assessed against this limit curve. To aid adoption and simplify assessment, the FIBP confirmed in Spring 2026 that a post-processor will soon be introduced in WUFI Pro. A future revision of relevant standards in Ireland and the UK, including BS 5250, would benefit from inclusion of this limit curve. It is important to note that hardwoods must also have temperature-dependent limit curves. That of oak would certainly be higher due to its greater durability; however, limit values have not yet been established. This means that, currently, risk assessment of hardwood structures, such as the installed oak window sub-frames, must be extrapolated from an overly onerous fixed value (i.e. 20 M-%) or a limit curve created for softwoods.

14.4.3.5 Establishing risk limits and resulting analysis

Given the German industry's slowness to adopt the latest guidance, when providing consultancy, the FIBP tends to assess timber first using the fixed risk limit of 20 M-% and, if that limit is breached, then apply the temperature-dependent risk limit curve. Their approach will be taken here to assess the risk of FSP being reached in the outer 20 mm of the oak sub-frame, accepting that oak, not a softwood, is being assessed.

Table 14.6 lists the density of 'oak, radial' as 685 kg/m^3. Multiplying this by 20 per cent identifies the moisture content of the traditional fixed risk limit (i.e. 20 M-%) as 137 kg/m^3. Using the moisture storage curve in the right graph of Figure 14.15 gives a relative humidity of 86 per cent, as shown by the orange line in the left graph. The moisture contents (kg/m^3) for Concretes A and B are established using the same relative humidity, giving values of 30 and 56 kg/m^3. As discussed, the WTA limit curve makes clear that higher temperatures lower the FSP at which

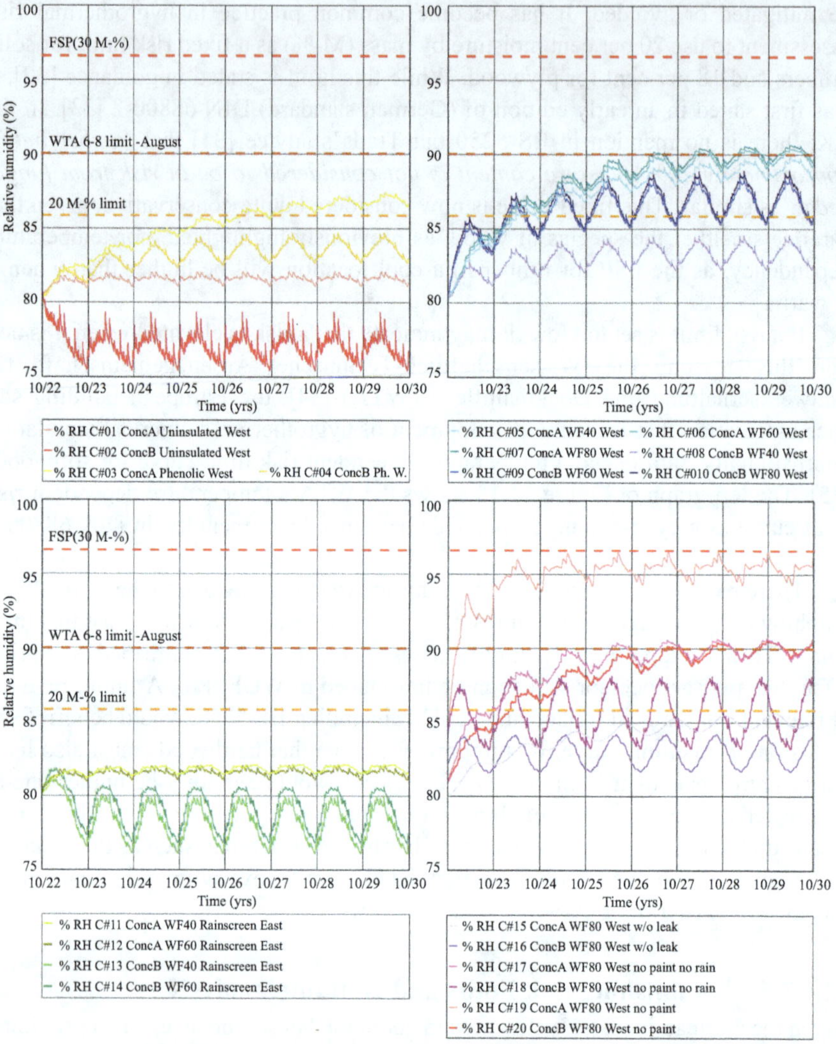

Figure 14.16 Relative humidity in the outer 20 mm of the window's wooden sub-frame

decay would commence. In the absence of the FIBP's post-processor, an assessment was made of the temperature in the outer 20 mm of the sub-frame. This was done by exporting the temperature file for that portion of the model and calculating an average daily temperature for the warmest day in August and the average temperature for all of August. Those temperatures were 18 and 16 °C, respectively. For this study, 16 °C will be considered significant and the 18 °C a '*short-term exceedance*'. The left graph of Figure 14.15 shows an arrow from 16 °C reaching the WTA 6–8 limit curve, then extending leftward as a brown line at 90 per cent relative humidity. As before, that value is then used in the right graph to establish moisture content for Concretes A and B, 34 and 63 kg/m^3, respectively. Lastly, the usually accepted FSP for oak, 30 M-%, is graphed in red on the right as 205 kg/m^3 and on the left as 97 per cent relative humidity.

The twenty cases were simulated for eight years using the HRY synthetic climate file with the masked horizon climate file described in Section 4.2.2. The relative humidity in the outer 20 mm of the sub-frame was then graphed and logically grouped in the four sets described in the sections that deal with cases and stressors above. Once graphed, the risk limits discussed were overlaid. The graphs show present conditions and risks in the outer portion of the timber sub-frame only. A stressed condition there does not necessarily indicate a stressed condition elsewhere, such as in the internal insulation. A fuller study would explore conditions at the interface of internal insulation with the plastered concrete wall.

The following issues have been highlighted by the analysis. The ten cases assessing the sub-frame embedded in Concrete A are all more stressed than their partner cases for Concrete B. In some cases, the variant featuring Concrete A breaches a threshold, where Concrete B does not. Remember, the material characteristics of the actual concrete are unknown: an assumption has been made that the real performance lies somewhere between them.

While Cases #1 and 2 (shown in top-left graph), representing the uninsulated west-facing walls, have the highest U-values (~3.0 W/m^{2}K at W_{80}), they are under the least moisture stress due to the optimal conditions for evaporation on both sides. Even a highly vapour-permeable insulation is more vapour-resisting than still air.

Cases #11–14 (bottom left graph) are the next group exhibiting low moisture stress at the timber sub-frame. These east-facing walls are modestly insulated (with U-values of 0.571 W/m^{2}K at W_{80} for a 40 mm wood fibre OSB composite system and 0.455 for 60 mm wood fibre). However, they are also sheltered by a ventilated timber rainscreen. This means the only liquid water stress is the 1 per cent driving rain deposited near the window-wall reveal junction. Dynamic equilibrium is reached quickly, the two A graphs and the two B graphs are closely grouped, and conditions are far below the conservative 20 per cent moisture by mass threshold. The real sub-frame embedded in this east-facing wall is experiencing conditions somewhere between Cases #11 and #13.

Cases #3 and #4 present relative humidity at the window sub-frame for a west-facing wall with a painted rendered finish where a conventional IWI system has been used, a 70 mm phenolic foam insulation board sheathed in vapour retarders, achieving a U-value of 0.27 W/m^{2}K at W_{80}. Despite the insulation being non-hygroscopic and the use of vapour retarders, the sub-frame is below the 20 M-% for Concrete B and just above that threshold for Concrete A. This indicates that the primary stressors in this location in the wall are not conditions in, or near, the room but the impact of the driving rain.

Cases #05–10 present relative humidity at the window sub-frame for a west-facing wall with a painted rendered finish and three thicknesses of the wood fibre OSB composite IWI system wall (with U-values of 0.728 W/m^{2}K at W_{80} for 40 mm wood fibre, 0.549 for 60 mm wood fibre and 0.44 for 80 mm wood fibre). Case #08 (Concrete B) featuring 40 mm wood fibre looks like Case #4 (Concrete B) featuring 70 mm phenolic foam and vapour retarders. Cases #09 and #10 (Concrete B) breach the 20 M-% risk limit but reach equilibrium in the fifth year and are well below the WTA temperature-dependent limit curve for August. Cases #06 and #07 featuring Concrete A and 60 or 80 mm wood fibre are more concerning. They both breach the WTA limit curve and are close to, but have not reached, dynamic equilibrium. This is a case where it would be useful to have the temperature-dependent limit curve for oak. Nonetheless, it shows that if the real concrete were like Concrete A, the timber sub-frame would be facing some level of moisture stress. The real sub-frame embedded in this west-facing wall is experiencing conditions somewhere between Cases #07 and #10.

The bottom right graphs explore risk further for the as-built west-facing wall. In Cases #15 and #16, the stressor of 1 per cent driving rain is removed. When compared with Cases #07 and #10, this greatly reduces stress by almost 5 per cent relative humidity annually in both cases. The value of careful application and maintenance of seals or other water ingress routes is clear.

In Cases #17 and #18, the paint finish is removed, exposing the capillary-active render and driving rain is 'turned off' in the model, although humidity generated from rainfall remains. Dynamic equilibrium is reached much quicker for Case #18 (Concrete B) and it sees a modest lowering of annual peaks and a significant improvement in drying capacity. Without the paint finish, Case #17 (Concrete A) absorbs vapour quicker and so reaches maximum values sooner. In both cases, the curve cumulatively shifts by a few months over several years.

Finally, in Cases #19 (Concrete A) and #20 (Concrete B), driving rain is restored but paint remains omitted. The results are the least acceptable cases in the sub-frame for either concrete type. Case #20 (Concrete B) is now performing like Case #06 (Concrete A), quickly growing to meet and slightly exceed the WTA limit curve. The sine curve of Case #19 (Concrete A) has jumped ~6 per cent above the equivalent Case #07 and is touching the generally accepted fibre saturation point for oak. This is the only case that is a clear failure.

14.4.3.6 Hygrothermal risk conclusions

If the real wall's concrete is more like the modelled Concrete B, the timber sub-frame is likely to be well below the WTA limit curve for softwood and, therefore, for hardwood too. If the actual concrete is the same as modelled Concrete A conditions, it becomes important to establish what the WTA limit curve for oak is. This research and supporting guidance suddenly take on a personal significance.

The retrofit condition that imposes the lowest risk on the oak sub-frame is the case where a ventilated rainscreen is used with internal wall insulation. This is substantially lower risk than the conditions where all driving rain is omitted. It is likely that ventilation in the rainscreen void enhances evaporation, giving added value beyond the removal of liquid water delivered via wind-driven rain. A further improvement would be to either externally insulate or internally *and* externally insulate, while maintaining the ventilated rainscreen. While this may seem costly, a non-proprietary mineral batt insulation and treated timber battens, taking a shiplap timber or cement board rainscreen, would be economic. This was what had been originally intended in the case study, but structural concerns prevented it.

The divergence found in the cases assessed makes clear the value of a bracketing approach for key materials in parametric assessment. Consider a scenario where only Concrete A or B, or worse still, a modern concrete, had been chosen. What would the assessor have learnt and what false certainty might have underpinned the poor guidance then given? While the state of the art has moved forward due to the range of historic Irish concrete data becoming available in 2026, it is hard not to conclude that key hygrothermal characteristics of historic concrete walls that are to be internally insulated should be measured.

Render is designed to protect walls and manage the uptake and removal of moisture. It is particularly striking, considering Cases #19 and #20, to see how much a paint finish can benefit a rendered wall. As paint manufacturers appear to be very reluctant to share data on their products' hygrothermal performance, some useful research could be undertaken.

14.5 Carbon and life cycle impacts

A formal life cycle assessment (LCA) of the case study cottage was carried out in accordance with EN 15978:2011 [36], concluding in May 2024. The assessment was carried out using One Click LCA software. The software complies with over 30 certifications and standards for Life Cycle Assessment and Life Cycle Costing, including all versions of LEED and BREEAM. The software includes curated and verified global and local databases. The quantities and volumes were taken from the *as-built* BIM model that also supported PHPP modelling.

The life cycle stages included in this assessment are indicated (by an 'X') in Figure 14.17. Stages B1–B3 (use, maintenance and repair) are excluded due to a lack of data on the One Click LCA platform and lack of a standardised methodology for Stages B2–B3. 'Stage D Benefits and Loads beyond the System Boundary' is reported in the assessment but excluded from the results, as per EN-15978. Stages A1–C4 focus on new rear extension and retrofit works to the existing retained structures, while Stages B1–C4 focus on the existing retained structures. As set out in EN 15978:2011, the system boundaries for Module B5 (Refurbishment) include several key elements:

1) Production of new building components.
2) Transportation of these components, accounting for any material losses during transport.
3) Construction activities associated with refurbishment, including the production of materials lost in the process.
4) Waste management generated during the refurbishment process.
5) End-of-life treatment for the replaced building components.

Product stage			Construction process stage		Use stage							End-of-life stage				Benefits and loads beyond the systems boundary		
Raw material supply	Transport	Manufacturing	Transport to building site	Installation into building	Use/application	Maintenance	Repair	Replacement	Refurbishment	Operational energy use	Operational water use	Deconstruction/demolition	Transport	Waste processing	Disposal	Reuse	Recovery	Recycling
A1	A2	A3	A4	A5	B1	B2	B3	B4	B5	B6	B7	C1	C2	C3	C4	D	D	D
	X		X	X				X	X	X	X		X				X	

Figure 14.17 Stages assessed in the LCA of case study

The renovation LCA stage and boundary diagram in Figure 14.18 [37] makes clear the additional complexity of assessing a retrofitted building. The strip-out, the new inclusions, how the continued existing elements are considered, and how end of life might look, are all crucial.

14.5.1 LCA results: mitigation and missed opportunities

The life cycle assessment results are summarised in Table 14.7. While all seven indicators were assessed, the primary focus was on Global Warming Potential (GWP, $kgCO_2e$) and Biogenic Carbon Storage ($kgCO_2e$). All building fabric and fitout were included. Energy and water use were included. The materials used for services, such as cabling, ducting, and drainage, were excluded.

The results represent the total life cycle impact per sqm over a 60-year service life for each impact category. GWP is the most common impact category covered by LCA. GWP describes changes in local, regional, or global surface temperatures caused by an increased concentration of GHGs in the atmosphere and is commonly referred to as the 'carbon footprint'. Table 14.7 shows the total GWP for the cottage is 225 tonnes of CO_2e, based on a whole-life calculation over 60 years. This equates to 98.8 $kgCO_2e/m^2$ per year and a total social carbon cost of €11,234 based on €50 per tonne of carbon.

Carbon dioxide emissions cannot be avoided but they can be mitigated. Overall, findings from the study demonstrate the value of the low-carbon and circular approach taken with the cottage renovation on mitigating production (A1–3) emissions, which is the dominant embodied carbon (EC) stage. The comparative analysis showed that, based on just a limited number of low-carbon and circular measures, the benefits of the approach taken over a 'mainstream' approach are clear. Findings from the study also demonstrate the following.

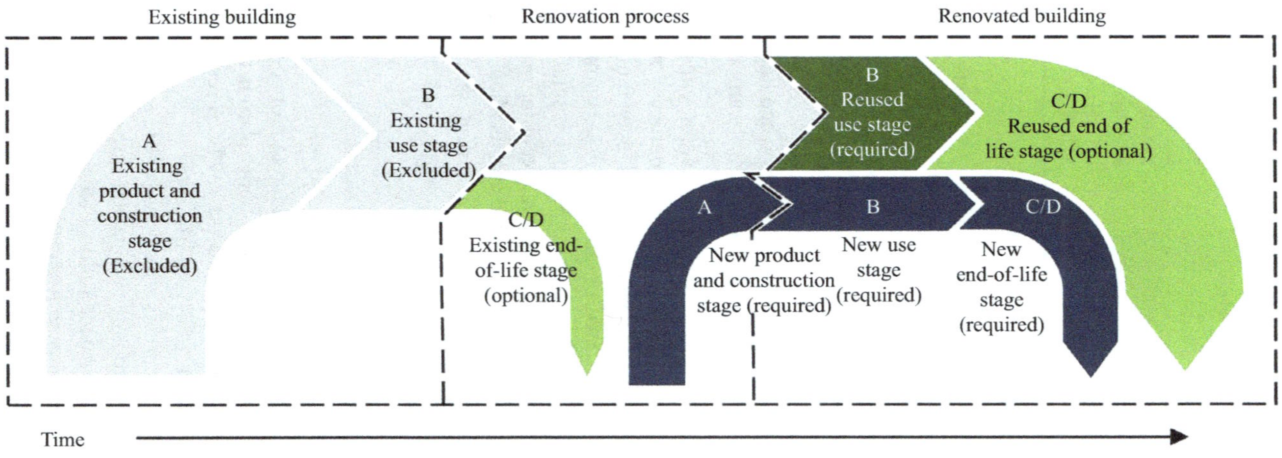

Figure 14.18 Renovation LCA stage and boundary diagram (courtesy of Hasik, [37])

Table 14.7 Total life cycle assessment results over 60 years per sqm

Impact category	Unit	Results
Global warming potential (greenhouse gases)	$kgCO_2eq$	224.7k
Biogenic carbon storage	$kgCO_2eq$ bio	9.2k
Ozone depletion potential	$kgCFC_{11}eq$	0.02
Acidification potential	$kgSO_2eq$	885
Eutrophication potential	$kgPO_4eq$	179.3
Formation of ozone in lower atmosphere	kgC_2H_4eq	37.8
Primary energy	MJ	4,002.7k

- The impact that short service life spans have on increasing replacement and refurbishment (B4–5) emissions. Therefore, for material longevity, avoiding or reducing these emissions is important.
- That materials should be retained and reused wherever possible and, while their greatest value is in avoided production (A1–3) emissions, construction (A4–5) emissions can be avoided too.
- That biogenic carbon storage potential increases when bio-based materials are used, which can have significant potential to offset carbon emissions.

While considerable effort was made to reduce the carbon emissions of the cottage renovation, it was not possible to take all opportunities. For example, for contractual reasons, the existing roof slats were replaced, even though 50 per cent were viable for reuse, resulting in increased production (A1–3) emissions. To facilitate the work practice of the installer of the blown cellulose insulation, a German paper cellulose was used instead of an Irish alternative, resulting in increased transport (A4) emissions. Finally, in uncomfortable conversations about reducing the size of the concrete foundations, the use of 50 or 70 per cent ground granulated blast furnace slag (GGBS) in the cement mix of the concrete was omitted.

14.5.2 A low-carbon and circular approach

A low-carbon, circular approach was taken in the energetic retrofit of the case study building in terms of material specification and retention of existing structures wherever possible. A key objective of the LCA undertaken is to assess the value of that approach. The following sections summarise key LCA results focused on GWP and biogenic carbon storage over a 60-year life cycle.

In general, LCA results for carbon emissions by life cycle stage show that the operational energy stage (B6) dominates whole-life carbon (WLC), followed by the production stage (A1–3). This project follows suit. Table 14.8 shows GWP and biogenic carbon storage results at each life cycle stage. Operational energy (B6) at 209,729 $kgCO_2e$ accounts for 93 per cent of WLC, shown by the value outlined in red in Table 14.8. This is a significant proportion attributable to several factors, including:

Table 14.8 GWP and biogenic carbon storage results

	Result category	Global warming kg CO_2eq	Biogenic carbon storage kg CO_2eq bio	Ozone depletion kg CFC_{11}e	Acidification kg SO_2eq	Eutrophication kg Po_4eq	Formation of ozone of lower atmosphere kg C2H4eq	Total use of primary energy ex. raw materials MJ
A1-A3	Construction Materials	7,375.22	9,239.18	0.00	35.59	8.19	3.08	204,481.69
A4	Transportation to site	520.88	0.00	0.00	2.33	0.49	0.04	13,738.82
A5	Construction / installation process	1,838.36		0.00	7.48	3.32	0.45	47,328.41
B1	Use phase							
B3	Repair	0.00		0.00	0.00	0.00	0.00	0.00
B4-B5	Material replacement and refurbishment	3,135.41		0.00	19.40	4.40	1.75	82,191.74
B6	Energy consumption	209,729.10		0.02	814.50	161.42	32.26	3,627,331.76
B7	Water use	2.50		0.00	0.02	0.00	0.00	41.06
C1-C4	End of life	2,078.65		0.00	5.67	1.49	0.19	27,300.28
D	External impacts (not included in totals)	−13807.81		0.00	3.57	0.71	0.15	12,758.76
	Total	**224,679.91**	**9,239.18**	**0.02**	**884.98**	**179.31**	**37.78**	**4,002,706.39**
	Results per denominator							
	Per gross internal floor area m^2/year	98.80	4.06	0.00	0.39	0.08	0.02	1,760.21
	Per gross internal floor area m^2	5,928.23	243.78	0.00	23.35	4.73	1.00	105,612.31

- The low-carbon and circular approach taken with the retrofit works has, as intended, reduced the associated levels of EC significantly. As a result, the impact of B6 appears *proportionately greater* than one would see if more common, higher carbon materials had been specified.
- No renewable energy sources are included, which would otherwise reduce operational energy (B6) and its proportion of whole-life carbon.
- Operational energy emissions (B6) are based on a Passive House Planning Package (PHPP) delivered energy value, which is approximately twice that of the Dwelling Energy Assessment Procedure (DEAP) equivalent value. If the DEAP value were used, the proportion of operational energy of WLC would be ~88 per cent.

Biogenic carbon storage accounts for the potential of any bio-based material, such as wood, to store carbon. When used in construction, these materials will be replaced with new planting, if harvested from sustainable forests, ensuring future carbon removal from the atmosphere as the next trees grow. The specification of bio-based materials, such as wood fibre insulation, cellulose, timber stud work, and doors, means the project has a high level of 'biogenic carbon storage', equating to ~9.2 tonnes of negative kgCO_2e. In the context of the renovation, this is a significant value. It more than offsets all production (A1–3) emissions and accounts for 62 per cent of embodied carbon.

The left image in Figure 14.19 shows GWP results by life cycle stage, excluding operational carbon (B6–7). While the production stage (A1–3) dominates embodied emissions at ~50 per cent overall, this value has been mitigated through the low-carbon material specification and retention of existing structures. The replacement and refurbishment stages (B4–5) also contribute significantly to embodied emissions. This is primarily due to the short service life of high-carbon materials and products that will need to be replaced within the 60-year life cycle of the cottage. Examples of such materials include the roof membrane (30 years – 1 replacement), windows and doors (40 years – 1 replacement) and the floor lacquer (10 years – 5 replacements).

Retention and reuse of existing structures, wherever possible, is central to the circular approach taken in the case study project. The right image in Figure 14.19

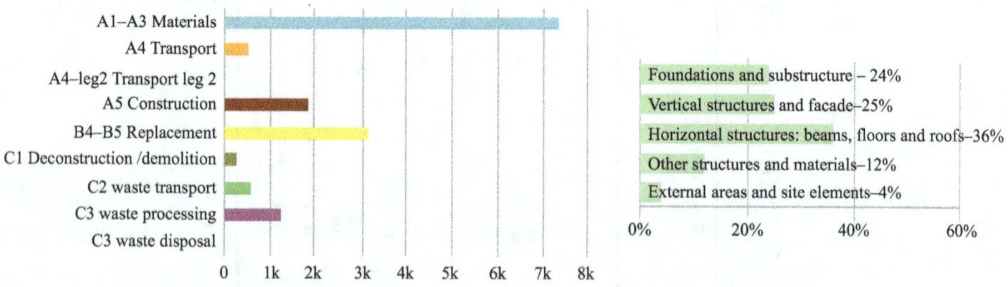

Figure 14.19 (left) GWP by life cycle stage, (right) embodied carbon by structure, stages A1–3

shows production (A1–3) emissions for the renovation by structure. The avoided production emissions from retaining and reusing the existing concrete walls mean that emissions for 'vertical structures' are relatively low in comparison to 'horizontal structures' despite the large external wall area of the cottage relative to its footprint.

14.5.3 Comparison to a 'mainstream' approach

The first step in reducing the environmental impacts in the built environment is to establish benchmarks [38]. However, methodological differences in the application of LCA, including energy consumption methodologies, system boundaries, scope, and differing datasets, make benchmarking and result comparison between studies challenging. Therefore, to better understand the value of the low-carbon and circular approach taken with the cottage renovation and provide context for the results, a partial comparative analysis was undertaken based on just three changes to the as-built renovation, described as a 'mainstream' approach, as follows:

- Cellulose and recycled cotton insulation materials were replaced with mineral wool.
- Wood fibre insulation boards were replaced with polyisocyanurate (PIR).
- The retained and reused concrete walls and footings were rebuilt with blockwork.

Figure 14.20 shows GWP and biogenic carbon storage results for all life cycle stages, excluding operational carbon (B6–7), for the as-built renovation and the 'mainstream' approach. The low-carbon and bio-based insulation materials, combined with the reuse of the concrete walls and footings, mean that the as-built renovation avoids more than half the carbon emissions (GWP) of the 'mainstream' approach, while also providing ~12 per cent more biogenic carbon storage.

Production (A1–3) and construction (A4–5) were the only stages affected by these illustrative changes. Figure 14.20 shows that:

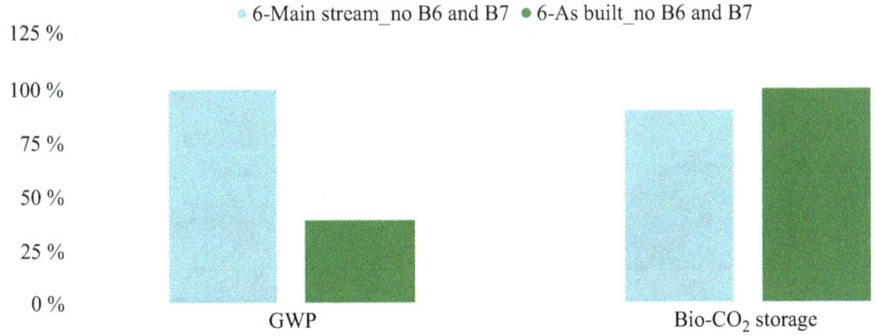

Figure 14.20 As-built construction and mainstream comparison, GWP and biogenic carbon storage for all stages, excluding B6 and B7

- The 'mainstream' approach would produce approximately 3.3 times more carbon emissions than the as-built renovation.
- The greatest saving is in the production stage (A1–3) due to the avoided production impacts of the retained concrete walls and footings and the low-carbon insulation materials used.
- Savings in transport (A4) are due to the reuse of the retained concrete walls and footings, which require no transport as they are already 'on site'.
- Savings in construction (A5) are due to the avoided construction impacts for the concrete walls and footings as these structures are already built.

14.5.4 Benchmarking embodied carbon and the Carbon Law

One of the outputs from One Click LCA is the bar chart for the Global Warming Potential (GWP) category, shown in Figure 14.21. It can be seen that the calculated WLC value of 225 $kgCO_2e/m^2$ for 'cradle to grave' emissions (A1–A5, B4–B7, C1–C4) gives this project a rating of 'A' by the standards prevalent in Western Europe in Quarter 3 of 2021. However, that does not make clear how the case study compares to the GWP of other retrofit projects, nor the extent to which the case study is aligned with the ambitions of the World Green Building Council, or the EU-27, to fully decarbonise the built environment by 2050.

There is a lack of data on all building types. For this reason, the IGBC received funding for the Indicate Project [39] to '*'increase the quality, quantity, and*

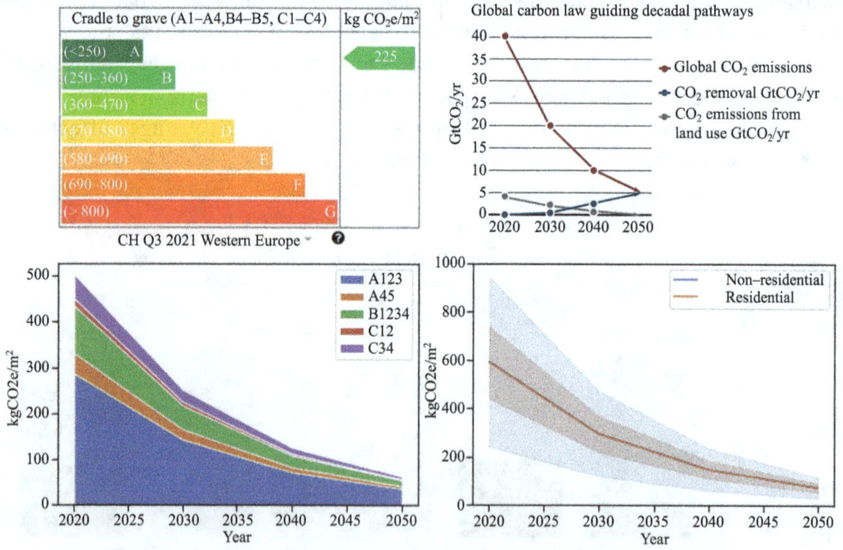

Figure 14.21 Benchmarking the carbon reductions achieved: (clockwise from top left) a bar chart from the case study assessment in One Click LCA; Rockström et al.'s Carbon Law; Röck et al.'s indicative embodied carbon reduction targets by decade and (at bottom left) their targets by life cycle stages and value (based on the Carbon Law)

application of building-level LCA data' with a primary focus on upfront embodied carbon (Stages A1–A5). The embodied carbon of fifty-two buildings was measured, representing a range of typologies, with a 50-year service life in all cases. The numbers in some of those groups are very small. For instance, there are only two retrofits of buildings but neither is a dwelling. There are three terraced houses assessed but all are new builds. Missing data or excluded stages, a common problem when compiling LCA data, also made comparisons and average values harder to establish. Nonetheless, this is an important start for Ireland and will support creation of benchmark values for the upcoming national LCA methodology. No doubt the data supporting the benchmarks will be expanded in the future.

The average treated floor area of Indicate's newly built terrace houses is 108.5 m^2. The average WLC is 81 tonnes of CO_2e, or 747 kgCO_2e/m^2. The average upfront EC (A1–A5) is 472.48 kgCO^2e/m^2. Röck *et al.* [40] measured the average WLC of new-built terrace houses as 400 kgCO_2e/m^2, using a 50-year service life. As such, they appear to significantly outperform the Irish terraced houses. On a brief review, the most relevant Irish project for which EC was measured is the nZEB retrofit of eight single-occupant terraced bungalows in Co. Wexford [41]. Similar to the case study cottage, the bungalows initially had very poor energy performance. Post-works, the average WLC value was calculated at 276 kgCO_2e/m^2, including PV and a heat pump. However, it is not clear what service life was allowed for. The WLC appears similar to the value achieved for the case study cottage, but this may be a coincidence if the boundary conditions are different. For now, it may not be possible to say how the case study compares to other Irish energetic retrofits. To aid future comparisons and contribute to the national dataset, the sharing of detailed case studies is essential.

To assess the extent to which the case study aligns with the ambitions of the World Green Building Council or the EU-27, it is useful to turn to the 'Carbon Law' first posited by Rockström *et al.* [42]. They proposed:

> '... *Framing the decarbonization challenge in terms of a global decadal roadmap based on a simple heuristic, a "carbon law", of halving gross anthropogenic carbon-dioxide (CO_2) emissions every decade. Complemented by immediately instigated, scalable carbon removal and efforts to ramp down land-use CO_2 emissions, this can lead to net-zero emissions around mid-century, a path necessary to limit warming to well below 2 °C ... An explicit carbon roadmap for halving anthropogenic emissions every decade, codesigned by and for all industry sectors, could help promote disruptive, nonlinear technological advances toward a zero-emissions world.*' [42]

Perhaps, what is most powerful about the Carbon Law is its simplicity in positing a loose structure for a dramatic change. The top right graph in Figure 14.21 was published in Rockström's paper. Given that CO_2 remains in the atmosphere for decades after emission, the graph appropriately shows that the steepest reduction must be made in the first of the three decades.

Röck *et al.* applied the Carbon Law to Europe's built environment [40]. They extensively reviewed the reported GWP, for both embodied and operational carbon, of diverse building types built between 2000 and 2020 to form a benchmark, harmonising the values to allow comparison with a notional 50-year life cycle. They then structured Rockström's graph line as the central line of a funnel, the breadth of which reflects the diverse GWP impacts of building types and how these must be reduced, shown in the bottom right graph in Figure 14.21. It is tempting to locate the value of 225 kgCO2e/m^2 achieved in 2023 on this funnel, near the bottom of the grey (non-residential) portion. However, the LCA of the cottage uses a 60-year, not 50-year, service life. Dividing it by 60 and multiplying by 50 years, resulting in 187.5 kgCO_2e/m^2, is a crude way to enable the comparison, though it ignores service life impacts. Nonetheless, the translated WLC value for the case study, once known, would likely be positioned in an *exemplary* position near the bottom of Röck *et al.*'s Carbon Law funnel.

Röck *et al.* took their funnel further by overlaying a notional decarbonising of each life cycle stage on their funnel graph (bottom left in Figure 14.21). The deepest and most urgent drop in carbon must be for Stages 1–3, materials supply, transport, and manufacturing, a change that can only happen if we build radically less, energetically retrofit far more, manufacture building materials with renewable energy, and greatly expand use of low-carbon and biogenic building materials.

> *'While this application of the carbon law approach is a simple exercise to give an indication of the drastic reduction of embodied carbon emissions from buildings, it provides a hint on the required trajectory. The definition of more specific, relative reduction targets is required and underway, but implementation of ambitious carbon reduction strategies must not wait another decade for methodological discussion to move forward and allocation issues to be settled.'* [40]

The choice to retrofit the case study building, rather than substantially rebuild, while doing so in a way that maximises the use of biogenic materials and post-industrial and post-consumer recycled materials, appears to align with this focus at Stages A1–3 also.

14.6 Does the case study project meet the challenges set?

In this next section, we consider whether the project meets the challenges that the team set out to achieve. These challenges include the CarbonLite Retrofit Standard, the RIAI Climate Challenge, and the World Green Buildings Council's Carbon Challenge. This section identifies the complexity of addressing the multiple parameters of the sustainable buildings challenges, while also addressing key elements of building performance.

14.6.1 The challenge of the CarbonLite Retrofit Standard

The case study is assessed against the operational energy challenge of the CarbonLite Retrofit Standard. Table 14.9 shows it achieves mixed results. Requirements for ventilation, airtightness, thermal bridging, and thermal comfort

Table 14.9 Analysis of compliance with CarbonLite Retrofit requirements

	Carbonlite retrofit requirements	Case study cottage	Judgement
Building Energy Modelling tool	PHPP	Yes	Yes
Space heating demand - kilowatt hours per m^2 per year	$\leq$50 kWh/m^2.a ($\leq$100 kWh/m^2.a with certifier-approved exemption)	89	Compliant, if certifier-approved exemption allowed
Delivered space cooling (kWh/m^2.a)	$\leq$15 kWh/m^2.a	n/a	
Either Primary Energy (PE, varies) or Renewable (PER) (kWh/m^2.a)	Report result	Future install of solar PV has been facilitated	Compliant, once planned PV array is installed in phase two
Windows	Not stated	$W_{W/D\ installed}$: 0.98 W/m^2.K $W_{W\ installed}$: 1.21 W/m^2.K Glazing (g-value): 0.49	EnerPHit compliant
Ventilation	DC-MEV or MVHR	DC-MEV, validated	Compliant
Air permeability (Q_{50}), airtightness (N_{50})	$\leq$2.0 m^3/m^2.hr (Q_{50})	1.51 m^3/m^2.hr	Compliant
Thermal Bridges	Assumed to be less than 0.01 W/mK	Modelled	Compliant. (Values are mostly negative psi-values when measured in accordance with PH Standard)
Surface condensation (f_{Rsi}) assessed	f_{Rsi} to meet criteria in PHPP, or 0.75 (as Building Regulations/PAS 2035)	Modelled	Mixed, but not contentious
Heating System	Existing heating systems may be retained, but a practical plan to allow for future low carbon heating supply must be in place	Wall-mounted direct electric heaters. Compact MVHR is considered as a Phase 2 alongside solar PV	**Not acceptable**
Thermal Comfort	PHPP modelled overheating risk <10 per cent acceptable	0 per cent	Compliant

are met. While most values achieved for thermal bridging (ψ) and surface risk of condensation (f_{Rsi}) are impressive, some do not meet the threshold values listed. Space heating demand is between the target of ≤ 50 kWh/m^2.a and the maximum value of 100 kWh/m^2.a, requiring an exemption to pass. However, given the high U-values for the internally insulated walls, the AECB certifiers' expert team, which manages the standard, is sympathetic in this regard. The sticking point turned out to be the direct electric energy supply.

The author discussed the project with the AECB certifiers' expert team and then identified a consultant to undertake a review of the project with a view towards certification. This led to internal conversations in February 2025. The following is an email extract.

> *'We've had two meetings of the AECB certifiers expert team now and the conclusion is that the standard does not allow for direct electric heating in dwellings such as yours. Only in multi-dwelling blocks of flats.*
>
> *There is an acknowledgement that it is common sense and maybe a conversation will start about some kind of cut-off of size for very small dwellings, but for now this is clearly the position.'*

In conclusion, then, the project does not meet the challenge of the CarbonLite Retrofit Standard as constituted but might do so in the future, if the AECB considers a wider range of energy supply approaches for very small dwellings. It is ironic that the grant now available for small heat pumps in Ireland, double what was available at the time, would likely have encouraged the installation and thus opened the door to certification to this standard. A future Phase 2 will include the installation of solar panels and may include the change from direct electric heating to a heat pump.

14.6.2 The RIAI Climate Challenge

The case study is then assessed against the RIAI 2030 Climate Challenge. Has the case study building met their three key challenges – a 33 per cent reduction in operational energy, a 40 per cent reduction in embodied carbon, and a 24 per cent reduction in potable water? In 2021, the RIAI's Sustainability Taskforce developed target metrics for the years 2021, 2025, and 2030 to show compliance [43]. These are listed in Table 14.10, alongside values achieved by the case study and a concluding judgement.

The case study cottage surpasses many targets significantly in terms of embodied carbon, potable water, overheating, and radon. It is unclear about volatile organic compounds (VOCs) and appears to fail the operational energy targets for dwellings, which include plug loads, despite having a measured usage 44 per cent lower than its current BER of C2. The extent of the overperformance of the embodied carbon figure and the underperformance of the operational energy figure suggest that the RIAI Climate Challenge currently relates only to new-build dwellings. If so, it would be useful for the RIAI to create a retrofit category. Without a clear category and appropriate challenge, the retrofit of dwellings is inadequately judged and may receive less attention. The RIAI could also define

Table 14.10 RIAI 2030 Climate Challenge target metrics

Sustainable outcome metrics	***Current benchmarks (Part L 2019)***	***2025 targets***	***2030 targets***	**2023 case study: values achieved**	**Conclusion**
Domestic buildings					
Operational Energy ($kWh/m^2/y$)	*90 (incl. plug loads)*	*60 (incl. plug loads)*	*35 (incl. plug loads)*	101 (based on first 12 months of recorded energy use, see Section 7.2)	Despite being 44 per cent lower than BER, value recorded is higher than targets given
Embodied Carbon ($kg\ CO_2e/m^2$)	*1200*	*<800*	*<625* *(A1-A5 < 400)* *Higher target* *<450*	225 over 60 years (as per Table 4.7)	Significantly surpasses all target values
Potable water (1/p/day)	*None required. DEAP credit where <125*	*<95*	*(A1-A5 < 300)* *<75*	77.5 (using DEAP Water-Efficiency-Calculator)	Exceeds 2025 value. Close to 2030 value
All buildings					
Overheating Daylighting	*25–28 C maximum for 1 per cent of occupied hours* *>2% av. daylight factor, 0.4 uniformity ratio*			0 per cent (as calculated in PHPP) 1.92 % DF	Surpasses target value Just below guided figure
CO_2 levels	*<900 ppm*			Unknown. $C0_2$ sensor in place	Unknown, but controls in place
Total VOCs	*<0.3 mg/m³*			Unknown	Unknown
Formaldehyde	*<0.1 mg/m³*			Unknown	Unknown
Radon (bq/m^3)	*<50*			27	Significantly surpasses target value

how operational energy (kWh/m^2/y) is to be gathered. Is the aggregation of a year's worth of meter readings, divided by floor area, the only acceptable approach?

14.6.3 The World Green Building Council's carbon challenge

How did the case study address foundational principles of the WGBC's Net Zero Carbon Buildings Commitment [44]? These include *prevention* of embodied carbon from the outset, *evaluating each design choice* to reduce and optimise, *planning* to avoid future embodied carbon, and *offsetting* as a last resort. Excluding the last resort of offsetting, the case study presents a series of focused attempts to actively engage with these principles, as evidenced by technical drawings, photographs, calculations, and measurements. Oversights or undershoots have also been disclosed in the spirit of 'no-fault' disclosure and community learning.

As both developer and designer, the author attempted to meet the WGBC's 'ambitious high-level pathways' for 2020 and 2025. These pathways, measures achieved, and conclusions are set out in Table 14.11.

The conclusions reported in Table 14.11 are all positive. Some flexibility can be granted given that the 'developer' is not commercial, nor acting at scale. However, the review prompts further questions. Should such results be audited and by whom? Who else is reporting such endeavours? Has this practice become normal amongst even a few developers and designers? It is hard to imagine that in 2026, *'all designers'* have now *'integrated low embodied carbon design at conceptual design stage'*. The targets for 2030 and 2035 need updating. Publicly sharing data that supports case studies like this is certainly best practice. Besides

Table 14.11 WGBC's high-level pathways for developers and designers [44]

Category of stakeholders	**2020**	**Achieved**	**Conclusion**
All developers by 2020	All developers commit to relevant industry roadmaps and require disclosure of supply chain data for structural elements.	Commitment to roadmaps made (AECB Carbonlite, RIAI, WGBC). Data disclosed & available on TU Dublin's open-access Arrow repository.	Compliance achieved
All developers by 2025	All developers set embodied carbon reduction targets and require mandatory disclosure of supply chain data and track construction site emissions.	Targets stated and met. Otherwise this appears to relate to large scale developers.	Compliance achieved
All designers by 2020	All designers commit to relevant industry roadmaps and have integrated low embodied carbon design at conceptual design stage.	Commitment to roadmaps made, principles followed through.	Compliance achieved
All designers by 2025	All design companies publicly share lifecycle assessment data Design companies propose best practice embodied carbon reduction targets and implement circularity principles	LCA data to be shared on TU Dublin's open-access Arrow repository. Circularity principles adopted	Compliance achieved

adding additional detail and colour to a case study, the data can be used to critique this, or other building projects, to support further research and the much-needed expansion of national and international LCA databases.

14.7 Post-occupancy and actual energy use

With any construction project, the success of the project can largely be determined by its use. In August 2023, the occupants, a young couple with a small boy, moved in. At the time of writing, 3 years later, the family has grown to four and are moving on, having purchased a new dwelling nearby. In this section, we reflect on two elements. Firstly, the retrofitted cottage's value from a user perspective, and secondly, its measured energy in use.

14.7.1 Occupant feedback

The occupants were asked to provide feedback 15 months after they moved in. In terms of the layout and design of the cottage, they identified the following features.

> *'There is lots of natural light in the house. I also really like the high ceilings, they make the rooms feel more spacious. All-in-all it's a very cosy cottage and we have lived here very comfortably. The house exudes a solid and homely feel, and once the main door is closed, it feels wonderfully self-contained.'*

They identified that the storage solutions provided by the house, although having some limitations, provided enough space when they considered what they did and did not need. They identified that the smaller size of the cottage encouraged the use of outdoor space.

In terms of the important issue of thermal comfort, in summer, the property's well-considered ventilation provided a flow of air that helped manage summer thermal comfort. The automated skylight was also highlighted as a key feature to help manage both thermal comfort and indoor air quality. In winter, the thermal comfort within the property was also considered to be easily managed.

> *'Generally, we use the wall heaters on a boost setting for 1 or 2 hours at a time. Our energy bills are averaged at €121 per month, though this might be based on the national average. For context, our highest two-month bill last year was €270, during a particularly cold and wet spell when the heating was used consistently alongside the washer-dryer—an outlier period for sure. The bedrooms can still be a little stuffy, even now with the temperature drop.'*

The management of the heating was interesting, with the occupants choosing manual control over the app-based approach. Overall, the design choices were well regarded. When asked to identify improvements, these were focused very much on practicality and durability, rather than building performance.

'Given what the cottage offers and the space available, it's hard to be critical. If I were to point out one area for improvement, it would be easier access to storage or perhaps additional storage higher up to make it simpler to rotate items as the seasons change. The [remanufactured pine] flooring, while undeniably beautiful, isn't particularly robust, so everyday activities can occasionally cause damage, which can be a bit anxiety-inducing.'

This suggests that the building, given the limitations of size, does perform well when introduced to real occupants living in it for a protracted period.

14.7.2 Energy in-use

While we have considered modelled energy use, actual in-use energy is useful to understand, due to the influence that individuals' needs and behaviours can have. The occupants shared their energy bills for the first 13 months of occupation to enable a comparison to be made between calculated energy use in the BER and actual energy use in the same units, kilowatt hours per square metre per year (kWh/m^2/y). The energy consumed includes unregulated plug loads, including intentional consumer usage, such as clothes washing, cooking, and phone charging, as well as unintended usage, such as standby power. These energy usages are excluded from BER assessments. SEAI estimates that ∼18% of total energy usage in Irish dwellings is typically due to unregulated uses. It would be reasonable to expect that the case study dwelling's measured energy use would be greater than the BER-calculated figure; however, this was not the case.

Figure 14.22 records the kilowatt hours consumed per electricity bill charged. The records graphed are not for uniform durations, due to suppliers changing and four bills being estimated. 3,738 kWh of electricity was used in the 12 months between 26th September 2023 and 25th September 2024.

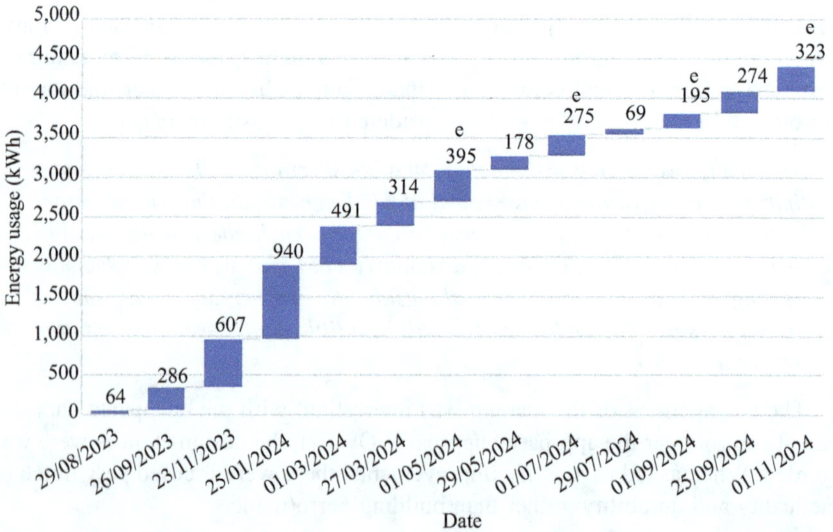

Figure 14.22 Case study electricity use, 28.09.23–01.11.24

If this figure is divided by the treated floor area of 37 sqm, it yields a value of 101 kWh/m^2/y. This is *44 per cent less* than the calculated BER value of C2 = 181.91 kWh/m^2/y. While the contribution of unregulated plug loads to the whole is unknown, the calculation suggests the building fabric and systems are performing far better than BER = C2 suggests.

14.8 Conclusion

14.8.1 A broad performance lens and climate challenges

This case study has examined the deep and circular retrofit of a small historic cottage in Dublin through a deliberately broad performance lens. Rather than defining retrofit success solely by operational energy, the chapter has assessed the project across interrelated domains: thermal performance, airtightness, radon, hygrothermal risk, life-cycle carbon, regulatory modelling, occupant experience, and measured energy use. It demonstrates that ambitious retrofit outcomes are achievable in modest historic dwellings, but also that they require careful negotiation among energy, moisture, carbon, heritage, comfort, and use.

The project largely meets the 2025-oriented challenges associated with the World Green Building Council, the RIAI Climate Challenge, and the CarbonLite Retrofit Standard, though with important qualifications. Most notably, the omission of a heat pump prevented CarbonLite certification under the present standard, despite the dwelling's strong fabric performance and low measured energy use. This suggests that some standards may not yet fully accommodate very small dwellings, phased retrofit strategies, or alternative low-demand approaches.

14.8.2 Systems, gaps in performance and policy, and carbon

A central finding is that retrofit must be understood as a systems problem. Measures that reduce heat loss may also alter moisture behaviour, ventilation effectiveness, radon pathways, and user comfort. The project achieved substantial improvements through fabric insulation, airtightness, and thermal bridge mitigation, while radon levels fell, despite a major reduction in air permeability. However, the hygrothermal analysis of embedded oak window sub-frames shows that apparently modest technical choices can carry complex moisture implications. A responsible retrofit process, therefore, requires more than generic guidance or simplified calculations: it must combine first-principles reasoning, measured evidence, parametric simulation, and ongoing monitoring.

The study also highlights persistent performance gaps in both practice and policy. These gaps occur not only between design intent and construction outcome, but also between modelling tools, regulatory assumptions, and actual use. The divergence between DEAP, PHPP, and measured energy consumption is particularly significant. The official BER for C2 does not adequately convey either the performance of the building fabric or the dwelling's actual energy use. Measured electricity consumption over the first year was 44 per cent lower than the calculated BER value, even though unregulated plug loads were included in the measured data and excluded from BER assessment. This raises questions about the use of BER as a proxy for real performance, as a basis for funding, and as a key metric in national retrofit policy. Asset-rating tools are useful, but their limitations must be clearly understood and communicated. TGD L would be improved by re-weighting its reference to IS EN 15026 and by including guidance from BS 5250.

The life cycle assessment confirms the value of a circular, low-carbon material approach to retrofit. Retaining existing structures, avoiding unnecessary demolition, using bio-based insulation, and specifying recycled or low-impact materials substantially reduced embodied carbon compared with a more conventional 'mainstream' alternative. The project avoided more than half of the embodied carbon emissions associated with that alternative, while also providing significant biogenic carbon storage. These findings reinforce a simple but important principle: the most effective embodied carbon reductions often begin with decisions not to demolish, not to replace, and not to over-specify. However, because embodied impacts were reduced, operational energy became a proportionally more dominant in whole-life carbon. Future installation of photovoltaic generation, and possibly a heat pump, would further strengthen the project's long-term decarbonisation performance.

14.8.3 Making quality outcomes normal

The research further identifies weaknesses in the knowledge infrastructure available to designers. Hygrothermal assessment was constrained by non-availability of climate files based on national climate data, measured hygric data for some historic materials, and temperature-dependent risk thresholds for hardwoods. The later availability of Irish historic concrete data significantly improved the post-construction hygrothermal assessment but it also illustrates the difficulty of making robust design decisions when critical data are unavailable. Where uncertainty remains, the bracketing approach used here offers a useful model: uncertainty is made visible, tested, and cautiously interpreted, rather than concealed.

The project also points to the need for new professional models. Although the cottage is only 37 sqm, its assessment required input from radon testers, airtightness testers, BER assessors, PHPP modellers, thermal bridge modellers, hygrothermal specialists, LCA consultants, LiDAR surveyors, and climate-file specialists (Table 14.12). The small scale of the dwelling did not reduce the

Table 14.12 Key professionals involved in building performance assessment

Radon tester	**David Doyle (Alpha Radon)**
Air permeability tester	Gavin Ó Sé (Greenbuild Services)
LIDAR scanner	Joe Mady (DCT)
BER assessors and DEAP building energy modellers	Michael Hanratty and Sheezan Salman (IHER)
BIM and PHPP building energy modeller	Shane Fenton (WMA)
Thermal bridge modeller	Shane Fenton (WMA)
Hygrothermal risk assessor	Joseph Little
Measurement of historic hygric datasets	TradFabs team, Dr Rosanne Walker, Dr Oliver Kinnane *et al.*
Creation of synthetic HRY climate file	John Morehead (WMA)
Life cycle assessors	Sara Carrigan and Alberta Congeduti

complexity of the risks being managed. If high-quality historic retrofit is to be delivered at scale, the sector will need interdisciplinary teams, stronger specialist networks, and pre-works assessment methods capable of supporting multiple dwellings, streets, or districts. The objective is not to make every modest retrofit exceptional, but to make rigorous assessment and informed decision-making ordinary.

14.8.4 Importance of post-occupancy review and a culture of measured performance

Finally, the occupants' experience confirms that performance must ultimately be judged in use. The dwelling (Figure 14.23), is reported as bright, comfortable, spatially efficient, and economical to run. Measured energy consumption suggests that compact living, improved fabric, effective ventilation, and user control can deliver outcomes that compare favourably with modelled expectations. Post-occupancy evidence should therefore be treated as an essential component of retrofit evaluation, not merely as anecdotal support.

Figure 14.23 Street view of completed cottage at sunset

In conclusion, the project demonstrates that low-carbon, moisture-conscious, and heritage-sensitive retrofit is achievable, even in a small historic dwelling with challenging geometry and material uncertainty. Yet, it also shows that success currently depends on unusual levels of expertise, persistence, disclosure, and assessment. The wider task is to transform these exceptional practices into dependable and scalable norms, supported by better data, improved standards, appropriate modelling tools, strong uptake of transformative technical upskilling, and a culture of measured performance and shared learning.

Acknowledgements and funding

The author is very grateful for the support and patience of this diverse group and many others beyond. Joe Mady of Digital Construction Technologies (DCT) provided LIDAR scanning free of charge. Almost everyone else provided support far beyond any fee paid. In particular, the skill, generosity, and patience of Michael Hanratty, Sara Carrigan, architect, and Shane Fenton and John Morehead, both of Wain Morehead Architects Ltd, WMA, must be noted. The author is also grateful to Ecological Building Systems Ltd, Acara Concepts Ltd, and Acoustic Materials Ltd for sharing the cost of purchasing MSR data loggers that are recording data on temperature and humidity in key locations within the case study's building envelope to support further research.

This case study project sits within Dublin City Council's long-running Built to Last project. Dublin City Council Heritage Office and the Department of Housing, Local Government and Heritage played a key role in enabling this research by part-funding the author's work, the life cycle assessment by Sara Carrigan with Alberta Congeduti, and the building energy modelling using PHPP by Shane Fenton. The author is very grateful for this funding.

References

[1] Dublin City Council. *Built to Last Case Studies* [online]. 2023. Available from http://www.dublincity.ie/builttolast [Accessed 12 March 2026]

[2] Swan W., Fitton R., Brown P. 'A UK practitioner view of domestic energy performance measurement'. *Proceedings of the Institution of Civil Engineers-Engineering Sustainability*. 2015;**168**(3):140–7

[3] Good Homes Alliance. *Performance Testing* [online]. 2011. Available from https://goodhomes.org.uk/performance-improvement [Accessed 12 March 2026]

[4] Bonfield P. *Each Home Counts: An Independent Review of Consumer Advice, Protection, Standards and Enforcement for Energy efficiency and Renewable Energy*. London: Department of Energy and Climate Change; 2016

[5] House of Commons Committee Report (UK). *Retrofitting Homes for Net Zero: First Report of Session 2024-25*. London: UK Parliament; 2025
[6] Department of Housing, Local Government, and Heritage (Ireland). *Technical Guidance Document C - Site Preparation and Resistance to Moisture*. Dublin: Government of Ireland; 2023
[7] International Standards Organisation. *EN ISO 10211:2017 Thermal Bridges in Building Constructions. Heat Flows and Surface Temperatures. Detailed Calculations*. Geneva: ISO; 2017
[8] Building Research Establishment. *BR497 Conventions for Calculating Linear Thermal Transmittance and Temperature Factors*. Watford: UK; 2007
[9] Department of Housing, Local Government, and Heritage (Ireland). *Technical Guidance Document L - Conservation of Fuel and Energy*. Dublin: Government of Ireland; 2022
[10] Little J., Arregi B. *Thermal Bridging - Understanding its Critical Role in Energy Efficiency* [online]. Available from https://arrow.tudublin.ie/cgi/viewcontent.cgi?article=1049&context=bescharcart [Accessed 12 March 2026]
[11] International Standards Organisation. *EN ISO 9972:2015 – Thermal Performance of Buildings – Determination of Air permeability of Buildings – Fan pressurization Method*. Geneva: ISO; 2015
[12] Ahern C., Norton B. 'Thermal energy refurbishment status of the Irish housing stock'. *Energy and Buildings*. 2019;**202**(1):109348
[13] Coyne B., Denny E. 'Mind the energy performance gap: Testing the accuracy of building Energy Performance Certificates in Ireland'. *Energy Efficiency*. 2021;14:57
[14] Gledhill T., Swan W., Fitton R. 'A practitioner study into the variability of UK domestic energy assessments'. *International Journal of Building Pathology and Adaptation*. 2023;**43**(5):1161–78
[15] Chartered Institute of Building. *Passivhaus vs SAP* [online]. 2023. Available from https://www.designingbuildings.co.uk/wiki/Passivhaus_vs_SAP [Accessed 12 March 2026]
[16] Little J., Ferraro C., Arregi B. *Technical Paper 15 - Assessing Risks in Insulation Retrofits Using Hygrothermal Software Tools: Heat and Moisture Transport in Internally Insulated Stone Walls*. Edinburgh: Historic Environment Scotland; 2015
[17] International Standards Organisation. *EN ISO 13788 (2012): Hygrothermal Performance of Building Components and Building Elements – Internal Surface Temperature to Avoid Critical Surface Humidity and Interstitial Condensation – Calculation Methods*. Geneva: ISO; 2012
[18] British Standards. *BS EN 15026:2023. Hygrothermal Performance of Building Components and Building Elements – Assessment of Moisture Transfer by Numerical Simulation*. London: BSI; 2023
[19] British Standards. *BS5250 (2021). The Code of Practice for Control of Moisture in Buildings*. London: BSI; 2021

[20] May N., Sanders C. *Moisture in Buildings: An Integrated Approach to Risk Assessment and Guidance*. London: BSI; 2017

[21] Department of Housing, Local Government, and Heritage (Ireland). *Section B3 Technical Guidance Document L - Conservation of Fuel and Energy*. Dublin: Government of Ireland; 2022

[22] S.R. 54:2014/A1:2019. *Code of Practice - Methodology for the Energy Efficient Retrofit of Existing Domestic Dwellings*. Dublin: NSAI; 2019

[23] Duarte R., Flores-Colen I., De Brito J., Hawreen A. 'Variability of in-situ testing in wall coating systems-Karsten tube and moisture meter techniques'. *Journal of Building Engineering*. 2020;27:100998

[24] Meteonorm. *Meteonorm* [online]. 2026. Available from https://meteonorm.com/en/ [Accessed 12 March 2026]

[25] Yu H., Zhang H. 'A new method for predicting wind-driven rain catch ratios on building facades in urban residential areas using machine learning models'. *Building and Environment*. 2025;270:112467

[26] Standard AS. *Standard 160–2021: Criteria for Moisture Control Design Analysis in Buildings*. Atlanta: American Society of Heating, Refrigerating and Air-Conditioning Engineers; 2021

[27] WUFI. *Dialog: Edit Hygrothermal Sources* [online]. Available from https://www.wufi-wiki.com/mediawiki/index.php?title=1D:Dialog_EditHygrothermalSources [Accessed 12 March 2026]

[28] Berry S.L., Roderick M.L. 'Plant–water relations and the fibre saturation point'. *New Phytologist*. 2005;168:25–37

[29] British Standards. *BS EN 942:2007. Timber in Joinery. General Requirements.* London: BSI; 2007

[30] British Standards. *BS EN 336:2013. Structural Timber - Sizes, Permitted Deviations*. London: BSI; 2013

[31] Trada Technology. *Moisture in Timber*. Wood Information Sheet. High Wycombe: Trada; 2011

[32] American Society of Heating, Refrigerating and Air-Conditioning Engineers. *ASHRAE Handbook: Fundamentals*. Atlanta: ASHRAE; 2021

[33] Deutsches Institut für Normung e.V. *DIN 68800-2:2012-02. Wood Preservation - Part 2: Preventive Constructional Measures in Buildings*. Berlin: DIN; 2012

[34] WTA. *Wissenschaftlich-Technische Arbeitsgemeinschaft für Bauwerkserhaltung und Denkmalpflege* [online]. Available from https://www.wta-international.org/de/ [Accessed 12 March 2026]

[35] Wissenschaftlich-Technische Arbeitsgemeinschaft für Bauwerkserhaltung und Denkmalpflege. *Merkblatt 6-8 - Assessment of Humidity in Timber Constructions – Simplified Verifications and Simulation.* Munich: WTA Publications; 2016

[36] European Committee for Standardisation. *EN 15978:2011 Sustainability of Constructions—Environmental Product Declarations—Core Rules for the Product Category of Construction Products*. Brussels: CEN; 2011

[37] Hasik V. 'Comparative whole-building life cycle assessment of renovation and new construction'. *Building and Environment*. 2019;161:106218

[38] Simonen K., Rodriguez B.X., De Wolf C. 'Benchmarking the embodied carbon of buildings'. *Technology Architecture and Design*. 2017;1(2):208–18

[39] Indicate. *Indicate (2024)* [online]. 2024. Available from https://www.indicatedata.com [Accessed 12 March 2026]

[40] Röck M., Allacker K., Auinger M., *et al.* 'Towards indicative baseline and decarbonization pathways for embodied life cycle GHG emissions of buildings across Europe'. *IOP Conference Series: Earth and Environmental Science*. 2022;1078(1):12055

[41] O'Hegarty R., Colclough S., Kinnane O., Lennon D., Rieux E. 'Operational and embodied energy analysis of 8 single-occupant dwellings retrofit to nZEB'. *Civil Engineering Research Association of Ireland (Ed.), Civil Engineering Research in Ireland2020: Conference Proceedings*; Cork, Ireland, Aug 2020. Dublin: CERAI, 2020:383–8

[42] Rockström J., Gaffney O., Rogelj J., Meinshausen M., Nakicenovic N., Schellnhuber H.J. 'A roadmap for rapid decarbonization'. *Science*. 2017;355 (6331):1269–71

[43] Royal Institute of Architects of Ireland. *2030 Climate Challenge (2021)* [online]. 2021. Available from https://www.riai.ie/whats-on/news/riai-sets-climate-challenge-for-architects [Accessed 12 March 2026]

[44] World Green Building Council (UK). *Bringing Embodied Carbon Upfront*. London: WGBC; 2019

Chapter 15

The materiality of housing insulation retrofit and implications for policy and practice

Hector Manuel Padilla Gil[1], *Trivess Moore*[2], *Niki Willand*[2] *and Ralph Horne*[3]

15.1 Introduction

Residential retrofit is taken here to mean the necessary and urgent task of reducing fossil energy demand of homes by making changes to the building fabric, design, and appliances. To date, support and delivery of retrofits have typically been framed through a techno-economic lens, where quantified thermal performance (actual and/or modelled) and cost-benefit analyses are central considerations. This framing has resulted in economic and policy approaches that focus on markets, costs, and activity choices. However, the slow uptake of retrofits suggests that other important dimensions of residential retrofit may be overlooked or ignored, which may mean opportunities for scaling up retrofit are being missed. This chapter explores materiality dimensions of retrofits. By materiality, we mean the aspects of retrofit practice that are shaped by the physical appearance and properties of retrofit materials and the networks and conditions of provisioning and installation. Analysing interviews with 45 industry stakeholders across insulation retrofit activities, we draw upon Actor Network Theory to bring to the fore the ways in which insulation technologies shape retrofit and the relations and networks of trust and conduct involved. The findings highlight the role of materiality in the retrofit industry, including through influencing selection, installation, and other associated retrofit practices. The chapter also discusses implications for policy and practice, including how supporting measures could circumvent market failures and vested interests and foster improved upscaling pathways. In so doing, the chapter makes the case for understanding factors beyond those foregrounded in dominant techno-economic frames that shape effective low-carbon housing retrofit policies and programmes.

[1]School of Global, Urban and Social Studies, RMIT University, Australia
[2]School of Property, Construction and Project Management, RMIT University, Australia
[3]College of Design and Social Context, RMIT University, Australia

The existing housing sector has a significant role to play in the transition to a low-carbon future [1,2]. Accordingly, policy and industry focus have typically sought to improve the quality and performance of new housing, and more recently, this focus has shifted to existing housing, much of which was built prior to modern energy efficiency standards [3]. There is also the question of what will be required for a low-carbon future under climate change [4,5].

Research from around the world has demonstrated that deep retrofit can significantly improve the quality and performance of existing housing [6,7]. Deep retrofit means the material improvement of the building envelope, for example, by retrofitting insulation, as well as the upgrade of space conditioning systems [8]. Benefits of deep retrofit include reduced environmental impact, reduced utility costs, improved thermal comfort, improved health and wellbeing, and improved market value [9–12]. There has been interest from policymakers to encourage homeowners to engage with retrofits as part of a wider approach to decarbonise the built environment [7,13–16]. However, despite the range of potential benefits from retrofitting and various financial and other support available in some locations, the uptake of retrofitting has been fragmented and is not yet at the scale or depth required [6].

Part of the challenge is that retrofit has largely been viewed through a techno-economic lens, where the belief has been that if the right market conditions are provided, retrofit will be desirable for households. However, this approach has been criticised for simplifying assumptions and overlooking key social understandings of retrofit [6,17,18]. This chapter explores this gap in knowledge through the lens of the materiality of insulation retrofits. Insulation refers to any product or material within a building element assembly that increases the thermal resistance through the assembly [19]. Insulation is used in new construction and retrofit projects; however, the choice of material and installation processes varies with its application. In particular, the practice of retrofitting insulation must contend with damp, inadequate ventilation, and cramped spaces and encounters risks associated with electric cabling, water pipes, and space conditioning ducts, to name a few [8,19]. Moreover, retrofitting insulation, along with double glazing, can reduce energy usage by 20–40 per cent, and, with a more nuanced approach, it can get as high as 80 per cent [19].

The research presented in this chapter is from the state of Victoria (Australia). The residential sector in Australia accounts for 8 per cent of energy use [19,20], of which 40 per cent is due to space cooling and heating [19]. Requirements for minimum insulation levels for new housing were first introduced in 1991 and have been increased several times as building codes have been updated [21]. However, there have been few requirements for improving the quality or performance of existing housing in Australia or Victoria. Retrofit has largely been left to the 'market' to undertake, although there has been some support provided at different points in time to scale up retrofit [22–24].

In Victoria, it is estimated that 69 per cent of existing dwellings were built prior to 1990, and 82 per cent before 2000 [25]. The majority of these existing dwellings are likely to remain in use for the next 50 years or more and so have an important role to play for Victoria (and Australia) to meet decarbonisation goals [14]. The existing housing stock in Victoria performs poorly, with houses built prior to 2004 estimated to have an energy rating of 2 stars out of 10 stars (best) [14]. The rating calculates heating and cooling energy loads for a proposed house design modelled within a specific local climate based on actual weather data [26–28]. In this heating-dominated climate, insulation retrofit has therefore been identified by researchers and policymakers as a key retrofit activity to reduce energy demand and carbon emissions [29].

In Victoria (and Australia), there is limited data on the number of insulated houses [30]. However, modelling has found that up to 1.65 tonnes of carbon dioxide equivalent (CO_2-e) may be prevented yearly for every dwelling that is insulated, demonstrating the importance of insulating existing dwellings [31,32].

15.2 Literature review

The role of markets in scaling up insulation retrofits is certainly constrained by the persistence of multiple market failures [24,33]. Other commonly cited issues that confound simple scaling up of insulation retrofits include weather, financial, and political conditions [24,34–36]. Local climate could encourage (or not) insulation take-up because the economic benefit varies accordingly [35,36].

However, other factors can be envisaged here, including the upfront energy/retrofit cost relative to other income/costs of living, the reliability/trust factors associated with different insulation technologies, the availability of trusted information and trades to install, and the level of social and political discourse supporting or conflating the logics of insulation. Thus, researchers have challenged simple techno-economic constructs of retrofit, showing that there are insufficient accounts of social, cultural and material considerations [17,18,37,38]. For example, Shove [17] looked at how social and local conditions shaped the practice of cavity wall insulation in the UK, Sweden, France, and Denmark. Through semi-structured interviews with insulation manufacturers, trade associations, and building regulators, Shove demonstrated that the provision of insulation often depended on the roles and influence of the people who delivered it. While Shove's research highlights how socio-technical conditions are influential in the delivery of insulation, there remains relatively little attention across the wider retrofit literature to the materiality of insulation and its effect on uptake.

In many locations, the delivery of insulation has been dependent on government support and political will [3,39,40]. This fits within the techno-economic approaches applied to retrofits more broadly in that the costs of retrofits influence the scale, speed, and quality of delivery [29,39,41,42]. Therefore, governments have been seen to play an important role in helping overcome market prices and other barriers through different types of support [39,41,42].

Governments rely on three generic types of support: regulation, subsidies, and education [41,43]. Regulation includes creating, maintaining, and improving building energy codes, and making certain retrofit activities compulsory for dwellings undergoing renovations, or at point of lease, if they fall below a certain performance standard [43]. Subsidies for retrofits can be interest-free loans, grants, and rebates that target homeowners to financially incentivise them to undertake retrofit activities [41,42]. Education efforts can include constructing low- or zero-carbon demonstration buildings, performing home energy audits, and providing information on matters such as the long-term benefits of energy retrofits, funding opportunities, and technical know-how [44].

Governments across much of the Westernised world have approached the scaling up of retrofits through these various approaches [39,44]. There are a range of market-based mechanisms that have been brought to bear that are variations on the concept of energy company obligation programmes [44,45], where energy suppliers are obliged to reduce fossil energy demand across consumers. In 2017, 47 energy company obligation programmes were in operation in 21 countries, including England and Australia [45]. By placing the responsibility on electricity suppliers to deliver retrofits through their need to purchase carbon certificates, the main assumption in such approaches is that price is central to stimulating the take-up of retrofits. This reduces the relative importance of social configurations in retrofits accordingly. It is also regressive in that the inevitable price increases in energy are borne disproportionately by the people on the lowest incomes, in the worst housing, and who are least likely to benefit from retrofits. The socio-technical literature has long recognised that techno-economic performance is important, but not sufficient, for successful diffusion or upscaling of retrofits [24,38,39].

However, the research also identifies that there are a range of other factors impacting the scaling up of retrofits [16,22,23,46–48]. For example, studies emphasise that retrofit installers are insufficiently prepared to develop and adapt new solutions to their current practices and lack the full set of skills and resources necessary to deal with increased opportunities [46,48]. This particular pattern of knowledge and skills across the retrofit industry has led to calls for training and reskilling [29,48,49]. It is not just the knowledge and skills of the retrofit process but also in relation to knowledge about alternative retrofit options, such as different materials or opportunities to benefit individual businesses [48,49]. Under pressure to deliver within tight time frames, retrofitters tend to offer their customers standardised solutions [47]. Even if a contractor applies a different solution for a specific client, it has been found that they will generally revert to established, conservative practices [47].

15.3 Actor Network Theory

Materiality is a concept used to examine the delivery of retrofit materials in ways that reveal more than techno-economic properties, including knowledge, networks, and relationships. Thus, by materiality, we mean the aspects of retrofit practice that are shaped by the physical appearance and properties of retrofit technologies and their associated networks and conditions which shape the retrofit industry, including through influencing selection, installation, and other associated retrofit practices [50]. Materiality has not previously been linked to exploring the role of insulation materials in shaping the delivery of retrofits. However, there is an emerging body of work that has applied Actor Network Theory (ANT), or related theories of socio-materiality, to highlight the human-technology interactions and challenges in low-carbon retrofits [38,47,50,51].

ANT is a post-structuralist approach that moves from essentialist explanations of the social by bringing sensitivity to unexplored entities. Researchers using ANT are interested in connections between humans and nonhumans because they subscribe to the notion that everything that exists in the world is the outcome of an interaction between two or more human and/or non-human agents [52–55]. The non-human is a term that includes artefacts, texts, markets, technology, and more. As Latour [56 p.34] stated, 'each and every thing can be nonhuman, excluding the human'. The main understanding of ANT is the associations and relationships between nonhumans that form networks to perform an action, as highlighted in the name, Actor Network Theory [52]. ANT aims to extend beyond conventional paradigms by embracing the agency of non-human entities, or materials, within intricate networks [53]. In its pursuit, ANT delves into the genesis of socio-material connections, their cohesive forces, and the outcomes they generate. By bringing attention to materiality, ANT and other socio-material approaches have provided a wider understanding of the challenges of retrofitting residential stock [38,47,50,51].

The foundation of ANT's transformative lens lies in its capacity to endow inanimate materials with agency, thereby challenging the primacy of human-centric explanations prevalent in other theoretical frameworks [53]. Here, every entity, whether human, artefact, or non-human, assumes the role of an actant, each wielding an agency that actively shapes the course of events. In essence, materials transcend their passive existence to become dynamic agents, capable of authorising, influencing, and guiding actions [53]. Through this departure from preconceived human-centredness, ANT unveils the socio-political dimensions that materials bring to the forefront – embedded with skills, biases, and limitations. This perspective provides a novel understanding of retrofits, as materials cease to be mere tools and evolve into influential actors shaping the delivery process. This understanding resonates with the idea that agency transcends human control, extending its reach to encompass a broader spectrum of entities [53,57]. It challenges us to probe into the socially constructed essence of all actants involved, be they human or material. This exploration into the realm of material agency unveils a landscape where materials, insulation in particular for this chapter, become orchestrators of change, forging relationships with installers, policymakers, and business owners. The interplay between materials and human stakeholders emerges as a dance of mutual influence, fundamentally altering the dynamics of insulation retrofit endeavours.

At the heart of ANT lies a potent tool for dissecting the interplay between human and non-human elements – an approach rooted in relational inquiry [55]. This perspective amplifies our comprehension of how insulation materials weave intricate relationships with key players in the retrofit network. ANT's lens compels us to unravel the threads of alliances that span across the socio-material landscape. Central to this discourse is the concept of 'enrolment', where actants, both human and non-human, forge new connections within the expansive network [53]. Enrolment unveils the intricate process by which elements align themselves, expanding the boundaries of their influence. Through this vantage point, we gain insights into the delicate choreography of alliances and the elements that breathe life into these dynamic networks. Through this chapter, we not only enrich our understanding of insulation retrofit processes but also unlock the potential of ANT as a transformative lens for deciphering complex socio-material phenomena.

15.4 Methods

The research presented in this chapter was undertaken as part of a wider project called Housing Energy Efficiency Transitions [58] and associated doctoral studies [50]. To capture the way the materiality of insulation retrofit affects processes (e.g. installing, promoting, selling), 45 semi-structured interviews were carried out with key stakeholders in 2020–2022. The interviews were conducted online or by telephone and lasted for around 60 minutes. Participants were from a diverse range of professions, including installers, business owners, managers, academics, contractors, Non-Government Organisations (NGOs), and incumbent actors in the insulation retrofit industry in Victoria, Australia. Participants had experience with different types of insulation retrofit; some had experience with all types of insulation retrofits of wall, ceiling, and underfloor insulation, and some had experience with one type.

Open-ended questions in semi-structured interviews allowed participants to express their beliefs and knowledge of the role of their relationships with human and non-human actors [54]. Furthermore, semi-structured interviews allow the recreation of networks to find the relationships between them [59]. Their personal and professional experiences shed light on the material aspects that affected the processes of delivering (sourcing, design, action, and key decision-making along the project chain) and how they interconnect and relate to the different types of insulation retrofit. The results were analysed using an abductive coding approach in which ANT was utilised to allow the iterative development of themes as the analysis developed.

15.5 Findings: invisibility, trust and lock-in

Analysis of the interviews revealed several key themes including the invisibility of insulation retrofit, a lack of trusted information, and material-specific training, and locked-in relationships. These are explored below.

15.5.1 The invisibility of insulation retrofits

The lack of visibility of ceiling and wall insulation when installed has hindered take-up and supervision of safety procedures. When stakeholders were asked about their experience working with insulation materials, they responded that the invisibility element of insulation retrofits is conditioning and shaping its take-up. Participants commented that once the insulation is installed in a ceiling or a wall, it is not visible to the human eye. Strategic decision-makers, as well as installers, agreed that this invisibility has made this type of home improvement less attractive to homeowners than changes that could be seen, such as solar panels. While for participants who work in managerial processes, this invisibility influenced quality processes and policy strategies, for installers, it affected the ability to sell and offer it to homeowners. For instance, participants 8 and 18 reported that the invisibility element prevented insulation from being a retrofit that is desirable, exciting, or appealing.

> *'Because the product is boring because you can't see it. It's not a prestige thing to have. I would rather have a big new colour TV than spend the money on insulation. You'd rather have solar panels on your roof so that people can see them. With insulation, you can't come in and say, "I've got wonderful insulation". It's all hidden.'* (Participant 8)

> *'It's not something you go up and touch every night, and it's hidden behind the walls.'* (Participant 18)

Moreover, respondents who work in the managerial part of programmes, research, or developing policies, reported that the lack of visibility hinders safety processes. Interviewees who have taken part in the development of regulation and quality processes reported poor installation and work practices in the industry. The retrofit's invisibility when it was installed made it hard to monitor or check and this was identified as a key challenge to ensuring quality control. Interviewees who have participated in safety and monitoring processes reported that the lack of visibility and monitoring has resulted in cases of ceiling insulation only being installed around the maintenance hole (the access point to the ceiling) and not across the entire ceiling as required. Despite the concern that poor installation practices have been prevalent across the industry, interviewees reported that tools that allow insulation to be detected once installed are not normalised in the industry. For instance, participant 1 reported that, in his experience of delivering insulation programmes, using a thermographic camera could help confirm the quality of installation:

> *'Once the wall gets, you know, done. You just don't really know how effective a job it is. Unless you have a certifier go out there with a tool, and they have tools, they can. I think they are heat-sensing tools, and they can just look at a wall and get an idea.'* (Participant 1)

Similarly, stakeholders working in policymaking and design reported that the invisibility of insulation retrofit was a challenge when trying to stimulate take-up. Policy designs depend upon neighbourhood effects, such as referrals. Interviewees often compared insulation with the high visibility of PV panels, which creates a fear of missing out for neighbouring households, but this did not occur with invisible insulation.

Furthermore, stakeholders reported that the lack of a visible factor necessitated more sophisticated marketing strategies around communicating the range of wider benefits from insulation, going beyond just reducing environmental impact to include improved comfort and financial aspects, which result in a package of benefits for households. There were some participants who mentioned that the best marketing was their own experience of living with insulation and the benefits it provided, which they could then discuss with potential clients:

> *'I've always understood the benefits technically, but now I can use it as a personal anecdote for people. When you're having a conversation with someone, you can say, Yes. I moved from an uninsulated house to an insulated one, and it made a massive difference to my comfort.'* (Participant 15)

The analysis depicts how ceiling and wall insulation retrofits condition its materiality and affect the social process of instalment, supervision, policymaking, and promotion. The analysis also demonstrates how the lack of attention to addressing this invisibility is affecting the ways in which the take-up of insulation can be promoted. Thus, this research sees that insulation retrofits have agency and how this is exerted on human factors [55]. This is an effect of the particularly invisible nature of the material and the current market structure.

15.5.2 Lack of trusted information and material-specific training for installers

While businesses and programme managers discussed that they could be more efficient in their work with insulation materials, there was no common consensus on the required training related to the materiality of insulation. Installers and business owners reported having a preferred insulation material due to different material aspects, such as texture, price, and size. Some of the interviewees contradicted each other in terms of the material characteristics of insulation, because they relied on experiential knowledge of the material, which may have been limited or partial. Installers reported that due to the lack of official information and material certifications, they would have to research and enquire about information themselves to ensure they had the best material. Such self-learning activities include looking on the Internet for information (e.g. Google, YouTube), joining Facebook groups of installers working overseas, relying on previous trade skills, knowledge acquired through social connections, or relying on information provided by manufacturers or other trades in the supply chain.

When installers and business owners were asked about their relationships with the materiality of insulation, respondents often identified training as a barrier to working with materials that they had not worked with previously. While some participants noted there was wider industry training available, there was a lack of trust in some of this due to a focus on a particular material type or support by particular suppliers; hence, the training was not always relevant or seen as impartial. Additionally, some training was aimed at new housing construction, with the nuances of retrofit often overlooked:

> *'[Referring to industry organizations] … They present themselves as a representative in the insulation industry, and they'll put out handbooks and information about how to install their material. There's nothing in the training that applies to the retrofit space, there is nothing there about how to identify faulty electrical connections in a roof.'* (Participant 20)

Additionally, when installers and business owners were asked about their relationship with different materials, they also mentioned that training affected the way they managed programmes. In particular, they mentioned that often installers working for them would have locked-in relationships with a specific material. In the following section, this is explored further.

15.5.3 Locked-in relationships

In Victoria, manufacturers and contractors provide training and information on a specific material. Installers often developed knowledge of one specific material, creating a path-dependent, locked-in, material–human relationship, where installers tended not to work with other materials. For instance, Participant 18, a business owner, reflected that the training and materials they use are aligned with their manufacturer and that, therefore, this was typically the product they would recommend:

> *'We are an approved (brand) grade distributor, so we tend to obviously promote their products predominantly, which, if it's underfloor or wall insulation or ceiling insulation, is a (brand) product.'* (Participant 18)

When asked about their reasons for relying on a locked-in relationship, Participant 18 reported that they rely on their manufacturers because they know the standards and requirements for insulation:

> *'Under our contract with [company] there is specific training required for that work. It tends to be driven by that contract that we've got there. They'll check for standards, and then we use that standard across all our work.'* (Participant 18)

When other installers were asked about the impact of this locked-in relationship, they often responded that they did not 'know' how to work with other ceiling insulation materials from other brands or types of ceiling insulation material.

While most installers and business owners reported being locked into a relationship, Participant 20, a retrofit-only business owner, reported that for him, it was important to remain 'material-agnostic'. He needed to be able to choose and find the best material and evaluate the pros and cons of each, and the locked-in relationships did not allow him to do so:

> *'I am very agnostic when it comes to products, I want to find what's best for an application. We do so many different things. And over the years of [company] and cellulose and all sorts of different products, we use polyester. And you basically just have to be very objective, looking at the pros and cons of each product.'* (Participant 20)

Similarly, programme managers reported that this locked-in relationship, between materials, installers, and suppliers, prevented the proper management of wall and ceiling insulation programmes. For instance, two interviewees reported that the material used in the programme they were managing was not familiar to their organisation. The interviewees said that the ongoing relationship between the installer and the material overruled their ability to choose another material that aligned better with the requirements of their organisation:

> *'And they all did training with the brand. When we spoke to the subcontractor, they said the process is that all the installers do training with the supplier of the brand to be. So, they were not familiar with the material we wanted to work with.'* (Participant 10)

> *'They are not familiar enough with the products, and they did a bad job. And sometimes they got the order wrong.'* (Participant 7)

These examples highlight how the different types of materials developed relationships with specific businesses and installers and how these existing relationships affected the management of insulation programmes.

15.6 Implications for policy and practice

The training and influence of manufacturers can limit installers' choices when they want to work with different insulation materials. Additionally, the locked-in relationships between insulation materials and installers, due to manufacturers' influence on the training, led some installers to look for training outside of manufacturers' influence. Often, the installers and business owners who wanted to remain, as they put it, 'material agnostic', expressed that the manufacturers' training often did not provide specific knowledge, or the insulation materials were not the best for retrofitting work. Implications of this are discussed below.

15.6.1 Making insulation visible

Enhancing visibility for insulation retrofit measures could mean literally, for example, by labelling, certifying, and aesthetically branding insulated places in the same way that energy efficiency labelling for appliances is used routinely in Australia. However, enhancing visibility in other ways links to measures that amplify awareness and lead to action. Common ground for collaboration among stakeholders, spanning policymakers, industry experts, and community advocates, is indispensable. This might include creating core language, discourse, and logic, pooling insights and resources to construct a comprehensive narrative and creating a shared mission around the significance of insulation retrofit in mitigating energy waste and bolstering sustainability efforts. Moreover, leveraging digital platforms and innovative communication channels offers a dynamic avenue to disseminate information effectively.

15.6.2 Training and locked-in relationships between installers and insulation materials

The findings above highlight how training was monopolised by manufacturers, and installers were often locked into working with one insulation material and did not know how to work with other materials. This builds upon earlier work where manufacturers have been recognised as influential and important in providing training on retrofit products [16,38,47]. When programme managers were asked about the locked-in relationships, they reported that often this deterred the capacity of installers to work with less carbon-intensive materials. The installers and business owners who did not develop a locked-in relationship needed to remain material-agnostic by actively looking for training online or having multiple manufacturers, but this required additional time and effort from them to stay on top of this.

While installers attend manufacturers' training courses because it gives them access to product-specific information [47,49], the findings reported above reveal that this training and influence of manufacturers could limit installers when they want to work with different insulation materials and could introduce important market failures into low-carbon retrofits. The installers and business owners who looked to remain, as they put it, 'material agnostic', expressed that often the manufacturers' training did not provide knowledge about the insulation materials that were the best for retrofitting work. Instead, this training (a) addressed only particular proprietary processes and products and (b) focussed on new build rather than retrofit applications.

Point (a) is critical as this positions installers as tied to individual manufacturers' product sales, rather than as independent advisors or experts on insulation and energy-efficient retrofits. However, this distinction is not overt in the market, and it is not visible to households. This information asymmetry presents a significant market failure. Point (b) is also important, as the requirements for the installation of ceiling and wall insulation retrofits are significantly different from new builds. For example, working within the existing housing structure presents additional logistical challenges for installing insulation.

The importance of the relationships between insulation materials, manufacturers, programme managers, and installers is not recognised in the wider techno-economic literature. This research has shed a light on the relevance of these relationships in Victoria but also has relevance for an international context. Evidence presented in this study suggests a need to attend to the influential role of manufacturers in the relationships between installers and their products. Several impacts on installers' experience and their ability to work with different insulation materials were identified, building on previous work [47,50].

The importance of neutral, independent, trusted advice, and information on different insulation technologies, processes, and performance is essential to underpin a functioning retrofit market in insulation. Moreover, the findings suggest that policymakers must pursue a cross-material, neutral, training approach that is free of potential vested interests. An avenue to achieve this is to partner with large retailers such as Bunnings, industry organisations and manufacturers, where independent industry experts, academics, researchers, and policy stakeholders are able to take advice and set curricula where trades are licensed according to their completion of such independently oriented training. Such training could be integrated into vocational training in Australia but also delivered by key industry peak bodies and other critical intermediaries.

15.6.3 Implications of visibility

Policies that support installers' and the public's understanding of the visible characteristics of insulation are important in increasing the current rate of installs. Business owners and industry organisations struggle to promote insulation measures due to the lack of visibility once it is installed. Therefore, insulation benefits should be communicated widely and clearly to the public to increase the appeal and retrofit uptake. Reality TV shows about renovation are quite common in various jurisdictions including Australia. Renovation shows present an opportunity to communicate the benefits of insulation (e.g. Dream Home Makeover, Open Homes Australia, The Block). Trusted intermediaries are essential in building a language, discourse, and practice of insulation retrofit.

The use of technologies, such as thermographic cameras, could assist in making insulation visible and also in quality control during insulation retrofits. Given that the lack of visibility when installed was mentioned as part of the material aspects that affect the scaling up of insulation, the Victorian Government could subsidise thermographic cameras for use across the industry. These cameras can identify areas where insulation might not be properly installed, leading to gaps or poor coverage. Subsidising the use of these cameras could help ensure that the retrofitting process is carried out effectively and efficiently. While such cameras are increasingly available on loan from libraries in some local government jurisdictions, their widespread use in reliable ways is still elusive. Contractors and inspectors could more easily ensure that insulation retrofits meet standards, avoiding potential issues and non-compliance penalties, if households were also equipped with the discourse and knowledge of/access to such technologies. Subsidised thermographic cameras could raise public awareness about the importance of insulation and energy-efficient retrofits. This can encourage more homeowners to invest in such projects, leading to broader energy savings, and environmental benefits.

15.7 Conclusion

The transition to lower-carbon housing is constituted by the rules, aspirations, and economic constraints that shape the retrofitting of existing homes. These are, in turn, shaped by materiality; the nature of non-human actors, networks, and relations. These, often overlooked, materialities are important and, indeed, the insights that can be drawn from materialities studies can, therefore, indicate directions through which upscaling of low-carbon retrofits might more readily be achieved.

It is within this context that the empirical study reported here has led to a set of implications for practice that otherwise would not have been revealed through a normative techno-economic lens. Namely, attention to visibility, to the networks and relations of training, and to the relative prevalence, availability, and independence of advice and information. The retrofitting of insulation is constituted through the practices by which stakeholders turn the objective of improving the energy efficiency of housing into action as part of a broader discourse and practice of home improvements. We have revealed how gaps in trust, knowledge, and the availability of training are filled by power-imbalanced social relations that can constrain insulation retrofits. Ultimately, these material circumstances that mark the practice of retrofitting insulation mediate policy and societal efforts to improve the thermal performance of housing.

References

[1] Fox-Reynolds K., Vadja K., Minunno R., Wilmot K. *Pathways to scale: Retrofitting One Million+ homes* [online]. 2022. Available from https://www.racefor2030.com.au [Accessed 16 June 2026]

[2] United Nations. 'United Nations UN climate change summit 2019'. *New Scientist*. 2019;**243**(3249):7

[3] Moore T., Berry S., Ambrose M. 'Aiming for mediocrity: The case of Australian housing thermal performance'. *Energy Policy*. 2019;132:602–10

[4] Moore T., Horne R., Gleeson B.J. (eds.). *Urban Sustainability Transitions: Australian Cases-International Perspectives*. Singapore: Springer Nature; 2018

[5] Brimblecombe P. 'Sustainability and heritage' in Saiz-Jimenez C. (ed.), *Air Pollution and Cultural Heritage*. London: Taylor Francis Group; 2004. pp. 275–80

[6] Wade F., Visscher H. 'Retrofit at scale: Accelerating capabilities for domestic building stocks'. *Buildings and Cities*. 2021;**2**(1):800–11

[7] Willand N., Maller C., Ridley I. 'Addressing health and equity in residential low carbon transitions–Insights from a pragmatic retrofit evaluation in Australia'. *Energy Research & Social Science*. 2019;53:68–84

[8] Hunt R., Suhr M. *Old House Eco Handbook: A Practical Guide to Retrofitting for Energy Efficiency and Sustainability*. London: Frances Lincoln; 2019

[9] Howden-Chapman P., Crane J., Matheson A., *et al.* 'Retrofitting houses with insulation to reduce health inequalities: Aims and methods of a clustered, randomised community-based trial'. *Social Science & Medicine*. 2005;**61** (12):2600–10

[10] Howden-Chapman P., Matheson A., Crane J., *et al.* 'Effect of insulating existing houses on health inequality: Cluster randomised study in the community'. *British Medical Journal*. 2007;**334**(7591):460

[11] Fuerst F., Warren-Myers G. 'Does voluntary disclosure create a green lemon problem? Energy-efficiency ratings and house prices'. *Energy Economics*. 2018;74:1–12

[12] Willand N., Moore T., Horne R., Robertson S. 'Retrofit poverty: Socioeconomic spatial disparities in retrofit subsidies uptake'. *Buildings and Cities*. 2020;**1**(1):14–35

[13] Brocklehurst F., Morgan E., Greer K., *et al.* 'Domestic retrofit supply chain initiatives and business innovations: An international review'. *Buildings and Cities*. 2021;**2**(1):533–53

[14] Moore T., Holdsworth S. 'The built environment and energy efficiency in Australia: Current state of play and where to next' in Rajagopalan P., Moore T., Andamon M. (eds) *Energy Performance in the Australian Built Environment*. Singapore: Springer Nature; 2019, pp. 45–59

[15] Smoleniec L., McManus P., Duncan E. 'Understanding the dynamics of sustainability transitions: The Home Insulation Program'. *Australian Geographer*. 2017;**48**(4):497–517

[16] Horne R., Maller C., Dalton T. 'Low carbon, water-efficient house retrofits: An emergent niche?'. *Building Research & Information*. 2014;**42**(4):539–48

[17] Guy S., Shove E. *A Sociology of Energy, Buildings and the Environment* Routledge Global Environmental Change Series, no. 5. Oxford: Routledge, 2000

[18] Wade F., Bush R., Webb J. 'Emerging linked ecologies for a national scale retrofitting programme: The role of local authorities and delivery partners'. *Energy Policy*. 2020;137:111179

[19] Thorpe D. *Sustainable Home Refurbishment: The Earthscan Expert Guide to Retrofitting Homes for Efficiency*. London: Routledge; 2010

[20] Australian Department of the Environment and Energy. *Bureau of resources and energy economics. Australian energy update* [online]. 2021. Available from https://www.energy.gov.au/sites/default/files/Australian%20Energy%

20Statistics%202021%20Energy%20Update%20Report.pdf [Accessed April 21 2024]

[21] Soebarto V., Williamson T. (eds). 'A critical review of home energy rating in Australia'. *Proceedings of the 34th Conference of the Australian and New Zealand Architectural Science Association*; Adelaide, Australia, 2000, pp. 101–10

[22] Horne R. *Housing Sustainability in Low Carbon Cities*. London: Routledge; 2017

[23] Horne R., Moloney S. 'Urban low carbon transitions: Institution-building and prospects for interventions in social practice'. *European Planning Studies*. 2019;**27**(2):336–54

[24] Killip G. 'Beyond the Green Deal: Market Transformation for Low-Carbon Housing Refurbishment in the UK'. Presented at Retrofit 2012; Salford, UK, 2012

[25] Australian Bureau of Statistics. *Census all persons QuickStats* [online]. Canberra: Australian Bureau of Statistics; 2021. Available from https://abs.gov.au/census/find-census-data/quickstats/2021/2 [Accessed 12 June 2026]

[26] Law T. 'Beyond minimum: Proposition for building surveyors to exceed the minimum standards of the construction code'. *Journal of Legal Affairs and Dispute Resolution in Engineering and Construction*. 2021;**13**(2):4521001

[27] Moore T., Horne R., Morrissey J. 'Zero emission housing: Policy development in Australia and comparisons with the EU, UK, USA and California'. *Environmental Innovation and Societal Transitions*. 2014;11:25–45

[28] Moore T., Sheriff G., Whaley D.M., Berry S.R. 'Does tenure matter for occupant experiences of low-energy housing?'. *Proceedings of the 10th International Conference in Sustainability on Energy and Buildings (SEB'18)*; Gold Coast, Australia, June 2018, pp. 453–63

[29] Forman T., Tweed C. 'Solid wall insulation retrofit in UK dwellings: Critical factors affecting management and quality'. *Proceedings of the 30th Annual ARCOM Conference*; Portsmouth, UK, September 2014, pp. 1367–76

[30] Judson E.P., Iyer-Raniga U., Horne R. 'Greening heritage housing: Understanding homeowners' renovation practices in Australia'. *Journal of Housing and the Built Environment*. 2014;29:61–78

[31] Insulation Council of Australia and New Zealand. *The value of insulation based residential energy savings measures in Australia* [online]. 2012. Available from http://icanz.org.au/government/key-issues/the-value-of-insulation-based-residential-energy-savings-measures/ [Accessed 4 April 2024]

[32] Washan P., Stenning J., Goodman M. *Building the Future: The Economic and Fiscal Impacts of Making Homes Energy Efficient*. Verco and Cambridge Econometrics; 2014

[33] Brown M.A. 'Market failures and barriers as a basis for clean energy policies'. *Energy Policy*. 2001;**29**(14):1197–207

[34] Dowling R., McGuirk P., Bulkeley H. 'Retrofitting cities: Local governance in Sydney, Australia'. *Cities*. 2014;38:18–24

[35] Rodríguez-Soria B., Domínguez-Hernández J., Pérez-Bella J.M., Del Coz-Díaz J.J. 'Review of international regulations governing the thermal insulation requirements of residential buildings and the harmonization of envelope energy loss'. *Renewable and Sustainable Energy Reviews*. 2014;34:78–90

[36] Kolaitis D.I., Malliotakis E., Kontogeorgos D.A., Mandilaras I., Katsourinis D.I., Founti M.A. 'Comparative assessment of internal and external thermal insulation systems for energy efficient retrofitting of residential buildings'. *Energy and Buildings*. 2013;64:123–31

[37] Sovacool B.K., Griffiths S. 'Culture and low-carbon energy transitions'. *Nature Sustainability*. 2020;**3**(9):685–93

[38] Banks N. 'Socio-technical networks and the sad case of the condensing boiler'. in Bertoldi P., Ricci A., de Almeida A. (eds) *Energy Efficiency in Household Appliances and Lighting*. Berlin, Heidelberg: Springer; 2001, pp. 141–55

[39] Hodson M., Marvin S. 'Mediating low-carbon urban transitions? Forms of organization, knowledge and action' in Priemus H., Davoudi S. (eds) *Climate Change and Sustainable Cities*. London: Routledge; 2016, pp. 145–63

[40] Hodson M., Marvin S. 'The mutual construction of urban retrofit and scale: Governing ON, IN and WITH in Greater Manchester'. *Environment and Planning C: Politics and Space*. 2017;**35**(7):1198–217

[41] Rosenow J., Eyre N. 'A post mortem of the Green Deal: Austerity, energy efficiency, and failure in British energy policy'. *Energy Research and Social Science*. 2016;21:141–44

[42] Rosenow J., Platt R., Demurtas A. 'Fiscal impacts of energy efficiency programmes – The example of solid wall insulation investment in the UK'. *Energy Policy*. 2014;**74**(C):610–20

[43] Weiss J., Dunkelberg E., Vogelpohl T. 'Improving policy instruments to better tap into homeowner refurbishment potential: Lessons learned from a case study in Germany'. *Energy Policy*. 2012;44:406–15

[44] Brown H.S., Vergragt P.J. 'Energy retrofitting of the residential housing stock' in Marletto G. (ed) *Creating a Sustainable Economy: An Institutional and Evolutionary Approach to Environmental Policy*. London: Routledge; 2013. p. 154.

[45] International Partnership for Energy Efficiency Cooperation. *Energy company obligations: Examples of delivering energy efficiency in buildings and future trends* [online]. 2019. Available from https://www.energy.gov.au/sites/default/files/buildings_energy_efficiency_task_group_beet_8-company_obligations_report_november2019.pdf [Accessed 4 April 2024]

[46] Killip G., Fawcett T., Janda K.B., Bellian V., Nosperger S. *Building expertise: Industry responses to the low-energy housing retrofit agenda in the UK and France* [online]. Available from https://ora.ox.ac.uk/objects/uuid:e125431e-1325-4040-8d73-992caa1a7633 [Accessed 12 June 2026]

[47] Killip G. 'Products, practices and processes: Exploring the innovation potential for low-carbon housing refurbishment among small and medium-sized enterprises (SMEs) in the UK construction industry'. *Energy Policy*. 2013;62:522–30

[48] Killip G., Owen A. 'The construction industry as agents of energy demand configuration in the existing housing stock'. *Energy Policy*. 2020; 147:111816

[49] Topouzi M., Killip G., Owen A. 'Learning from 'horror' Stories: A plan of work to reduce the performance gap in deep retrofit'. *Proceedings of the 33rd PLEA International Conference: Design to Thrive*. Edinburgh, UK, July 2017, pp. 550–57

[50] Padilla Gil H.M. 'The Pink elephant in the room: The materiality, networks and conditions of home insulation'. Doctoral dissertation, RMIT University, Melbourne, Australia; 2023

[51] Buser M., Carlsson V. 'What you see is not what you get: Single-family house renovation and energy retrofit seen through the lens of socio-materiality'. *Construction Management and Economics*. 2017;**35**(5):276–87

[52] Latour B. 'On actor-network theory: A few clarifications'. *Soziale Welt*. 1996:369–81

[53] Latour B. *Reassembling the Social: An Introduction to Actor-Network-Theory*. Oxford: Oxford University Press; 2007

[54] Law J. 'Actor network theory and material semiotics' in Turner B.S. (ed) *The New Blackwell Companion to Social Theory*. Oxford: Wiley-Blackwell; 2008. pp. 141–58

[55] Law J., Hassard J. (eds.). *Actor Network Theory and After*. Oxford: Blackwell; 1999

[56] Latour B., Woolgar S. *Laboratory Life: The Construction of Scientific Facts*. Princeton, NJ: Princeton University Press; 1986

[57] Latour B. 'Where are the missing masses? The sociology of a few mundane artifacts' in Law J., Bijker W. (eds) *Shaping Technology/Building Society: Studies in Sociotechnical Change*. Cambridge, MA: MIT Press; 1992. pp. 225–58

[58] RMIT University. *HEET (Housing Energy Efficiency Transitions). Sustainable building innovation laboratory* [online]. n.d. Available from https://www.rmit.edu.au/about/schools-colleges/property-construction-and-project-management/research/research-centres-and-groups/sustainable-building-innovation-laboratory/projects/heet-housing-energy-efficiency-transitions [Accessed 12 June 2026]

[59] Decuypere M. 'Visual network analysis: A qualitative method for researching sociomaterial practice'. *Qualitative Research*. 2020;**20**(1):73–90

Index

www.ingramcontent.com/pod-product-compliance
Lightning Source LLC
LaVergne TN
LVHW010948191026
850388LV00003B/25
* 9 7 8 1 8 3 9 5 3 4 0 4 1 *